Methods in Enzymology

Volume 405
MASS SPECTROMETRY: MODIFIED PROTEINS
AND GLYCOCONJUGATES

METHODS IN ENZYMOLOGY

EDITORS-IN-CHIEF

John N. Abelson   Melvin I. Simon

DIVISION OF BIOLOGY
CALIFORNIA INSTITUTE OF TECHNOLOGY
PASADENA, CALIFORNIA

FOUNDING EDITORS

Sidney P. Colowick and Nathan O. Kaplan

Methods in Enzymology

Volume 405

Mass Spectrometry: Modified Proteins and Glycoconjugates

EDITED BY

A. L. Burlingame

DEPARTMENT OF PHARMACEUTICAL CHEMISTRY
UNIVERSITY OF CALIFORNIA
SAN FRANCISCO, CALIFORNIA

AMSTERDAM • BOSTON • HEIDELBERG • LONDON
NEW YORK • OXFORD • PARIS • SAN DIEGO
SAN FRANCISCO • SINGAPORE • SYDNEY • TOKYO
Academic Press is an imprint of Elsevier

ELSEVIER

Elsevier Academic Press
525 B Street, Suite 1900, San Diego, California 92101-4495, USA
84 Theobald's Road, London WC1X 8RR, UK

This book is printed on acid-free paper.

For all information on all Elsevier Academic Press publications
visit our Web site at www.books.elsevier.com

ISBN-13: 978-0-12-182810-3
ISBN-10: 0-12-182810-7

PRINTED IN THE UNITED STATES OF AMERICA
05   06   07   08   09   9   8   7   6   5   4   3   2   1

Working together to grow
libraries in developing countries

www.elsevier.com | www.bookaid.org | www.sabre.org

ELSEVIER   **BOOK AID** International   **Sabre Foundation**

Table of Contents

Contributors to Volume 405

Article numbers are in parentheses following the name of contributors.
Affiliations listed are current.

REUDI AEBERSOLD (4), *The Institute for Systems Biology, Seattle, Washington*

ROLAND S. ANNAN (5), *GlaxoSmith Kline Pharmaceuticals, King of Prussia, Pennsylvania*

MICHAEL A. BALDWIN (8), *Mass Spectrometry Research Resource, Department of Pharmaceutical Chemistry, University of California, San Francisco, San Francisco, California*

A. L. BURLINGAME (9), *Department of Pharmaceutical Chemistry, University of California, San Francisco, San Francisco, California*

ODILE BURLET-SCHILTZ (11), *Institut de Pharmacologie et de Biologie Structurale, CNRS, Toulouse, France*

STEVEN A. CARR (5), *GlaxoSmithKline Pharmaceuticals, King of Prussia, Pennsylvania*

STÉPHANE CLAVEROL (11), *Pôle Protéomique, Pateforme Génomique Fonctionnelle Bordeau, Université Victor Segalen Bordeaux, France*

GARRY L. CORTHALS (4), *Turku Centre for Biotechnology, University of Turku, Åbo Akademi University, Turku, Finland*

A. D. COX (13), *Institute for Biological Sciences, National Research Council of Canada, Ottawa, Ontario, Canada*

JEAN EDOUARD GAIRIN (11), *Institut de Sciences et Technologies du Médicament de Toulouse, CNRS – Pierre Fabre, Toulouse, France*

DAVID R. GOODLETT (4), *Department of Medicinal Chemistry, University of Washington, Seattle, Washington*

LAN HUANG (9), *Science I, University of California, Irvine, California*

MICHAEL J. HUDDLESTON (5), *GlaxoSmithKline Pharmaceuticals, King of Prussia, Pennsylvania*

CONNIE R. JIMÉNEZ (2), *Mass Spectrometry Resource, Department of Pharmaceutical Chemistry, University of California, San Francisco, San Francisco, California*

STEVEN B. LEVERY (12), *Department of Chemistry, University of New Hampshire, Durham, New Hampshire*

J. LI (13), *Institute for Biological Sciences, National Research Council of Canada, Ottawa, Ontario, Canada*

A. MARTIN (13), *Institute for Biological Sciences, National Research Council of Canada, Ottawa, Ontario, Canada*

KATALIN F. MEDZIHRADSZKY (3, 6), *Department of Pharmaceutical Chemistry, School of Pharmacy, University of California, San Francisco, San Francisco, California*

BERNARD MONSARRAT (11), *Institut de Pharmacologie et de Biologie Structurale, CNRS, Toulouse, France*

E. R. MOXON (13), *Institute for Molecular Medicine, John Radcliffe Hospital, Headington, Oxford, United Kingdom*

GITTE NEUBAUER (10), *European Molecular Biology Laboratory, Meyerhofstrasse 1, Heidelberg, Germany*

JASNA PETER-KATALINIĆ (7), *Institute for Medical Physics und Biophysic, University of Muenster, Muenster, Germany*

J. C. RICHARDS (13), *Institute for Biological Sciences, National Research Council of Canada, Ottawa, Ontario, Canada*

KERSTIN STRUPAT (1), *Thermo Electron GmbH (Bremen), Bremen, Germany*

P. THIBAULT (13), *Institute for Research in Immunology and Cancer, Université de Montréal, Montreal, Quebec, Canada*

Preface

Unparalleled advances in macromolecular mass spectrometry continue at a rapid pace, and are providing ever more powerful tools to unravel and define the protein and glycolipid composition of cells on a global scale. Indeed, complex suites of proteins and glycolipids can be studied in concert and even in their functional biological context.

There is accelerating awareness in the biological research community that mass spectral strategies can provide crucial information to new areas, such as the challenging problems associated with the elucidation of protein signaling cascades and networks, temporal modulation of protein machine assembly and disassembly, and their posttranslational regulation. The broad utility and accessibility of this technology has begun to influence the design of experimental protocols in fields in which previously mass spectrometry had been only minimally involved. Since the continuing introduction of better technology platforms breed even higher quality experimentation, this trend will certainly continue but at an accelerated pace.

The more challenging biological problems drive tailoring and development of new methodologies, techniques, and instruments. Such development usually represents a mixture of multi-disciplinary, inter-dependent efforts involving sample handling, adaptation of techniques, which require gaining experience in their optimal use, and development of new software to enable the analysis of data and facilitate its comprehension.

This volume is a companion to Biological Mass Spectrometry in this series (Burlingame, 2005) and covers the additional topics and knowledge base essential for the understanding and practice of peptide and protein mass spectrometry, and glycolipidomics. These include protocols for sample preparation and clean up, proteolytic and chemical digests, factors essential for measurement of the molecular weight accurately, and then introduction into the mass spectrometer.

In addition, this volume describes the strategies employed for structural analyses of the two most pervasive classes of protein posttranslational modification, namely phosphorylation and the three major types of protein glycosylation. Many of these experiments utilize capillary HPLC coupled with electrospray ionization and tandem mass spectrometry. These treatments are followed by case studies of two protein machines, the proteasome and spliceosome, and the identification of MHC antigens. These examples reveal archetypal insights into how and why mass spectrometry revolutionized our global understanding of the functional modulation of protein assemblages, complexes, and organelles so quickly. Finally, comprehensive treatments of both endogenous and bacterial glycolipid analyses are presented. All of these contributions are written by authorities who are world leaders in their fields. These authors have provided a detailed view into the classes of biopolymers that are readily tractable together with our state of knowledge about the biological context. In addition, projections of the technical developments and challenging problems ahead are indicated.

I am indebted to all of my colleagues who have participated in this work, to Candy Stoner for her assistance and talents during the manuscript organization and preparation phase, and to Raisa Talroze for the completion of both volumes. I would like to acknowledge the NIH, National Center for Research Resources, for generous financial support (Grant RR 01614).

A. L. BURLINGAME

REFERENCE

Burlingame, A. L. (2005). Biological Mass Spectrometry. *Meth. Enz.* 402.

METHODS IN ENZYMOLOGY

VOLUME 72. Lipids (Part D)
Edited by JOHN M. LOWENSTEIN

VOLUME 73. Immunochemical Techniques (Part B)
Edited by JOHN J. LANGONE AND HELEN VAN VUNAKIS

VOLUME 74. Immunochemical Techniques (Part C)
Edited by JOHN J. LANGONE AND HELEN VAN VUNAKIS

VOLUME 75. Cumulative Subject Index Volumes XXXI, XXXII, XXXIV–LX
Edited by EDWARD A. DENNIS AND MARTHA G. DENNIS

VOLUME 76. Hemoglobins
Edited by ERALDO ANTONINI, LUIGI ROSSI-BERNARDI, AND EMILIA CHIANCONE

VOLUME 77. Detoxication and Drug Metabolism
Edited by WILLIAM B. JAKOBY

VOLUME 78. Interferons (Part A)
Edited by SIDNEY PESTKA

VOLUME 79. Interferons (Part B)
Edited by SIDNEY PESTKA

VOLUME 80. Proteolytic Enzymes (Part C)
Edited by LASZLO LORAND

VOLUME 81. Biomembranes (Part H: Visual Pigments and Purple Membranes, I)
Edited by LESTER PACKER

VOLUME 82. Structural and Contractile Proteins (Part A: Extracellular Matrix)
Edited by LEON W. CUNNINGHAM AND DIXIE W. FREDERIKSEN

VOLUME 83. Complex Carbohydrates (Part D)
Edited by VICTOR GINSBURG

VOLUME 84. Immunochemical Techniques (Part D: Selected Immunoassays)
Edited by JOHN J. LANGONE AND HELEN VAN VUNAKIS

VOLUME 85. Structural and Contractile Proteins (Part B: The Contractile Apparatus and the Cytoskeleton)
Edited by DIXIE W. FREDERIKSEN AND LEON W. CUNNINGHAM

VOLUME 86. Prostaglandins and Arachidonate Metabolites
Edited by WILLIAM E. M. LANDS AND WILLIAM L. SMITH

VOLUME 87. Enzyme Kinetics and Mechanism (Part C: Intermediates, Stereo-chemistry, and Rate Studies)
Edited by DANIEL L. PURICH

VOLUME 88. Biomembranes (Part I: Visual Pigments and Purple Membranes, II)
Edited by LESTER PACKER

VOLUME 89. Carbohydrate Metabolism (Part D)
Edited by WILLIS A. WOOD

VOLUME 352. Redox Cell Biology and Genetics (Part A)
Edited by CHANDAN K. SEN AND LESTER PACKER

VOLUME 353. Redox Cell Biology and Genetics (Part B)
Edited by CHANDAN K. SEN AND LESTER PACKER

VOLUME 354. Enzyme Kinetics and Mechanisms (Part F: Detection and Characterization of Enzyme Reaction Intermediates)
Edited by DANIEL L. PURICH

VOLUME 355. Cumulative Subject Index Volumes 321–354

VOLUME 356. Laser Capture Microscopy and Microdissection
Edited by P. MICHAEL CONN

VOLUME 357. Cytochrome P450, Part C
Edited by ERIC F. JOHNSON AND MICHAEL R. WATERMAN

VOLUME 358. Bacterial Pathogenesis (Part C: Identification, Regulation, and Function of Virulence Factors)
Edited by VIRGINIA L. CLARK AND PATRIK M. BAVOIL

VOLUME 359. Nitric Oxide (Part D)
Edited by ENRIQUE CADENAS AND LESTER PACKER

VOLUME 360. Biophotonics (Part A)
Edited by GERARD MARRIOTT AND IAN PARKER

VOLUME 361. Biophotonics (Part B)
Edited by GERARD MARRIOTT AND IAN PARKER

VOLUME 362. Recognition of Carbohydrates in Biological Systems (Part A)
Edited by YUAN C. LEE AND REIKO T. LEE

VOLUME 363. Recognition of Carbohydrates in Biological Systems (Part B)
Edited by YUAN C. LEE AND REIKO T. LEE

VOLUME 364. Nuclear Receptors
Edited by DAVID W. RUSSELL AND DAVID J. MANGELSDORF

VOLUME 365. Differentiation of Embryonic Stem Cells
Edited by PAUL M. WASSAUMAN AND GORDON M. KELLER

VOLUME 366. Protein Phosphatases
Edited by SUSANNE KLUMPP AND JOSEF KRIEGLSTEIN

VOLUME 367. Liposomes (Part A)
Edited by NEJAT DÜZGÜNEŞ

VOLUME 368. Macromolecular Crystallography (Part C)
Edited by CHARLES W. CARTER, JR., AND ROBERT M. SWEET

VOLUME 369. Combinational Chemistry (Part B)
Edited by GUILLERMO A. MORALES AND BARRY A. BUNIN

VOLUME 370. RNA Polymerases and Associated Factors (Part C)
Edited by SANKAR L. ADHYA AND SUSAN GARGES

[1]   Molecular Weight Determination of Peptides and Proteins by ESI and MALDI

By Kerstin Strupat

Abstract

Several topics are covered, namely, general aspects important for mass determination of peptides and proteins, sample preparation for both ESI and MALDI, and various mass analyzers coupled to these ionization techniques. Finally, the discussion is carried out on peptide and protein mass analysis as related to accuracy and precision of mass determination for both ESI–MS and MALDI–MS.

Introduction

The techniques of electrospray/ionization (ESI) and matrix-assisted laser desorption/ionization (MALDI) have revolutionized biological mass spectrometry (MS). All state-of-the-art biochemistry and biology laboratories possess at least one of these ionization techniques and, in general, have access to both of them. Parallel to and motivated by the development of these techniques, a diminution of chromatographic separation techniques and purification techniques has taken place, and the success of ESI and MALDI in biochemistry and biology is also due to the possibility of direct or indirect coupling of the ionization techniques to appropriate separation and purification techniques. Both ionization techniques are applicable to peptides and proteins (Yates, 1998), DNA and RNA (Gross, 2000), glycoconjugates, and synthetic polymers (Nielen, 1999).

Considerable information can be derived from mass spectra of biological samples such as peptides and proteins. Besides the determination of the molecular mass M of a given compound (often named molecular weight M_r, which is not quite correct because no weight or force, respectively, is measured by MS) and the identification of proteins by accurate mass determination of their proteolytic fragments, mass spectrometry is capable of providing structural information. (i.e., sequences) for peptides. The elucidation of post-translational modifications of peptides or proteins is an important branch of mass spectrometry. In combination with chromatography, known compounds can be determined quantitatively. Together with enzymatic degradation, the carbohydrate content of a glycoprotein

METHODS IN ENZYMOLOGY, VOL. 405   0076-6879/05 $35.00
Copyright 2005, Elsevier Inc. All rights reserved.   DOI: 10.1016/S0076-6879(05)05001-9

can be evaluated at least semiquantitatively. Furthermore, the quaternary structure of protein complexes, the interaction between proteins and ligands (Rogniaux *et al.*, 1999) or metal ions (Strupat *et al.*, 2000), as well as protein folding (Yao, 2005) can be studied by mass spectrometry.

In the rapidly expanding field of proteomics, high quality mass data—namely accurate mass determination—represent the key to unambiguous protein identification by peptide mass mapping and to determination of posttranslational modifications of proteins harvested from cells grown under different conditions (Stults, 2005). The accuracy of mass determination in peptide mass analysis obtained by ESI–MS and MALDI–MS has increased from less than $5 \cdot 10^{-4}$ or 500 ppm (0.5 u at 1000 u) 10 years ago to $0.5 - 2 \cdot 10^{-6}$ or 0.5 – 2 ppm (0.0005 – 0.002 u at 1000 u) at present (Senko, 2004). The striking advantage of an improved mass accuracy is the dramatic reduction of false hits in database interrogation for protein identity (Clauser *et al.*, 1999; Jensen *et al.*, 1996; Shevchenko *et al.*, 1996). It has to be noted, however, that such a high accuracy of mass determination is typically not achievable for analyte molecules of higher masses, such as proteins. This is due to both practical and fundamental reasons (see following paragraphs).

This paper is divided into several parts for ease of understanding. First, general aspects important for the understanding of the context are explained. This rather theoretical part is followed by an explanation of practical features, such as preparative steps and mass analyzers coupled to the ionization techniques, and, more importantly, results in a discussion of peptide and protein mass analysis with respect to accuracy and precision of mass determination for both ESI–MS and MALDI–MS. The performance of a Fourier transform ion cyclotron resonance (FTICR) mass analyzer with its ultra-high mass resolving power is discussed in a separate chapter.

Some General Aspects Important for Mass Determination of Peptides and Proteins

Isotopic Distribution of Peptide and Protein Signals

When discussing the topic of mass determination of biological molecules, one main aspect that should be kept in mind is the naturally occurring isotope distribution. It is necessary to distinguish between nominal mass, monoisotopic mass, and the average mass of a molecule (Yergey *et al.*, 1983). The nominal mass of a molecule is calculated by using the most abundant isotope without regard of mass defect/excess (i.e., H = 1, C = 12, N = 14, O = 16, etc.). The monoisotopic mass (M_{MONO}) of a molecule again refers to the most abundant isotope, but the exact mass is used (i.e., $^1H = 1.007825$, $^{12}C = 12.000000$, $^{14}N = 14.003074$, $^{16}O = 15.994915$, $^{31}P = 30.973762$, $^{32}S = 31.972070$, etc.). The average mass (M_{av}) of a molecule is calculated from the average masses of the elements weighted for abundance (i.e., $H = 1.00794$, $C = 12.011$, $N = 14.00674$, $O = 15.9994$, $P = 30.97376$, $S = 32.066$, etc.) (Price, 1991; Winter, 1983; Yergey et al., 1983). Mainly, due to the naturally occurring $^{12}C/^{13}C$ ratio (98.89% ^{12}C, 1.11% ^{13}C), biological molecules of mass 1000 units and larger show a considerable contribution of the ^{13}C isotope (see Fig. 1). The difference between monoisotopic mass and average mass increases with increasing mass and is larger for molecules containing a larger number of atoms with mass excess (such as $^1H = 1.007825$ for hydrogen) or a relatively abundant heavy isotope (Biemann, 1990). The difference is largest for peptides and proteins because of their relatively high content of carbon and hydrogen; the difference is less for carbohydrates and least for oligonucleotides because of their low carbon and hydrogen content and the large number of (monoisotopic and mass-deficient) phosphorus atoms (Biemann, 1990).

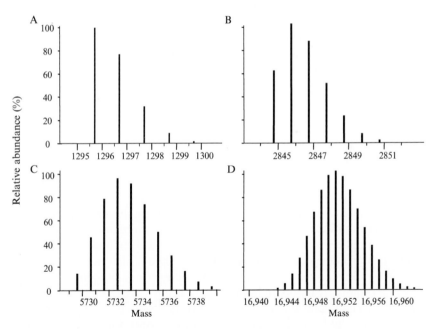

FIG. 1. Isotope distributions of selected peptides and one protein taking into account the naturally occurring $^{12}C/^{13}C$ ratio. (A) Angiotensin I, human ($M_{MONO} = 1295.6775$ u, $C_{62}H_{89}N_{17}O_{14}$). (B) Melittin, honeybee ($M_{MONO} = 2844.7542$ u, $C_{131}H_{229}N_{39}O_{31}$). (C) Insulin, bovine ($M_{MONO} = 5729.6009$ u, $C_{254}H_{377}N_{65}O_{75}S_6$). (D) apomyoglobin, horse ($M_{MONO} = 16940.9650$ u, $C_{769}H_{1212}N_{210}O_{218}S_2$).

In the mass range below 1500 u the monoisotopic mass of peptides is the most abundant mass with an asymmetry to larger masses. The isotope distribution loses its asymmetry with increasing mass. Particularly for proteins, the isotope distribution follows a Gaussian distribution (bell-shaped curve). For proteins above $\approx$15 ku, the abundance of the monoisotopic peak is negligible; for example, the monoisotopic peak of apomyoglobin (horse), a protein of mass 16951.49 u (average mass), is of low abundance (0.03%) and occurs at 16940.97 u, about 10.5 u beyond the average molecular mass of apomyoglobin. Figure 1 gives the isotope distributions of angiotensin I, melittin, insulin, and apomyoglobin. Sum formulae and organisms are given together with monoisotopic masses and average (molecular) masses, respectively, in the legend for Fig. 1. For apomyoglobin, (see Fig. 1D), the monoisotopic mass cannot be shown due to its low abundance.

Precision and Accuracy of Mass Determination

Precision of mass determination describes to what extent a mass measurement of a given compound can be reproduced. The precision is expressed by a statistic distribution (mean variation) of several independent mass determinations of a given analyte molecule; the average value of these independent mass measurements is given together with its standard deviation (σ_n). The accuracy of mass determination describes the accuracy of the measurement (i.e., the accuracy of mass determination expresses the deviation between the measured mass and the theoretical mass of the compound under investigation).

Both accuracy and precision are important characteristics of a mass spectrometric technique and account for its reliability. Achieving both high precision and high accuracy of mass determination are necessary to satisfy the demands on biological mass spectrometry, such as in protein identification by peptide mass mapping (Yates, 1998). Put simply, a high precision of mass determination allows reliance on a limited number of mass spectra of the same peptide map (high reproducibility of determined mass means a small standard deviation of measurement), while a high mass accuracy of mass determination results in a high probability of unambiguous identification of a protein by database interrogation using the masses of the peptide mass map. The higher the accuracy of mass determination of peptides in a peptide map, the fewer peptide masses are necessary for this approach.

Precision and accuracy of mass determination of peptides (1 to 5 ku) using ESI and MALDI with state-of-the-art mass analyzers are in the low parts per million range. For example, insulin β-chain (3494.651 u) was measured by Edmondson and Russell with a precision of 7.3 ppm

(0.026 u) and an accuracy of -2.6 ppm (-0.009 u) using MALDI–MS and applying an internal mass calibration (see the following paragraphs) (Edmondson and Russell, 1996; Russell and Edmondson, 1997). The same authors describe how accurate mass measurement can aid correct assignment of proteolytic fragments while also taking into account patterns of isotope distributions: Patterns can be uncommon compared to simple peptides of the same size, such as if a prosthetic group is bound to the peptide. Russell and Edmondson (1997) report of MALDI mass spectra obtained for a proteolytic digestion of cytochrome c with its covalently bound heme group. The heme-containing tryptic fragment of cytochrome c (Cys14 – Lys22, m/z 1633.620) can clearly be assigned by the relative abundances of the resolved isotopes and, therefore, distinguished from another potentially occurring fragment (Ile9 – Lys22, m/z 1633.820). The prosthetic group contains iron, and the observed (uncommon) isotope distribution is due to the iron-containing heme group (Edmondson and Russell, 1996; Russell and Edmondson, 1997).

A frequently discussed challenge for peptide mass analyses is to distinguish, for example, between Lys and Gln in a tryptic fragment of typical mass (1000–5000 u) or in a small protein only from the determination of its absolute molecular mass (and not by acetylating the compounds and deducing the number of Lys by the mass differences of 42 u). Because Lys and Gln differ by 0.0364 u, an accuracy of 36.4 ppm at mass 1000 u is required, whereas an accuracy of 3.6 ppm is required at mass 10,000 u. There are many other examples for amino acid mass coincidences requiring high accuracy of mass determination to distinguish different amino acid compositions just from peptide mass.

Mass Resolution and Resolving Power

A peak width definition as well as a 10% valley definition exist (Price, 1991). The *peak width definition* considers a single peak in a mass spectrum made up of singly charged ions at mass m. The resolution R is expressed as $m/\Delta m$, where Δm is the width of the peak at a height that is a specific fraction of the maximum peak height. A common standard is the definition of resolution R based upon Δm being full width at half maximum (FWHM). Considering a signal at mass $m = 1000$ u and a peak width of this signal of $\Delta m = 0.5$ u FWHM, mass resolution is $R = m/\Delta m = 2000$. The *10% valley definition* considers two peaks of equal height in a mass spectrum at mass m and $(m - \Delta m)$ that are separated by a valley that at its lowest point is just 10% of the height of either peak. The resolution is then $R = m/\Delta m$. It is usually a function of m, and therefore $m/\Delta m$ should be given for a number of different values of m (Price, 1991).

The ability of a mass analyzer to distinguish between ions differing slightly in mass-to-charge ratio is expressed by its resolving power. The resolving power is characterized by the peak width (in mass units) for at least two points on the peak (50% and 5% of the maximum peak height) (Price, 1991). High resolving power of a mass analyzer is neither a necessary prerequisite for accurate mass determination nor a sufficient one because accurate mass determination depends also on well-defined calibration standards of known mass, correct calibration procedures (see the following paragraphs), and precise (i.e., reproducible) signal peak shapes. However, high mass resolution helps to obtain high mass accuracy if the latter conditions are optimized and adapted to the analytical problem. As a rough guideline, if mass resolution is $R = 10,000$ at mass 1000 u, that is the peak width $\Delta m = 0.1$ u FWHM, the achievable accuracy of mass determination is about or better than one-tenth of the peak width Δm. This means that the mass of an ion at mass 1000 u can be determined with an accuracy of at least 0.01 u at 1000 u (10 ppm).

Calibration of Mass Spectra

The calibration of mass spectra can be performed internally or externally (i.e., the calibration peptides or proteins can either be within the sample containing the analyte, or the calibration compound and the analyte can be measured separately from each other using the same instrument configuration and voltages). It should be noted that a given mass calibration is only valid for the mass range covered by the standards.

In general, high precision of mass determination is achievable for both MALDI and ESI because triggering of electronics, stability of power supplies (AC and DC voltages), and other such factors are reliable (precise). The achievable mass accuracy, therefore, depends strongly on the quality of standards used for mass calibration. In the mass range below 30 ku a number of homogeneous peptides and proteins of known mass exist that can be used for mass calibration. Table I gives an overview of commonly used standard peptides and proteins. The average mass of a peptide or protein is abbreviated by M_{av}, while the monoisotopic mass of a compound is abbreviated by M_{MONO}.

Due to a typically occurring sample heterogeneity of proteins increasing with mass (a fundamental problem) and a heterogeneity induced by impurities and/or by the ionization technique itself (a practical problem), the accuracy of mass determination is limited at high mass range. An intrinsic protein heterogeneity typical for large proteins (posttranslational modifications such as phosphorylation and glycosylation) and sample impurities also in the standards (such as sodium and potassium) do not allow

TABLE I
MASSES, SUM FORMULAE, AND SWISSPROT REFERENCES OF SOME PEPTIDES AND PROTEINS
FREQUENTLY USED FOR MASS CALIBRATION

Peptide/Protein	Sum formula	Monoisotopic mass, M_{MONO} Average mass, M_{av}, u	SwissProt access
Angiotensin II, human, free acid	$C_{50}H_{71}N_{13}O_{12}$	$M_{MONO} = 1045.5345$ $M_{av} = 1046.19$	P01019
Bradykinin, human, free acid	$C_{50}H_{73}N_{15}O_{11}$	$M_{MONO} = 1059.5614$ $M_{av} = 1060.22$	P01042
Angiotensin I, human, free acid	$C_{62}H_{89}N_{17}O_{14}$	$M_{MONO} = 1295.6775$ $M_{av} = 1296.50$	P01019
Substance P, human, free acid, ... Met–OH	$C_{63}H_{97}N_{17}O_{14}S_1$	$M_{MONO} = 1347.7122$ $M_{av} = 1348.81$	P20366
Substance P, human, ... Met–NH$_2$	$C_{63}H_{98}N_{18}O_{13}S_1$	$M_{MONO} = 1346.7281$ $M_{av} = 1347.65$	P20366
Neurotensin, human, free acid	$C_{78}H_{121}N_{21}O_{20}$	$M_{MONO} = 1671,9097$ $M_{av} = 1672.95$	P30990
ACTH (CLIP), 18–39, human, free acid	$C_{112}H_{165}N_{27}O_{36}$	$M_{MONO} = 2464.1911$ $M_{av} = 2465.70$	P01189
Melittin, honeybee, ... Gln–NH$_2$	$C_{131}H_{229}N_{39}O_{31}$	$M_{MONO} = 2844.754$ $M_{av} = 2846.50$	P01501
ACTH (CLIP), 1–39, human, free acid	$C_{207}H_{308}N_{56}O_{58}S$	$M_{MONO} = 4538.2594$ $M_{av} = 4541.13$	P01189
Insulin, bovine, β-chain, oxidized, sulphated C (R-SO$_3$H)	$C_{157}H_{232}N_{40}O_{47}S_2$	$M_{MONO} = 3493.6435$ $M_{av} = 3495.94$	P01317
Insulin, bovine	$C_{254}H_{377}N_{65}O_{75}S_6$	$M_{MONO} = 5729.6009$ $M_{av} = 5733.58$	P01317
Ubiquitin, bovine	$C_{378}H_{629}N_{105}O_{118}S_1$	$M_{MONO} = 8559.6167$ $M_{av} = 8564.86$	P02248
Lysozyme, hen egg, oxidized form	$C_{613}H_{951}N_{193}O_{185}S_{10}$	$M_{MONO} = 14295.8148$ $M_{av} = 14305.14$	P00698
Apomyoglobin, horse	$C_{769}H_{1212}N_{210}O_{218}S_2$	$M_{MONO} = 16940.9650$ $M_{av} = 16951.49$	P02188
Carbonic anhydrase, bovine[a]	$C_{1312}H_{1998}N_{358}O_{384}S_3$	$M_{MONO} = 29005.6750$ $M_{av} = 29023.65$	P00921

[a] Assuming that amino acids 10, 100, and 101 assigned as Asx are actually Asp and that residue at position 13 designated Glx is actually Glu. See Senko, M. W., Beu, S. C., and McLafferty, F. W. (1994). *Anal. Chem.* **66**, 415.

reliance upon calibrations based on masses given in databases. Bovine serum albumin (BSA, 66 ku) used as a standard has often led to controversy due to its intrinsic sample heterogeneity (more than one sequence, ragged ends) and its high affinity for cations, both causing an inaccurate mass determination.

Nevertheless, peptides and proteins used for mass calibration should be used in a highly purified form to produce clean, well-resolved peaks in the mass spectrum and should be freshly prepared to obtain optimal results. If available, recombinant proteins are recommended for mass calibration of proteins; however, it should be remembered that such proteins are still sensitive to oxidation and other modifications as well as proteolysis during storage in solution. The content of alkali cations or inorganic anions can also vary among suppliers and vials. Both modifications and adduct formation result in moving of the peak centroid and peak height, respectively. Especially in ESI mass analysis, a minor amount of salts and detergents can weaken protein ion intensities by adduct formation and can degrade spectra quality. Independent of these arguments, the question remains whether the protein under investigation behaves in the same way as the calibration protein with respect to both adduct formation and decay induced by the ionization technique (for the latter, see upcoming text).

Although it is preferable to use peptides or proteins to calibrate peptide or protein signals to obtain high mass accuracy, mixtures of inorganic salts, such as CsI/NaI, can be used as calibration compounds in ESI. Signals from these calibration compounds are not affected by the effects previously discussed and allow for determination of a precise peak centroid because these salts are monoisotopic. It should be noted, however, that the use of inorganic salts often leads to source contamination, especially in ESI under flow rates in the microliters per minute range, and internal mass calibration is not recommended due to the cation formation already discussed.

Factors Affecting Mass Resolution and Mass Accuracy

Besides the problem of intrinsic mass heterogeneity, a signal heterogeneity can be induced by the ionization technique itself (a practical problem): adduct ion formation (e.g., cations, solvents, or matrix molecules getting attached to the analyte molecule) or small neutral loss (loss of H_2O or NH_3) from the calibration compound or from the compound of interest are the most important factors that can deteriorate mass calibration in MALDI–MS and ESI–MS. Adduct ion formation and small neutral loss can cause severe problems if mass resolution is insufficient and accurate signal assignment is no longer possible. The former leads to a shift of the peak centroid toward higher masses, while the latter leads toward lower masses. In addition, adduct ion formation and small neutral loss can cause particularly severe problems when the calibration compound and the compound of interest behave differently in adduct ion formation or decay.

In this context, explaining sample purification seems to be appropriate because purified calibration standards and analyte molecules are required for best results. Typically, sample cleanup procedures are required prior to mass analysis to obtain high mass accuracy. Samples must be freed from sodium and potassium as much as possible; often—in protein identification by peptide mass mapping—sample cleanup is essential to get rid of detergents and other biochemical additives that were required for proteolytic degradation in steps prior to mass analysis. MALDI–MS is known to be more tolerant toward common impurities than ESI–MS using flow rates in the microliters per minute range. In comparison to ESI, nanoelectrospray is significantly more tolerant toward common contaminants than ESI due to the formation of smaller, more highly charged droplets undergoing fissions at earlier stages (Juraschek et al., 1999). For both the MALDI–MS and ESI–MS techniques, sample cleanup procedures were established, including cation or anion exchange procedures and purification using reversed phase surfaces, which are on a microliter scale and are easy to use (Gobom et al., 1999; Kussmann et al., 1997).

Several other parameters can influence the achievable mass resolution of a given system; these parameters include sample preparation, purity of the ion source, density of ions (e.g., in the mass separating device, such as an ion trap or FTICR), and other such factors. The detection system can also limit mass resolution; for example, the postacceleration of large mass ions onto a conversion dynode followed by a secondary electron multiplier produces a variety of (nonresolved) secondary ions as well as electrons, and mass dispersion leads to a significant peak broadening (Spengler et al., 1990). Multichannel plate detectors (MCP) are therefore typically used for both ESI and MALDI mass analysis (Baldwin, 2005), and the conversion dynode approach appears superior only for very high m/z values produced by MALDI.

ESI–MS

In this section, the results and discussion are based mainly on quadrupole mass filters and orthogonal acceleration TOF (oa-TOF) mass analyzers for electrospray/ionization–mass spectrometry (ESI–MS) applications.

ESI: Sample Preparation

The general idea of an ESI sample preparation is to dissolve the analyte molecules (peptides or proteins) in a 1/1 (v/v) water/methanol or water/acetonitrile mixture, typically containing 0.5–1% acetic or formic acid (*denaturing conditions*). As a rough rule, the sample concentration is in the 10^{-6} M range (1 pmol/μl). The sample is continuously injected into an electric field via a metal capillary (Fenn *et al.*, 1989) (atmospheric pressure ESI) with flow rates of 1–5 μl per minute or via a metal-coated glass capillary (Wilm and Mann, 1994, 1996) (nanoelectrospray source) with flow rates in the low nanoliters per minute range. The electric field generates a mist of highly charged droplets containing analyte molecules. The droplets move down a potential gradient ΔU and a pressure gradient Δp—which both finally liberate analyte molecules from solvent molecules—toward the mass analyzer (Kebarle, 2000).

While the preparation conditions just described are appropriate for the determination of molecular masses of single peptide or protein chains, ESI–MS is being used more for the mass analysis of noncovalently bound complexes, such as multimeric proteins or protein/ligand complexes (Strupat *et al.*, 2000). To guarantee a complex in solution, mild solvent conditions must be chosen. Typically, the multimeric protein or protein/ligand complex is dissolved in an aqueous, buffered solution adjusted to appropriate pH-values (*native conditions*). Many noncovalently bound complexes are dissolved in and measured directly from aqueous solutions containing, for example, 5–25 mM $_{NH4Ac}$ with pH values between 5 and 8.5, depending on the analyte molecules under investigation (Loo, 1997; Smith *et al.*, 1997; Strupat *et al.*, 2000; Yao, 2005).

Mass Analyzers

The continuity of the spray in electrospray/ionization enables a scanning device, such as the quadrupole mass filter, to be conveniently used as a mass analyzer (Fenn *et al.*, 1989). Quadrupole mass filters (scanning filters) coupled to ESI sources, therefore, were the workhorses for ESI for many years in both off-line and on-line couplings (liquid chromatography [LC] or capillary electrophoresis [EC]–ESI–MS) approaches (Fenn *et al.*, 1989; Voyksner, 1997).

Triple quadrupole instruments offer the opportunity to obtain tandem MS (MS/MS) data; the first quadrupole serves as precursor ion selector, followed by a collision cell (hexapole in rf-only mode) in which collision-induced dissociation (CID) between analyte molecules and Argon (low-energy collision) are performed. The collision cell is followed by another quadrupole mass filter that separates the fragment ions by the m/z ratio. The coupling of ESI sources to other scanning devices, such as ion traps (van Berkel *et al.*, 1990) and magnetic sectors (Meng *et al.*, 1990a,b), which are also able to perform MS/MS measurements, is described in the literature.

The coupling of ESI sources to (pulsed) time-of-flight (TOF) analyzers in so-called orthogonal acceleration TOF setups (oa-TOF) is routinely used after the introduction of the oa-TOF by the groups of Guilhaus (Dawson and Guilhaus, 1989) and Dodonow (Dodonow, 1991). A recently published review by Guilhaus *et al.* reports on the principle instrumentation and different applications of the oa-TOF instrument (Guilhaus *et al.*, 2000). The main advantages of an oa-TOF over a quadrupole are the higher mass resolution, a higher achievable mass range, and a higher sensitivity (Guilhaus *et al.*, 2000). The higher sensitivity is achieved by the orthogonal accelerator, which is a highly efficient device for sampling ions from an ion beam into a TOF mass analyzer. A higher mass range is of interest because proteins or noncovalent complexes measured under native conditions (see previous paragraphs) require a high mass-to-charge range of the mass analyzer. The excellent performance of the oa-TOF mass analyzer for very high masses (noncovalent complexes) was first realized by the Manitoba group (Chernushevich *et al.*, 1999; Werner, 2005) and by Robinson *et al.* (Rostom and Robinson, 1999; Yao, 2005).

A further development of the oa-TOF instrumentation was made by the introduction of a hybrid instrument (quadrupole-TOF combinations, Q-TOF or qQ-TOF (Chernushevich *et al.*, 1999) that enabled MS/MS applications (Werner, 2005). ESI-produced ions are transferred to an analytical quadrupole, where a specific precursor ion is selected. The selected ion can be fragmented in the succeeding collision cell. Mass analysis of the fragment ions is performed in the adjacent oa-TOF.

Appearance of ESI–MS

ESI–MS generates highly charged ion species of peptides or proteins. An ESI mass spectrum is characterized by a number of signals that each differ by one charge. Such a distribution of charge states is typically produced by multiple protonation (positive ion mode) or deprotonation (negative ion mode) of the species, but cation formation is also known to occur. An ion signal of a species of mass M produced by n-fold protonation (mass of a proton m_H) has the mass $M + n \cdot m_H$; this signal is assigned as $(M + n H)^{n+}$ or M^{n+} (positive ion mode, $n \in N$) in the mass spectrum. Note that the signal occurs at mass-to-charge ratios $m/z = (M + n \cdot m_H)/n$, $(n \in N)$. Depending on the purity of the sample, undesired though typically less pronounced, sodium and potassium adduct ions, such as $(M + nH + mNa)^{(m+n)+}$ $(n + m > 0, n \in \mathbf{Z}, m \in N)$, also appear. An ion signal of an n-fold deprotonated species (of mass M) possesses a mass $M - n \cdot m_H$, which is assigned as $(M - n H)^{n-}$ or M^{n-} in the mass spectrum (negative ion mode).

Dissolving peptides and proteins under denaturing conditions (organic solvent plus water containing acid), as already described, results in a more or less pronounced unfolding of the analyte molecule in solution. The degree of unfolding depends on the features of the individual analyte; unfolding is typically limited for proteins with intramolecular bonds (disulfide bonds). Acetic or formic acid in the solvent ensures protonation of basic amino acids, and unfolding of the analyte molecule by organic solvent increases the number of achievable basic amino acids. As a rough rule, the number of basic amino acids determines the number of charges observed in positive ESI–MS.

Peptides are typically observed with two to five protons, $(M + 2H)^{2+}$, ..., $(M + 5H)^{5+}$, in the mass spectrum depending on the mass of the peptide and the number of its basic residues. The m/z ratios are typically between $m/z = 300$ and $m/z = 1500$. ESI–MS data of an equimolar mixture of angiotensin I, substance P, and neurotensin (1 pmol/μl each peptide, all peptides of human source) are shown in Fig. 2. The triply charged species $(M + 3H)^{3+}$ are observed predominantly for these three peptides; angiotensin I is also observed as fourfold-charged species ($m/z = 325$). The charge state information observed for the three peptides, therefore, is in agreement with the number of basic amino acids plus one charge for the N terminus. Angiotensin I has four likely protonation sites, while substance P and neurotensin have three potential protonation sites (i.e., the sum of basic amino acid residues (Lys, Arg, His) plus one protonation site at the amino terminus). Data shown in Fig. 2 were obtained using an ESI–oa-TOF mass analyzer (Mariner, Applied Biosystems, Framingham, MA, www.appliedbiosystems.com) equipped with a nanoelectrospray source. Typically, a mass resolution of $R = 6000$ FWHM for each peptide is observed as shown in the insets of Fig. 2. If the triply charged species of angiotensin I and neurotensin are used for calibration of the mass scale 300 $\leq m/z \leq 1000$, then the mass of substance P is determined with an accuracy of $\approx$10 ppm or 10^{-5} (Fig. 2, internal calibration). Using the calibrated mass spectrum for further mass analysis in subsequent spectra (external calibration), the accuracy of mass determination is still better than 100 ppm. For peptides, common mass analyzers enable the resolution of the corresponding isotope distribution (Fig. 2). To resolve adjacent isotopes is of particular interest because it allows the assignment of the charge state of an ion only by the spacing between adjacent signals. The spacing between the monoisotopic peak and the peak including one ^{13}C isotope equals to (charge state)$^{-1}$. In the example shown in Fig. 2, the spacing between adjacent signals of angiotensin equals to $m/z = 0.\overline{3}$; this corresponds to a charge state of 3. Knowing the charge state of a given peptide signal (see the following paragraphs) enables the determination of the monoisotopic mass of a peptide M_{MONO}.

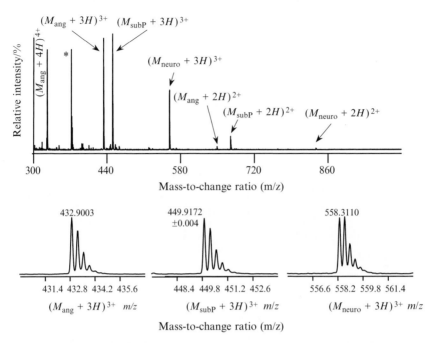

FIG. 2. ESI (nanoelectrospray) mass spectrum of three peptides: equimolar mixture of angiotensin I (M_{ang}), substance P (M_{subP}, Met–NH$_2$), and neurotensin (M_{neuro}) ([angiotensin] = [substance P] = [neurotensin] = 1 μM; oa-TOF mass analyzer). The triply charged species of the three peptides are enlarged, and the monoisotopic signals are assigned. Note that the spacing between adjacent isotopes ($m/z = 0.\bar{3}$) equals to the (charge state)$^{-1}$. Calibrating the spectrum internally with angiotensin and neurotensin as calibration peptides, the mass of substance P can be measured with a precision better than ±10 ppm. The peak labeled with the symbol * is due to an impurity (no peptide).

For proteins, a broader charge state distribution and more highly charged species are observed. As a rough guide, one charge per 1000 u is observed using denaturing solvent conditions, and the mass-to-charge range covered by the distribution is between $m/z = 500$ and $m/z = 2000$. The width of this charge state distribution is often about half that of the highest charge state (Smith *et al.*, 1990). For a protein such as myoglobin (apo-form, horse, 16951.49 u) (Zaia *et al.*, 1992), signals will appear between $m/z = 680$ and $m/z = 1700$ with 24 ($m/z = 679.06$, $(M + 25H)^{25+}$) to 10 ($m/z = 1696.15$, $(M + 10H)^{12+}$) charges (see Fig. 3).

The average mass M_{av} of the molecule is "reconstructed" from the ESI mass spectrum while taking into account the assumptions that (i) adjacent

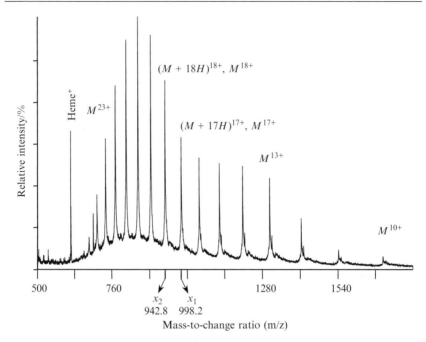

FIG. 3. ESI (nanoelectrospray) mass spectrum of apomyoglobin (horse heart) measured from a water/acetonitrile mixture containing 1% formic acid ([apomyoglobin] = 2 μM; oa-TOF mass analyzer). Some charge states are assigned. For x_1 and x_2, see text.

peaks differ by one charge and (ii) charging is due to the adduction of the same species (e.g., proton H^+). The basic equations for this reconstruction are given below although average masses or monoisotopic masses of analyte molecules are calculated easily by appropriate software provided with the mass spectrometer. The procedure described in the upcoming paragraphs is slightly simplified because it takes into account protonated ion species only; a more complete description is given by the "averaging algorithm procedure" described by Mann and coworkers that also takes into account cations as adducts and generalizes for negative ion mode (Mann *et al.*, 1989).

For reconstruction of the molecular mass of a protein, any two adjacent peaks are sufficient to determine the molecular mass of a species (Edmonds, 1990; Mann *et al.*, 1989); in this context, the redundancy of mass and charge state information contained in ESI mass spectra is notable. Referring to Fig. 2, the mass-to-charge ratios $m/z = x_1$ and $m/z = x_2$ are two adjacent members of an ion series obtained by protonation (m_H = 1.00794 u). The m/z ratios of the two ions can be expressed by the following:

$$x_1 = (M + z \cdot m_H)/z. \tag{1}$$

$$x_2 = (M + (z + 1) \cdot m_H)/(z + 1). \tag{2}$$

The factors in equations (1) and (2) can be defined as:

   M : molecular mass of the analyte molecule, M_{av} or M_{MONO}

   x_1 : m/z ratio of the analyte molecule with z protons

   x_2 : m/z ratio of the analyte molecule with $(z + 1)$ protons

Equations (1) and (2) require that x_1 is greater than x_2. Combining these equations and solving them for charge z allows the determination of the charge state z of the ion signal at $m/z = x_1$, as shown in the following equation:

$$z = x_2 - m_H/x_1 - x_2. \tag{3}$$

The number of charges z of the ion signal of $m/z = x_1$ allows the determination of the molecular mass M of the analyte molecule:

$$M = z \cdot x_1 - z \cdot m_H = (z + 1) \cdot x_2 - (z + 1) \cdot m_H \tag{4}$$

Taking the spectrum of Fig. 3 as an example, two adjacent charge states might be considered: $x_1 = 998.2$ and $x_2 = 942.8$; the charge state of x_1 is easily calculated to $z = 17$ from these numbers. The molecular mass of apomyoglobin (16951.49 u) can be determined with a precision of better than ± 50 ppm or $\pm 5 \cdot 10^{-5}$ (± 0.85 u) using external mass calibration.

Common mass analyzers (such as quadrupole mass filters or oa-TOF-mass analyzers) enable the observation of the envelope of the isotope distribution of proteins as in the case of apomyoglobin (Fig. 3), but these analyzers do not allow resolution of certain isotopes of the distribution (see Fig. 1). Therefore, the values of peak height or, better yet, of peak centroid (averaging over the peak area) of a given charge state are taken into account, and the experimentally determined value is the average mass, M_{av}.

Calibration of Mass Spectra

Denaturing Conditions. A standard mixture containing some peptides is typically used for mass calibration of peptides as analyte molecules. Due to the complexity of an ESI mass spectrum and of peptide mixtures (a number of different charge states show up for each peptide), external mass calibration might be preferred; that is, the standard mixture is run, the mass spectrum is then calibrated with the corresponding monoisotopic masses of the peptides, and, finally, the sample of interest is run under the same instrumental conditions. A similar procedure is performed for proteins. Apomyoglobin from horse heart is used as a calibration protein in many applications because it is well-defined (one amino acid sequence, no heterogeneity due to posttranslational modifications) and because it results in a wide charge state distribution (see Fig. 3). Bourell and colleagues report a mass determination accuracy of less than or equal to 200 ppm for purified antibody fragments of 100 ku that are engineered in a recombinant manner using a triple quadrupole mass analyzer and performing peak height measurement (Bourell *et al.*, 1994). Such an accuracy will reflect whether correct translation and proper posttranslational modification of the proteins are achieved (Bourell *et al.*, 1994).

Native Conditions. For the analysis of noncovalently bound multimeric proteins or protein-ligand complexes, samples are dissolved in aqueous, buffered solutions (see previous paragraphs) with the aim of preserving the (native) solution state of the complex. If compounds are mass analyzed under these conditions, mass calibration is typically a little more tedious because charge state distributions are narrower and shifted to higher m/z values and because well-defined standards that give intense signals at high m/z- values are rare. In addition, the quality of ion signals may be degraded with respect to mass resolution due to unresolved solvent and/or buffer molecules (e.g., acetate, ammonia, water) attached to the protonated ion species (Potier *et al.*, 1997; Rogniaux *et al.*, 1999). Lysozyme (14305.14 u, chicken egg white), when dissolved in pure water and measured from a 5 μM solution, produces nicely resolved peaks and a relatively wide charge state distribution from $z = 4$ to $z = 11$, which covers the mass range between $1300 < m/z < 3600$. It is, therefore, a reasonable calibration protein for medium-sized proteins measured under native conditions.

The noncovalent complex between the protein apomyoglobin and its heme group (616 u) is observed (myoglobin) if the protein is dissolved in an aqueous buffered solution (5–10 mM NH$_4$Ac). The intact complex is observed largely with two charge states at $m/z = 1953$, $(M + 9H)^{9+}$, and $m/z = 2197$, $(M + 8H)^{8+}$ (as shown in Fig. 4). Emphasized are the differences in the width of charge state distribution, most abundant charge states, and last, but not least, the determined molecular mass of apomyoglobin (16951 u, Fig. 3, denaturing conditions) and myoglobin (17568 u, Fig. 4, native conditions).

MALDI–MS

In this section, results and discussion are based mainly on axial acceleration (aa-TOF mass analyzers) for matrix-assisted laser desorption/ionization–mass spectrometry (MALDI–MS) applications (Gross and Strupat, 1998). The general idea of a MALDI sample preparation is to mix a solution of analyte molecules of relatively low concentration with a

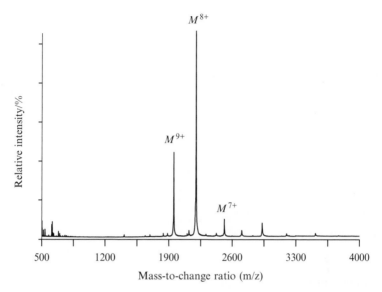

FIG. 4. ESI (nanoelectrospray) mass spectrum of myoglobin (horse heart) measured from an aqueous, buffered solution (10 mM NH$_4$Ac, pH 7) ([myoglobin] = 10 μM; oa-TOF mass analyzer). The noncovalent complex between the protein chain and the prosthetic group (heme) survives the transfer from the liquid to the gas phase.

so-called matrix solution of a relatively high concentration. The matrix— typically a small organic molecule—has mainly three tasks: (**i**) to strongly and resonantly absorb the irradiated laser wavelength, (**ii**) to force separation of analyte molecules from each other (matrix isolation), and (**iii**) to help or to initiate analyte ionization (Karas and Hillenkamp, 1988; Karas *et al.*, 1985). The matrix finally enables a desorption independent of the individual features of the investigated analyte molecule. The requirements for the matrix are as follows: First, to absorb the laser light strongly and resonantly, the matrix compound is an aromatic system in the case of ultraviolet-MALDI (UV-MALDI, electronic excitation) or an aromatic or aliphatic system in the case of infrared-MALDI (IR-MALDI, rotational-vibration excitation). Specific matrix compounds can be derived by various functional groups (-OH, -NH$_2$, OCH$_3$, etc.) to accommodate the absorption of the matrix to the irradiating laser wavelength. Second, the required analyte separation (matrix isolation) is achieved by a large molar excess of the matrix and a molar matrix-to-analyte ratio between 10^3 and 10^6 in the final sample preparation. Therefore, the matrix concentration is $\approx 10^{-1}$ M, while the analyte concentration is in the range between 10^{-5} M and 10^{-7} M, depending on the size of the analyte molecule and its purity. In other words, a peptide or protein concentration of 0.1 g/l (10^{-4} M for mass 1000 u and 10^{-6} M for mass 100,000 u) is sufficient for a successful MALDI mass analysis. Third, it is assumed that the choice of the matrix plays an important role in analyte ionization, and a proton transfer from the electronically excited matrix compound to the analyte molecules in the expanding plume might be responsible for ionization; however, a model taking into account the different physical properties of UV– and IR–MALDI has yet to be explored (Ehring et al., 1992; Karas et al., 2000; Zenobi and Knochenmuss, 1999).

MALDI Sample Preparation

Two different main sample preparation techniques should be distinguished and can lead to very different sample morphologies, which influence the achievable precision and accuracy of mass determination. The fast evaporation or thin layer preparation introduced by Vorm et al. (Vorm and Mann, 1994; Vorm et al., 1994) results in a very homogeneous sample morphology. This preparation technique is suited to matrix compounds that are almost water insoluble, such as α-cyano-4-hydroxy cinnamic acid (ACCA) (Beavis et al., 1992). The MALDI sample is prepared by first producing a thin matrix layer on the target and afterward spotting the analyte solution on top of this matrix layer. For this purpose, a saturated matrix solution dissolved in acetone is spread over the target; the solvent evaporates quickly, leaving a thin, dry matrix layer behind. The analyte solution (preferentially in a slightly acidic solution containing 0.1% trifluoroacetic acid or TFA) is added on top of the matrix layer that is not dissolved completely by the solvent. Whether this results in analyte incorporation into the ACCA matrix crystals or the analyte molecules are only attached to the matrix surface is still debated (Horneffer et al., 1999). The thin layer preparation technique using the ACCA matrix is preferentially used for peptide mass mapping with the aim of protein identification. The ACCA matrix is not a matrix of choice for high mass compounds, such as proteins analyzed in a reflector TOF, because the ACCA matrix tends to induce a considerable amount of metastable fragmentation of analyte molecules (Karas et al., 1995).

The dried droplet preparation is best suited for water-soluble matrix compounds, such as 2,5-dihydroxybenzoic acid (2,5-DHB), or is best suited for mixtures with 2-hydroxy-5-methoxy benzoic acid (DHBs) (Karas et al., 1993), 3-hydroxy picolinic acid (both UV–MALDI), or succinic acid (IR–MALDI) and results in a more heterogeneous sample morphology.

The MALDI sample is prepared by mixing analyte and matrix solution directly on the target and air-drying the sample. The cocrystallization of matrix and analyte results in matrix crystals with dimensions of the 100-μm range or larger. Crystals can tower into the acceleration region of the ion source, which limits precision of mass determination, if mass spectra are taken from different spots. Typically a dried droplet preparation leads to a more or less pronounced hot spot phenomenon (i.e., signal intensity can differ quite dramatically within one given sample preparation). This is especially pronounced for the 3-high-performance addressing (HPA) matrix.

Several helpful hints about sample preparation, including sample purification steps on a microliter scale prior to matrix incubation and on-target reactions such as dithiothreitol (DTT)-reduction, are described in the literature (Gobom et al., 1999; Kussmann et al., 1997).

Last, but not least, some comment about the liquid IR matrix glycerol seems appropriate. Glycerol acts as an IR matrix due to its three hydroxyl groups and can be used at a laser wavelength of 3 μm as well as of 10.6 μm (Berkenkamp et al., 1997; Menzel et al., 1999). Glycerol has the advantage of also being a liquid in a vacuum, and no cocrystallization with the analyte molecules occurs. The laser beam irradiates the liquid sample and always finds a "healed" surface from which analyte molecules can be desorbed. In practice, the observed ions—singly to more highly charged monomeric ions or singly charged oligomers of the analyte molecules—depend on the molar matrix-to-analyte ratio chosen (Berlenpamp, 2000; Berkenkamp et al., 1997; Menzel et al., 1999). Molar matrix-to-analyte ratios are between 10^4 and 10^7. The lower the molar glycerol excess, the more pronounced are the oligomeric states (most likely gas-phase induced) and the less pronounced become doubly or triply charged ions of the analyte molecule.

Mass Analyzers. The laser-pulsed MALDI source is most suitable for TOF mass analyzers. Therefore, TOF mass analyzers in an axial acceleration (aa-TOF) geometry were the first to be employed in this ionization technique (Hillenkamp et al., 1991). Peptide mass mapping is straightforward and the ease of interpretation of MALDI mass spectra (see *appearance of mass spectra*) often avoids the need to couple to the chromatographic separation of peptides prior to mass analysis.

A state-of-the-art MALDI–TOF instrument is equipped with both a linear port (ion source, field-free drift region, and detector are in a linear row) and a reflector port (which divides the field-free drift region by an electrostatic mirror that compensates for energy deficits of ions of the same m/z ratio) and, most importantly, with the possibility for delayed ion extraction in the MALDI source. In particular, the introduction of delayed ion extraction has improved the quality of MALDI mass spectra significantly in terms of mass resolution R (Brown and Lennon, 1995; Colby et al., 1994; Vestal et al., 1995; Whittal and Li, 1995). Basically, delayed ion extraction (DE–MALDI) compensates for the initial velocity distribution (Beavis and Chait, 1991) of MALDI-produced ions (Juhasz et al., 1997).

Mass resolution of more than $R = 10,000$ is achievable for peptides in the mass range up to 5000 u (Vestal et al., 1995), and DE–MALDI-produced ions of a peptide mass map can be determined with an accuracy of 10 to 50 ppm (0.01 u to 0.05 u in 1000 u) or better. Such a high mass accuracy dramatically increases the specificity of database interrogation, and identification of proteins can be achieved unambiguously if at least five peptide masses are determined with better than 50 ppm accuracy (Clauser et al., 1999; Jensen et al. 1996; Shevchenko et al., 1996).

In contrast to the first conclusions made after the introduction of the technique—that MALDI-produced ions would be extremely stable and that no fragment ions would be observed—postsource decay (PSD) analysis allows investigation and identification of structural fragment ions of peptides up to 3000 u that result from decay taking place in the field-free drift region of the mass spectrometer after leaving the ion source (Kaufmann et al., 1996; Spengler et al., 1992). The complexity of PSD spectra, the relatively low abundance of fragment ions, and, most importantly, the limited mass accuracy still make PSD-based peptide sequencing for protein identification difficult, at least for high-throughput analysis (Spengler, 1997).

A MALDI source has been coupled to an oa-TOF arrangement (Krutchinsky et al., 1998) and to a qQ-TOF setup (Krutchinsky, 1998; Loboda, 1999). The oa-TOF geometry substantially decouples the desorption process of MALDI (ions with large initial velocity distribution) from the subsequent mass analysis in the TOF. This facilitates mass calibration for both MS and MS/MS applications. MALDI-produced ions are cooled in the collisional damping interface (q) and transferred to the analytical quadrupole (Q), which is operated to transmit the ions to the oa-TOF (peptide mass map) or a precursor ion is selected that is fragmented in the succeeding collision cell (MS/MS) (Werner, 2005). Details on the design and performance of the qQ-TOF equipped with a MALDI-source have been published by Loboda et al. (2000). The power of this instrumentation and its promising impact on proteomics by MALDI-produced ions were described recently by Shevchenko et al. (2000). With high mass resolution R ($R = 10000$ FWHM) and high mass accuracy (10 ppm) of both, MS and MS/MS spectra are provided by this approach to peptide mass analysis (Shevchenko et al., 2000).

MALDI sources have also been coupled to other mass analyzers, such as ion traps (Doroshenko et al., 1992; Qin et al., 1996), magnetic sectors

(Hill *et al.*, 1991), and double-focusing instruments (combined with an oa-TOF) (Bateman *et al.*, 1995), allowing the performance of MS and MS/MS applications.

Appearance of MALDI Mass Spectra

When accumulating spectra in the positive ion mode, single protonation of peptide and protein species is the most frequent mechanism. The most abundant ion signal is, therefore, assigned as $(M + H)^+$ or M^+. Depending on sample purity, less pronounced sodium and potassium adduct ions $(M + Na)^+$ or $(M - H + Na + K)^+$ occur and broaden the signal to the higher mass side. This becomes a severe problem with increasing mass of the protein, because cation adducts cannot be resolved any more from the protonated species.

The singly charged (protonated) ion species M^+ is accompanied by less intense doubly and more highly charged ions (M^{n+}) and some still less abundant singly or more highly charged (most likely) gas phase-induced oligomers (mM^{n+}) of the species. The fact that MALDI produces predominantly singly charged ions significantly facilitates mass spectra interpretation compared to what is possible with ESI if more than one species is present in the sample. Even in the analysis of mixtures, signals corresponding to the same analyte molecule are easy to assign, making MALDI mass analysis most straightforward for peptide mass mapping (Clauser *et al.*, 1999; Jensen *et al.*, 1996; Shevchenko *et al.*, 1996).

In peptide mass analysis and positive ion mode, signals corresponding to singly charged protonated species $(M + H)^+$ are almost the only signals obtained. The information in Fig. 5 shows the same peptide mixture as shown in Fig. 2, but the data in Fig. 5 result from using an ACCA matrix prepared as a thin layer analyzed by MALDI–MS; angiotensin I, substance P, and neurotensin are added on top of the dried matrix layer. Each peptide has a concentration of 1 μM. Interestingly, only singly charged species appear in the mass spectrum (as protonated species and, although less abundant, as sodium and potassium attached species) together with matrix signals. Signals corresponding to dimers $(2M^+)$ and doubly charged species (M^{2+}) are not observed. This finding is very typical for the mass analysis of peptides by MALDI–MS.

With increasing mass, more highly charged signals, such as $(M + 2H)^{2+}$ or $(M + 3H)^{3+}$, and singly or doubly charged oligomers, such as $(2M + H)^+$ or $(3M + 2H)^{2+}$, may show up in the mass spectrum. To provide an example of a medium-sized protein, Fig. 6 shows apomyoglobin (3 μM) prepared with DHBs matrix (Bahr *et al.*, 1997). The inset shows the singly charged ion species; a peak width at half maximum of $\Delta m = 9.6$ u is

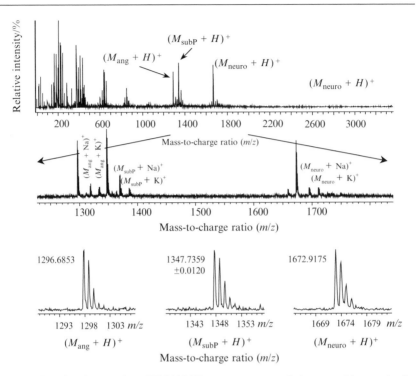

FIG. 5. Delayed extraction UV–MALDI mass spectrum of three peptides: equimolar mixture of angiotensin I (M_{ang}), substance P (M_{subP}), and neurotensin (M_{neuro}) ([angiotensin] = [substance P] = [neurotensin] = 1 μM). ACCA thin-layer preparation; aa-TOF mass analyzer. The singly charged species of the three peptides are enlarged, and the monoisotopic signals are assigned. Calibrating the spectrum internally with angiotensin and neurotensin as calibration peptides, the mass of substance P can be measured with a precision better than ±10 ppm.

obtained (this corresponds to a mass resolution of $R = m/\Delta m = 1765$), which is still broader than the envelope of the isotope distribution of apomyoglobin (see Fig. 1). A mass resolution of $R = 3500$ would be required to obtain the envelope of the isotope distribution of apomyoglobin (Bahr *et al.*, 1997).

Figure 7 shows various data, taken from the literature, obtained by UV– and IR–DE–MALDI of the four compounds simulated in Fig. 1. The agreement of the isotopic distributions of angiotensin I (Fig. 7A, mass resolution $R = m/\Delta m = 8600$), melittin (Fig. 7B, mass resolution $R = m/\Delta m = 9500$), and insulin all obtained by UV–DE–MALDI (Fig. 7C, mass resolution $R = m/\Delta m = 12500$) is obvious. For apomyoglobin obtained by UV–DE– MALDI (Fig. 7D), a mass resolution of $R = m/\Delta m = 1765$ is obtained (Bahr *et al.*, 1997).

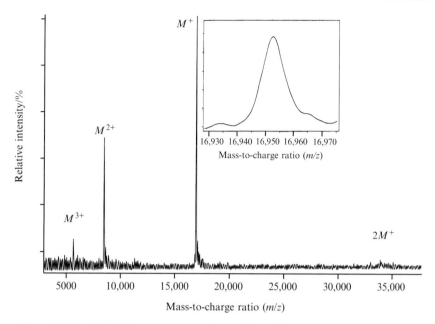

FIG. 6. Delayed extraction UV–MALDI mass spectrum of apomyoglobin (horse heart) ([apomyoglobin] = 3 μM; aa-TOF mass analyzer). Use of DHBs matrix. Singly, doubly, and triply protonated species are shown together with a very low abundant dimer, and the singly charged species is enlarged. (Refer to acknowledgments).

Calibration of Peptide Mass Spectra

For the most frequent application—peptide mass mapping by UV–DE–MALDI–MS—a standard containing peptides is used for external mass calibration, or the calibration is performed internally using autolysis products of the enzyme (e.g., 2163.057 u, autolysis product of trypsin) and matrix signals (Jensen *et al.*, 1996). The ACCA matrix, using a thin-layer preparation technique, is often used for this kind of analysis because it introduces a more homogeneous sample morphology than matrices better suited to dried droplet preparation, such as the 2,5-DHB or DHBs matrices. Therefore, the ACCA matrix is best suited for rapid mass finger printing.

To determine how accurately masses of peptides can be measured over a wide mass range, Takach *et al.* (1997) investigated a peptide mixture containing 12 standard peptides in the mass range between 900 u and 3700 u. Mass resolution of each peptide signal is between $R = 7500$ and $R = 10,000$ in the reported UV–MALDI mass measurements. Using two of the peptides (904.4681 u and 2465.1989 u) to calibrate the mass scale

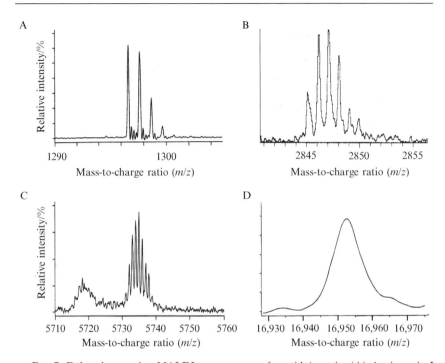

FIG. 7. Delayed extraction MALDI mass spectra of peptide/protein. (A) Angiotensin I (human), ACCA matrix. (B) Melittin (honeybee), succinic acid. (C) Insulin (bovine), sinapic acid matrix. (D) Apomyoglobin (horse), DHBs matrix. The singly protonated species are shown ([peptide/protein] = 1 - 5 μM; axial-TOF as mass analyzer). For graphs A and C, the spectra are according to Vestal *et al.* (1995); for graph B, the spectra are according to Berkenkamp *et al.* (1997); and for graph D, the spectra are according to Bahr *et al.* (1997). Reprinted with permission from John Wiley & Sons, New York. Graph B.

internally results in mass errors of less than 6 ppm for all peptides in a single mass spectrum (Takach *et al.*, 1997).

High accuracy of mass determination in peptide mass analysis allows one to distinguish between Gln and Lys in peptides on the basis of peptide mass. This is demonstrated by Takach *et al.* (1997) for the renin inhibitor K10 (1318.6737 u) and the renin inhibitor Q10 (1318.6773 u), respectively. The mass accuracy required to distinguish the chosen peptide masses is 27.5 ppm; the experimental results differ from the theoretical values by 1.5 ppm or less, using internal mass calibration (Takach *et al.*, 1997).

Mass calibration can be performed externally; that is, the standard is run, the mass spectrum is then calibrated with the corresponding mono-isotopic masses, and, finally, the sample of interest is run under the same instrumental conditions. Consequences and limitations with respect to mass accuracy resulting from external mass calibration of MALDI data are mainly due to the heterogeneity of sample preparation both from spot-to-spot and from preparation-to-preparation. Russell and Edmondson discuss the influence of mass resolution and peak shapes on accurate mass assignment in the mass analysis of peptides; they have obtained an accuracy better than 5 ppm for internal and 10 to 15 ppm for external mass calibration in the mass range between 1 and 4 ku while achieving a mass resolution of $R = 10,000 - 15,000$ in the mass range of peptides (Edmondson and Russell, 1996; Russell and Edmondson, 1997).

Calibration of Protein Mass Spectra

In protein mass analysis, precision and accuracy of mass determination drops to values of 100 to 1000 ppm when using static ion extraction. Beavis and Chait reported results of UV–MALDI mass analysis of proteins in the mass range up to 30 ku and determined the mass of bovine pancreatic trypsinogen to 23980.3 ± 2.6 u, which is equal to a precision of 110 ppm and an accuracy of -29 ppm; the mass of protease subtilisin Carlsberg (*Bacillus subtilis*) was measured to (27288.2 ± 1.7 u), which is equal to a precision of 62 ppm and an accuracy of -7.3 ppm (Beavis and Chait, 1990). However, both the accuracy and precision drop down for proteins with increasing mass due to their intrinsic heterogeneity and their decay and adduct formation. For monoclonal antibodies (150 ku), Siegel *et al.* (1991) demonstrated that a precision between 100 and 700 ppm is achievable using nicotinic acid as UV–MALDI matrix.

Due to the introduction of delayed ion extraction and the higher mass resolution thereby obtained, precision of mass determination of proteins above 20 ku has been increased to ±50 ppm for successive measurements from a given spot and is still ±200 ppm for several spots in between one preparation of a DHBs matrix rim, as shown by Bahr *et al.* (1997). The accuracy of mass determination is in the 100 ppm range for proteins exceeding 25,000 u. The best conditions were explored for weak extraction field strengths and long delay times (Bahr *et al.*, 1997). By far the best results with respect to matrix choice were obtained using a DHBs matrix; the strength of this matrix for high mass compounds is due to the fact that it does not transfer much energy into the analyte molecules (as the ACCA matrix does, for example) and therefore prevents or reduces decay of analyte molecules in the field-free drift region of the mass analyzer. The softness of the DHBs matrix (Karas *et al.*, 1993) compensates for the heterogeneous sample morphology obtained using this type of matrix. Bahr *et al.* (1997) investigated α-amylase (from

Bacillus amyloliquefaciens) by DE–MALDI–MS using the DHBs matrix. A sharp peak of the singly protonated species is obtained and is accompanied by matrix adducts attached to the protein (see Fig. 8). From the peak shape obtained it is obvious that several adducts contribute to the shoulder at the high mass side of the peak. The average mass of α-amylase was determined from six measurements on different spots performing an external calibration using bovine carbonic anhydrase as a calibration protein (M^+, $2M^+$). A mass of 54850 ± 9 u was determined for α-amylase, which is in very good agreement with literature data of 54851 u (accuracy −18 ppm, precision ±164 ppm).

There may be several reasons why mass accuracy and precision of DE–MALDI data are still limited for proteins above 50 ku (Bahr *et al.*, 1997). These reasons include an increase in protein heterogeneity with increasing mass (covalent modifications such as glycosylation or phosphorylation, ragged ends, etc.), ion formation between analyte molecules and

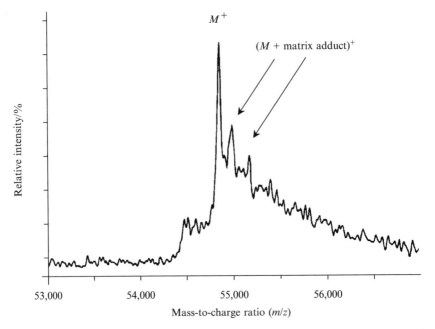

FIG. 8. Delayed extraction MALDI mass spectrum of α-amylase (*B. amyloliquefaciens*) (DHBs matrix; [α-amylase] = 3 μM). Use of aa-TOF mass analyzer. The singly protonated species is shown together with matrix adducts attached to the molecular ion. Spectrum according to Bahr *et al.* (1997). Reprinted with permission from John Wiley & Sons, New York.

matrix molecules (matrix adducts), and an increasing contribution of the initial energy and energy distribution with increasing analyte mass (Bahr et al., 1997).

Calibration of Mass Spectra Using IR–MALDI

In static extraction using a reflectron TOF mass analyzer, the precision of mass determination of IR–MALDI is typically 400 to 700 ppm for molecular masses up to 150 ku using solid matrices, while the precision is a little better, at 200 to 500 ppm, using the liquid matrix glycerol. Using solid matrices precision is mainly limited by the need for frequent changes of desorption location, as is typical for IR–MALDI. Precision in IR–MALDI is, therefore, one order of magnitude worse for analyte molecules below 30 kDa using static extraction compared to UV–MALDI, while it is better by a factor of 2 for larger analyte molecules. This finding might be explained by the typically better mass resolution of IR–MALDI ion signals compared to UV–MALDI ion signals of high mass compounds, which is a consequence of the lower yield of metastable fragmentation in IR–MALDI compared to UV–MALDI. For proteins up to 40 ku, mass accuracy is about 100 to 500 ppm, while it is limited to 1000 to 5000 ppm for analytes exceeding 40 ku (Berkenkamp et al., 1997).

Mass calibration of high mass compounds in IR–MALDI can be performed conveniently by using lysozyme (chicken egg white) desorbed from glycerol matrix. Depending on the mass range of interest, slightly different molar lysozyme-glycerol ratios are prepared, resulting in more or less pronounced oligomeric signals of lysozyme of the type $(nM + H)^+$ up to 200,000 u. This performance of IR–MALDI using a glycerol matrix permits calibration of higher mass ranges; however, high accuracy of mass determination is still desirable: lysozyme was used to calibrate two different chondroitinase enzymes that digest a polysaccharide part of the eye's proteoglycan. The molecular masses of these two enzymes were calculated from their c-DNA-derived sequences; chondroitinase I has a molecular mass of 112,508 u, while chondroitinase II has a molecular mass of 111,713 u. IR–MALDI determined the molecular masses with an accuracy of -1600 ppm (-185 u) using static ion extraction (Berlenpamp, 2000; Kelleher et al., 1997).

Using delayed extraction, the achievable accuracy of mass determination of a mixture of peptides was tested for succinic acid matrix (2.94 μm) (Berkenkamp et al., 1997), glycerol matrix (2.94 μm) (Berkenkamp et al., 1997), and fumaric acid (10.6 μm) (Menzel et al., 1999). All three matrices resulted in an accuracy of 10 ppm or better using internal mass calibration. With this respect, accuracy of mass determination of peptides achieved by

IR–MALDI–MS is in excellent agreement with values obtained by UV–MALDI–MS and ESI–MS.

FTICR Mass Analyzers Coupled to ESI and MALDI

The underlying physical principle of mass determination by a FTICR mass analyzer is the relationship between the cyclotron frequency (v_c) of an ion in a magnetic field of the magnetic field induction (B) and the mass-to-charge ratio (m/z) of the ion: $v_c \propto (m/z)^{-1}$ (Marshall and Grosshans, 1991). The accuracy of mass determination by FTICR mass analyzers is potentially ultra-high because frequencies can be measured more accurately than any other physical property. Mass calibration is performed by the determination of the cyclotron frequency of a calibration peptide or protein. The cyclotron frequency of the analyte compound under investigation is compared to that of the calibration molecule. As a rough guide, mass accuracy performing an external mass calibration is about 10 ppm or better (Li et al., 1994), while applying mass calibration is better by a factor of 3 (Wu et al., 1995). The masses of the most abundant isotopes of the two chondroitinases (112,508 u and 111,713 u) mentioned earlier could be determined with an accuracy of ± 3 u (± 2.6 ppm, external mass calibration) by Kelleher et al. (1997).

Although ESI is the more frequently used ionization technique employed with FTICR mass analyzers, MALDI can also be applied with this kind of mass analyzer. The challenge to couple MALDI-produced ions to FTICR mass analyzers lies in the relatively broad mass-independent velocity distribution of MALDI ions (Beavis and Chait, 1991). The kinetic energy dependent on velocity distribution makes an efficient trapping of ions in the FTICR cell more difficult (Hettich and Buchanan, 1991; Li et al., 1996).

The striking feature of FTICR mass analyzers is their ultra-high mass resolution ($R > 10^5$ at 1000 u). The mass resolution increases with the applied magnetic field induction B, which motivates the performance of experiments with higher and higher values of B. The high mass resolution achievable with FTICR mass analyzers allows the determination of the charge state of high molecular compounds, such as proteins, directly from one single charge state: the distance between isotopic signals is now also resolved for larger compounds (compare to Fig. 2 and Fig. 3), and adjacent ion signals differing by one mass unit (1 u) have a distance of (charge state)$^{-1}$.

This feature greatly simplifies the mass assignment of an ion in ESI (Beu et al., 1993; Senko et al., 1996; Speir et al., 1995). Figure 9 shows the

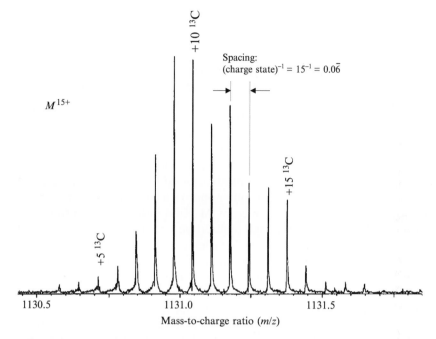

FIG. 9. Inset of an ESI mass spectrum of the 15-fold charged apomyoglobin (horse heart), m/z 1131.1, $(M + 15H)^{15+}$, using an FT–ICR mass analyzer. Adjacent isotopes of the distribution can be distinguished; the spacing correlates to (charge state)$^{-1}$. Figure according to Beu *et al.* (1993). Reprinted with permission from Elsevier Science.

signal of the 15-fold charged apomyoglobin, $(M + 15H)^{15+}$, obtained by an FTICR mass analyzer (data taken from the literature). Mass resolving power is $9 \cdot 10^5$ (Beu *et al.*, 1993). The irregular shape of the isotopic distribution is due to a small ion population because a few thousand ions only contribute to the entire isotope profile shown in Fig. 9 (Beu *et al.*, 1993).

Because ions can be stored in the trap, MS^n investigations can be performed in an FTICR analyzer using a variety of different approaches, such as SORI–CAD (Senko *et al.*, 1994), BIRD (Price *et al.*, 1996), IR–MPD (Little *et al.*, 1994), and ECD (Zubarev *et al.*, 1998), which are applicable to multiply charged ions obtained by an ESI source (Williams, 1998). The possibilities and features mentioned in the previous paragraphs make ESI–FTICR–MS an important, valuable tool in biological mass spectrometry and, moreover, in life sciences.

Addendum

Mass Analyzers

A variety of different analyzers or combinations of analyzers (hybrid instruments) have been applied very successfully for both ESI and MALDI ionization techniques, and most of them are commercially available. These include quadrupole time-of-flight instruments available with exchangeable ion sources, making them applicable to a large spectrum of analytical questions (Chernushevich et al., 2001). The TOF–TOF instrument equipped with a MALDI source and a collision cell for high energy collisions allows the identification of up to 10,000 proteins per day (Medzihradszky et al., 2000). Another approach to identifying proteins via a sequencing of specific peptides of MALDI-induced PSD products is opened by the LIFT technology realized on an aa-reflectron TOF; PSD-spectra can be accumulated much faster because all fragment ions are found for one given reflectron potential (La Rotta, 2001). In addition, a hybrid instrument that is a combination of a linear ion trap followed by a FTICR mass spectrometer has also been introduced. The LTQ–FT is capable of detecting fragment ions in the linear trap (low mass resolution and accuracy) of a given precursor, while (in parallel) the corresponding precursor is detected with high mass resolution and ultra-high mass accuracy (0.5–2 ppm) (Olsen and Mann, 2004). The future will show how these analyzers will answer the variety of analytical questions in life sciences with respect to factors such as speed, sensitivity, dynamic range, and mass accuracy.

MALDI Sample Preparation

When addressing the question whether analyte molecules should be incorporated into the solid matrix to enable a successful desorption/ionization event of analyte molecules by MALDI, it should be mentioned that results indicate the necessity of a chemisorption of analyte molecules (i.e., only partly incorporated) into the matrix surface to enable MALDI. A physisorption (a simple deposition of analyte molecules onto the matrix surface) is not sufficient (Gluckmann, 2001; Horneffer, 2001).

Acknowledgments

ESI mass spectra of the peptide mixture, of apomyoglobin and myoglobin, were accumulated using a Mariner instrument (oa-TOF) equipped with a nanoelectrospray source. The instrument was kindly given to chapter author Kerstin Strupat for research work; this support by Applied Biosystems is highly appreciated. Strupat wants to thank Dr. Jonathan

Butler (Laboratory of Molecular Biology, Cambridge, United Kingdom) for proofreading the manuscript and helping with the English language. The MALDI mass spectrum of apomyoglobin (Fig. 6) was kindly donated by Dr. Ute Bahr and Dr. Michael Karas from the University of Frankfurt, Germany. Dr. Michael Mormann (Institute for Medical Physics and Biophysics, Biomedical Analytics, University of Münster) is also thanked for the discussion about FTICR instrumentation.

References

Bahr, U., Stahl-Zeng, J., Gleitsmann, E., and Karas, M. (1997). Delayed extraction time-of-flight MALDI mass spectrometry of proteins above 25,000 Da. *J. Mass Spectrom.* **32,** 1111–1116.

Baldwin, M. L. (2005). Mass spectrometers for the analysis of biomolecules. *Methods in Enzymology* **405,** 172–187.

Bateman, R. H., Green, M. R., Scott, G., and Clayton, E. (1995). A combined magnetic sector-time-of-flight mass spectrometer for structural determination studies by tandem mass spectrometry. *Rapid Commun. Mass Spectrom.* **9,** 1227–1233.

Beavis, R. C., and Chait, B. T. (1990). High-accuracy molecular mass determination of proteins using matrix-assisted laser desorption mass spectrometry. *Anal. Chem.* **62,** 1836–1840.

Beavis, R. C., and Chait, B. T. (1991). Velocity distributions of intact high mass polypeptide molecule ions produced by matrix assisted laser desorption. *Chem. Phys. Letters* **181,** 479–484.

Beavis, R. C., Chaudhary, T., and Chait, B. T. (1992). A-Cyano-4-hydroxycinnamic acid as a matrix for matrix-assisted laser desorption mass spectrometry. *Org. Mass Spectrom.* **27,** 156–158.

Berkenkamp, S., Menzel, C., Karas, M., and Hillenkamp, F. (1997). Performance of infrared matrix-assisted laser desorption/ionization mass spectrometry with lasers emitting in the 3 mm wavelength range. *Rapid Commun. Mass Spectrom.* **11,** 1399–1406.

Berlenpamp (2000). Dissertation. University of Munster, Germany.

Beu, S. C., Senko, M. W., Quinn, J. P., Wampler, F. M., III, and McLafferty, F. W. (1993). Fourier-transform electrospray instrumentation for tandem high-resolution mass spectrometry of large molecules. *J. Am. Soc. Mass Spectrom.* **4,** 557–565.

Biemann, K. (1990). *In* "Methods in Enzymology" (J. McCloskey, ed.). Academic Press, New York.

Bourell, J. H., Clauser, K. P., Kelley, R., Carter, P., and Stults, J. T. (1994). Electrospray ionization mass spectrometry of recombinantly engineered antibody fragments. *Anal. Chem.* **66,** 2088–2095.

Brown, R. S., and Lennon, J. J. (1995). Mass resolution improvement by incorporation of pulsed ion extraction in a matrix-assisted laser desorption/ionization linear time-of-flight mass spectrometer. *Anal. Chem.* **67,** 1998–2003.

Chernushevich, I. V., Ens, W., and Standing, K. G. (1999). Orthogonal injection TOFMS for analyzing biomolecules. *Anal. Chem.* **71,** 452A–461A.

Chernushevich, I. V., Loboda, A. V., and Thomson, B. A. (2001). An introduction to quadrupole-time-of-flight mass spectrometry. *J. Mass Spectrom.* **36,** 849–865.

Clauser, K. R., Baker, P., and Burlingame, A. L. (1999). Role of accurate mass measurement (±10 ppm) in protein identification strategies employing MS or MS/MS and database searching. *Anal. Chem.* **71,** 2871–2882.

Colby, S. M., King, T. B., and Reilly, J. P. (1994). Improving the resolution of matrix-assisted laser desorption/ionization time-of-flight mass spectrometry by exploiting the correlation between ion position and velocity. *Rapid Commun. Mass Spectrom.* **8**, 865–868.

Dawson, J. H. J., and Guilhaus, M. (1989). Orthogonal-acceleration time-of-flight mass spectrometer. *Rapid Commun. Mass Spectrom.* **3**, 155–159.

Dodonow, J. H. J., Chernushevich, I. V., and Laiko, V. V. (1991). *Proceed. 12th Intern. Mass Spectrom. Conf.*, p. 153.

Doroshenko, V. M., Cornish, T. J., and Cotter, R. J. (1992). Matrix-assisted laser desorption/ionization inside a quadrupole ion-trap detector cell. *Rapid Commun. Mass Spectrom.* **6**, 753–757.

Edmonds, C. G., and Smith, R. D. (1990). Methods in Enzymology. Academic Press, New York.

Edmondson, R. D., and Russell, D. H. (1996). Evaluation of matrix-assisted laser desorption ionization-time-of-flight mass measurement accuracy by using delayed extraction. *J. Am. Soc. Mass Spectrom.* **7**, 995–1001.

Ehring, H., Karas, M., and Hillenkamp, F. (1992). Role of photoionization and photochemistry in ionization processes of organic molecules and relevance for matrix-assisted laser desorption ionization mass spectrometry. *Org. Mass Spectrom.* **27**, 472–480.

Fenn, J. B., Mann, M., Meng, C. K., Wong, S. F., and Whitehouse, C. M. (1989). Electrospray ionization for mass spectrometry of large biomolecules. *Science* **246**, 64–71.

Gluckmann, M., Pfenninger, A., Karas, M., Horneffer, V., Hillenkamp, F., and Strupat, K. (2001). Poster presentation, part 1. *Proceed. 34th Ann. Meeting German Mass Spectrom. Conf.*, p. 70. Leipzig, Germany.

Gobom, J., Nordhoff, E., Mirgorodskaya, E., Ekman, R., and Roepstorff, P. (1999). Sample purification and preparation technique based on nano-scale reversed-phase columns for the sensitive analysis of complex peptide mixtures by matrix-assisted laser desorption/ionization mass spectrometry. *J. Mass Spectrom.* **34**, 105–116.

Gross, J., and Strupat, K. (1998). Matrix-assisted laser desorption/ionization-mass spectrometry applied to biological macromolecules. *TrAC* **17**, 470–484.

Gross, J., and Hillenkamp, F. (2000). *In* "Encyclopedia of Analytical Chemistry" (R. A. Meyers, ed.), pp. 225–275. John Wiley & Sons. In press.

Guilhaus, M., Selby, S., and Mlynski, V. (2000). Orthogonal acceleration time-of-flight mass spectrometry. *Mass Spectrom. Rev.* **19**, 65–107.

Hettich, R. L., and Buchanan, M. V. (1991). Matrix-assisted laser desorption Fourier transform mass spectrometry for the structural examination of modified nucleic acid constituents. *Inter. J. Mass Spectrom. Ion Proc.* **111**, 365–380.

Hill, J. A., Annan, R. S., and Biemann, K. (1991). Matrix-assisted laser desorption ionization with a magnetic mass spectrometer. *Rapid Commun. Mass Spectrom.* **5**, 395–399.

Hillenkamp, F., Karas, M., Beavis, R. C., and Chait, B. T. (1991). Matrix-assisted laser desorption/ionization mass spectrometry of biopolymers. *Anal. Chem.* **63**, 1193A–1203A.

Horneffer, V., Dreisewerd, K., Ludemann, H. C., Hillenkamp, F., Lage, M., and Strupat, K. (1999). Is the incorporation of analytes into matrix crystals a prerequisite for matrix-assisted laser desorption/ionization mass spectrometry? A study of five positional isomers of dihydroxybenzoic acid. *Intern. J. Mass Spectrom.* **185/186/187**, 859–870.

Horneffer, V., Hillenkamp, F., Strupat, K., Gluckmann, M., Pfenninger, A., and Karas, M. (2001). Poster, part 2, presented at conference. *Proceed. 34th Ann. Meet. German Mass Spectrom. Conf.*, p. 72. Leipzig, Germany.

Jensen, O. N., Podtelejnikov, A., and Mann, M. (1996). Delayed extraction improves specificity in database searches by matrix-assisted laser desorption/ionization peptide maps. *Rapid Commun. Mass Spectrom.* **10**, 1371–1378.

Juhasz, P., Vestal, M. L., and Martin, S. A. (1997). On the initial velocity of ions generated by matrix-assisted laser desorption ionization and its effect on the calibration of delayed extraction time-of-flight mass spectra. *J. Am. Soc. Mass Spectrom.* **8**, 209–217.

Juraschek, R., Dulcks, T., and Karas, M. (1999). Nanoelectrospray—more than just a minimized-flow electrospray ionization source. *J. Am. Soc. Mass Spectrom.* **10**, 300–308.

Karas, M., Bachmann, D., and Hillenkamp, F. (1985). Influence of the wavelength in high-irradiance ultraviolet laser desorption mass spectrometry of organic molecules. *Anal. Chem.* **57**, 2935–2939.

Karas, M., Bahr, U., Strupat, K., Hillenkamp, F., Tsarbopoulos, A., and Pramanik, B. N. (1995). Matrix dependence of metastable fragmentation of glycoproteins in MALDI TOF mass spectrometry. *Anal. Chem.* **67**, 675–679.

Karas, M., Ehring, H., Nordhoff, E., Stahl, B., Strupat, K., Hillenkamp, F., Grehl, M., and Krebs, B. (1993). Matrix-assisted laser desorption/ionization mass spectrometry with additives to 2,5-dihydroxybenzoic acid. *Org. Mass Spectrom.* **28**, 1476–1481.

Karas, M., Gluckmann, M., and Schafer, J. (2000). Ionization in matrix-assisted laser desorption/ionization: Singly charged molecular ions are the lucky survivors. *J. Mass Spectrom.* **35**, 1–12.

Karas, M., and Hillenkamp, F. (1988). Laser desorption ionization of proteins with molecular masses exceeding 10,000 daltons. *Anal. Chem.* **60**, 2299–2301.

Kaufmann, R., Chaurand, P., Kirsch, D., and Spengler, B. (1996). Post-source decay and delayed extraction in matrix-assisted laser desorption/ionization-reflectron time-of-flight mass spectrometry. Are there trade-offs? *Rapid Commun. Mass Spectrom.* **10**, 1199–2208.

Kebarle, P. (2000). A brief overview of the present status of the mechanisms involved in electrospray mass spectrometry. *J. Mass Spectrom.* **35**, 804–817.

Kelleher, N. L., Senko, M. W., Siegel, M. M., and McLafferty, F. W. (1997). Unit resolution mass spectra of 112 kDa molecules with 3 Da accuracy. *J. Amer. Soc. Mass Spectrom.* **8**, 380–383.

Krutchinsky, A. N., Loboda, A. V., Spicer, V. L., Dworschak, R., Ens, W., and Standing, K. G. (1998). Orthogonal injection of matrix-assisted laser desorption/ionization ions into a time-of-flight spectrometer through a collisional damping interfacet. *Rapid Commun. Mass Spectrom.* **12**, 508–518.

Krutchinsky, A. N., Loboda, A. V., Bromirski, M., Standing, K. G., and Ens, W. (1998). *Proc. 46th ASMS Conf.* p. 794. Orlando, FL.

Kussmann, M., Nordhoff, E., Rahbek-Nielsen, H., Haebel, S., Rossel-Larsen, M., Jakobsen, L., Gobom, J., Mirgorodskaya, E., Kroll-Kristensen, A., Palm, L., and Roepstorff, P. (1997). Matrix-assisted laser desorption/ionization mass spectrometry sample preparation techniques designed for various peptide and protein analytes. *J. Mass Spectrom.* **32**, 593–601.

La Rotta, A., Holle, A., and Hillenkamp, F. (2001). *Proc. 49th ASMS Conf.*, p. 1240. Chicago, IL.

Li, Y., Hunter, R. L., and McIver, R. T., Jr. (1996). Ultrahigh-resolution Fourier transform mass spectrometry of biomolecules above m/z 5000. *Inter. J. Mass Spectrom. Ion Proc.* **157/158**, 175–188.

Li, Y., Hunter, R. L., and McIver, R. T., Jr. (1994). High-accuracy molecular mass determination for peptides and proteins by Fourier transform mass spectrometry. *Anal. Chem.* **66**, 2077–2083.

Little, D. P., Speir, J. P., Senko, M. W., O'Connor, P. B., and McLafferty, F. W. (1994). Infrared multiphoton dissociation of large multiply charged ions for biomolecule sequencing. *Anal. Chem.* **66**, 2809–2815.

Loboda, A. V., Krutchinsky, A. N., Bromirski, M., Ens, W., and Standing, K. G. (2000). A tandem quadrupole/time-of-flight mass spectrometer with a matrix-assisted laser desorption/ionization source, design and performance. *Rapid Commun. Mass Spectrom.* **14**, 1047–1057.

Loboda, A. V., Krutchinsky, A. N., Spicer, V. L., Ens, W., and Standing, K. G. (1999). *Proc. 47th ASMS Conf.*, p. 1956, Dallas, TX.

Loo, J. A. (1997). Studying noncovalent protein complexes by electrospray ionization mass spectrometry. *Mass Spectrom. Rev* **16**, 1–23.

Mann, M., Meng, C. K., and Fenn, J. B. (1989). Interpreting mass spectra of multiply charged ions. *Anal. Chem.* **61**, 1702–1708.

Marshall, A. G., and Grosshans, P. B. (1991). Fourier transform ion cyclotron resonance mass spectrometry: The teenage years. *Anal. Chem.* **63**, 215A–229A.

Medzihradszky, K. F., Campbell, J. M., Baldwin, M. A., Falick, A. M., Juhasz, P., Vestal, M. L., and Burlingame, A. L. (2000). The characteristics of peptide collision-induced dissociation using a high-performance MALDI–TOF/TOF tandem mass spectrometer. *Anal. Chem.* **72**, 552–558.

Meng, C. K., McEwen, C. N., and Larsen, B. S. (1990a). Electrospray ionization on a high-performance magnetic-sector mass spectrometer. *Rapid Commun. Mass Spectrom.* **4**, 147–150.

Meng, C. K., McEwen, C. N., and Larsen, B. S. (1990b). Peptide sequencing with electrospray ionization on a magnetic sector mass spectrometer. *Rapid Comm. Mass Spectrom.* **4**, 151–155.

Menzel, C., Berkenkamp, S., and Hillenkamp, F. (1999). Infrared matrix-assisted laser desorption/ionization mass spectrometry with a transversely excited atmospheric pressure carbon dioxide laser at 10.6 mm wavelength with static and delayed ion extraction. *Rapid Commun. Mass Spectrom.* **13**, 26–32.

Nielen, M. W. F. (1999). MALDI time-of-flight mass spectrometry of synthetic polymers. *Mass Spectrom. Rev.* **18**, 309–344.

Olsen, J. V., and Mann, M. (2004). Improved peptide identification in proteomics by two consecutive stages of mass spectrometric fragmentation. *Proc. Natl. Acad. Sci. USA* **101**, 13417–13422.

Potier, N., Barth, P., Tritsch, D., Biellmann, J. F., and Van Dorsselaer, A. (1997). Study of noncovalent enzyme-inhibitor complexes of aldose reductase by electrospray mass spectrometry. *Eur. J. Biochem.* **243**, 274–282.

Price, P. (1991). Standard definitions of terms relating to mass spectrometry. A report from the committee on measurements and standards of the american society for mass spectrometry. *J. Amer. Soc. Mass Spectro.* **2**, 336–348.

Price, W. D., Schnier, P. D., and Williams, E. R. (1996). Tandem mass spectrometry of large biomolecule ions by blackbody infrared radiative dissociation. *Anal. Chem.* **68**, 859–866.

Qin, J., Ruud, J., and Chait, B. T. (1996). A practical ion trap mass spectrometer for the analysis of peptides by matrix-assisted laser desorption/ionization. *Anal. Chem.* **68**, 1784–1791.

Rogniaux, H., van Dorsselaer, A., Barth, P., Biellmann, J. F., Barbanton, J., van Zandt, M., Chevrier, B., Howard, E., Mitschler, A., Potier, N., Urzhumtseva, L., Moras, D., and Podjarny, A. (1999). Binding of aldose reductase inhibitors: Correlation of crystallographic and mass spectrometric studies. *J. Am. Soc. Mass Spectrom.* **10**, 635–647.

Rostom, A. A., and Robinson, C. V. (1999). Disassembly of intact multiprotein complexes in the gas phase. *Curr. Opin. Struct. Biol.* **9**, 135–141.

Russell, D. H., and Edmondson, R. D. (1997). High-resolution mass spectrometry and accurate mass measurements with emphasis on the characterization of peptides and proteins by matrix-assisted laser desorption/ionization time-of-flight mass spectrometry. *J. Mass Spectrom.* **32**, 263–276.

Senko, M., Zabrouskov, V., Lange, O., Wieghaus, A., and Horning, S. (2004). LC/MS with external calibration mass accuracies approaching 100 ppb. *Proc. 52nd ASMS Conf. Mass Spectrom. Allied Top*, p. 735. Nashville, TN.

Senko, M. W., Hendrickson, C. L., Pasa-Tolic, L., Marto, J. A., White, F. M., Guan, S., and Marshall, A. G. (1996). Electrospray ionization Fourier transform ion cyclotron resonance at 9.4 T. *Rapid Comm. Mass Spectrom.* **10,** 1824–1828.

Senko, M. W., Speir, J. P., and McLafferty, F. W. (1994). Collisional activation of large multiply charged ions using Fourier transform mass spectrometry. *Anal. Chem.* **66,** 2801–2808.

Shevchenko, A., Jensen, O. N., Podtelejnikov, A. V., Sagliocco, F., Wilm, M., Vorm, O., Mortensen, P., Boucherie, H., and Mann, M. (1996). Linking genome and proteome by mass spectrometry: Large-scale identification of yeast proteins from two dimensional gels. *Proc. Natl. Acad. Sci. USA* **93,** 14440–14445.

Shevchenko, A., Loboda, A., Shevchenko, A., Ens, W., and Standing, K. G. (2000). MALDI quadrupole time-of-flight mass spectrometry: A powerful tool for proteomic research. *Anal. Chem.* **72,** 2132–2141.

Siegel, M. M., Hollander, IJ., Hamann, P. R., James, J. P., Hinman, L., Smith, B. J., Farnsworth, A. P., Phipps, A., King, D. J., and Karas, M. (1991). Matrix-assisted UV-laser desorption/ionization mass spectrometric analysis of monoclonal antibodies for the determination of carbohydrate, conjugated chelator and conjugated drug content. *Anal. Chem.* **63,** 2470–2481.

Smith, R. D., Bruce, J. E., Wu, Q., and Lei, Q. P. (1997). New mass spectrometric methods for the study of noncovalent associations of biopolymers. *Chem. Soc. Rev.* **26,** 191–202.

Smith, R. D., Loo, J. A., Edmonds, C. G., Barinaga, C. J., and Udseth, H. R. (1990). New developments in biochemical mass spectrometry: Electrospray ionization. *Anal. Chem.* **62,** 882–899.

Speir, J. P., Senko, M. W., Little, D. P., Loo, J. A., and McLafferty, F. W. (1995). High-resolution tandem mass spectra of 37–67 kDa proteins. *J. Mass Spectrom.* **30,** 39–42.

Spengler, B. (1997). Post-source decay analysis in matrix-assisted laser desorption/ionization mass spectrometry of biomolecules. *J. Mass Spectrom.* **32,** 1019–1036.

Spengler, B., Kirsch, D., Kaufmann, R., and Jaeger, E. (1992). Peptide sequencing by matrix-assisted laser-desorption mass spectrometry. *Rapid Commun. Mass Spectrom.* **6,** 105–108.

Spengler, B., Kirsch, D., Kaufmann, R., Karas, M., Hillenkamp, F., and Giessmann, U. (1990). The detection of large molecules in matrix-assisted UV-laser desorption. *Rapid Commun. Mass Spectrom.* **4,** 301–305.

Strupat, K., Rogniaux, H., van Dorsselaer, A., Roth, J., and Vogl, T. (2000). Calcium-induced noncovalently linked tetramers of MRP8 and MRP14 are confirmed by electrospray ionization-mass analysis. *J. Am. Soc. Mass Spectrom.* **11,** 780–788.

Stults, J. T., and Arnott, D. (2005). Proteomics. *Meth. Enzym.* **402,** 245–289.

Takach, E. J., Hines, W. M., Patterson, D. H., Juhasz, P., Falick, A. M., Vestal, M. L., and Martin, S. A. (1997). Accurate mass measurements using MALDI–TOF with delayed extraction. *J. Protein Chem.* **16,** 363–369.

van Berkel, G. J., Glish, G. L., and McLuckey, S. A. (1990). Electrospray ionization combined with ion trap mass spectrometry. *Anal. Chem.* **62,** 1284–1295.

Vestal, M. L., Juhasz, P., and Martin, S. A. (1995). Delayed extraction matrix-assisted laser desorption time-of-flight mass spectrometry. *Rapid Commun. Mass Spectrom.* **9,** 1044–1050.

Vorm, O., and Mann, M. (1994). Improved mass accuracy in matrix-assisted laser desorption/ionization time-of-flight mass spectrometry of peptides. *J. Amer. Soc. Mass Spectrom.* **5,** 955–958.

Vorm, O., Roepstorff, P., and Mann, M. (1994). Improved resolution and very high sensitivity in MALDI–TOF of matrix surfaces made by fast evaporation. *Anal. Chem.* **66,** 3281–3287.

Voyksner, R. D. (1997). ESI–MS *In* "Electrospray ionization mass spectrometry" (R. B. Cole, ed.). John Wiley & Sons, New York.

Werner, E., and Standing, K. G. (2005). Hybrid quadrupole/time-of-flight mass spectrometers for analysis of biomolecules. *Meth. Enzym.* **405.** In press.

Whittal, R. M., and Li, L. (1995). High-resolution matrix-assisted laser desorption/ionization in a linear time-of-flight mass spectrometer. *Anal. Chem.* **67,** 1950–1954.

Williams, E. R. (1998). Tandem FTMS of large biomolecules. *Anal. Chem.* **70,** 179A–185A.

Wilm, M., and Mann, M. (1996). Analytical properties of the nanoelectrospray ion source. *Anal. Chem.* **68,** 1–8.

Wilm, M. S., and Mann, M. (1994). Electrospray and Taylor-Cone theory, Dole's beam of macromolecules at last? *Intern. J. Mass Spectrom. Ion Proc.* **136,** 167–180.

Winter, M. (1983). WebElements.

Wu, J., Fannin, S. T., Franklin, M. A., Molinski, T. F., and Lebrilla, C. B. (1995). Exact mass determination for elemental analysis of ions produced by matrix-assisted laser desorption. *Anal. Chem.* **67,** 3788–3792.

Yao, Z., Tito, P., and Robinson, C. V. (2005). Site-specific hydrogen exchange of proteins: Insights into the structures of amyloidogenic intermediates. *Meth. Enzym.* **405.** In press.

Yates, J. R., 3rd. (1998). Mass spectrometry and the age of the proteome. *J. Mass Spectrom.* **33,** 1–19.

Yergey, J., Heller, D., Hansen, G., Cotter, R. J., and Fenselau, C. (1983). Isotopic distributions in mass spectra of large molecules. *Anal. Chem.* **55,** 353–356.

Zaia, J., Annan, R. S., and Biemann, K. (1992). The correct molecular weight of myoglobin, a common calibrant for mass spectrometry. *Rapid Commun. Mass Spectrom.* **6,** 32–36.

Zenobi, R., and Knochenmuss, R. (1999). Ion formation in MALDI mass spectrometry. *Mass Spectrom. Rev.* **17,** 337–366.

Zubarev, R. A., Kelleher, N. L., and McLafferty, F. W. (1998). Electron capture dissociation of multiply charged protein cations. A nonergodic process. *J. Amer. Chem. Soc.* **120,** 3265–3266.

[2] Batch Introduction Techniques

By Connie R. Jiménez

Abstract

Mass spectrometry (MS) is widely used as a rapid tool for peptide profiling and protein identification. However, the success of the method is compromised by dirty and contaminated samples. Moreover, analysis from a small sample volume with a relative high concentration is usually required. In this chapter, different microscale sample preparation methods are discussed for off-line, matrix-assisted laser desorption/ionization (MALDI) and nanoelectrospray ionization (nanoESI) MS analysis.

METHODS IN ENZYMOLOGY, VOL. 405   0076-6879/05 $35.00
DOI: 10.1016/S0076-6879(05)05002-0

Introduction

Biomolecules form a heterogeneous group; therefore, multiple different purification schemes and sample preparation protocols exist for mass spectrometry (MS) (Fig. 1). In this era of proteomics, proteins are the most widely studied biomolecules. Both matrix-assisted laser desorption/ionization (MALDI) and electrospray ionization (ESI) MS have become very successful in detecting and identifying proteins at increasingly lower amounts (Aebersold and Mann, 2003; Pandey and Mann, 2000). Appropriate sample preparation is crucial for attaining low detection levels. Therefore, this chapter emphasizes sample preparation (including cleaning of the sample) for off-line analysis of proteins/peptides using MALDI and nanoelectrospray ionization (nanoESI).

In most laboratories, MALDI–MS is being used for rapid identification of proteins using tryptic peptide mass mapping of gel-purified proteins as well as for subsequent characterization of post-translational modifications. Other applications of MALDI–MS include its use as a tool to visualize native peptides directly in tissues and even single cells (Jimenez and Burlingame, 1998) as well as its use in biomedical/clinical applications for proteomics pattern analysis of human body fluids (Villanueva et al., 2004). The application of direct peptide profiling of a single cell will be discussed in this chapter. In addition, nanoESI tandem MS (MS/MS) has become very important for the generation of peptide sequence tags that allow for even more specific identification of proteins. For a successful nanoESI analysis, in-gel digests need to be desalted using microtip extraction columns. The following paragraphs discuss the requirements and strategies of sample preparation for MALDI–MS and nanoESI analyses.

Sample Preparation for MALDI–MS

Samples for MALDI–MS are typically prepared off-line. Many samples can be deposited on a single sample stage and measured in a single session. The MALDI process produces ions from the analyte cocrystallized with a matrix that absorbs at the wavelength of laser irradiation. Matrices are low molecular weight, aromatic organic acids such as α-cyano-4-hydroxy-*trans*-cinnamic acid or 2,5-dihydroxybenzoic (2,5-DHB) acid. The analyte is desorbed and ionized by the incident laser pulse.

Sample preparation may be the most crucial step in MALDI mass analysis of peptides and proteins. Subtle variations of the experimental parameters (e.g., matrix, matrix solvent, laser power, analyte-to-matrix ratio, etc.) can dramatically affect the result of the measurement. Equally important, detection sensitivity is not usually determined by the absolute amount of analyte

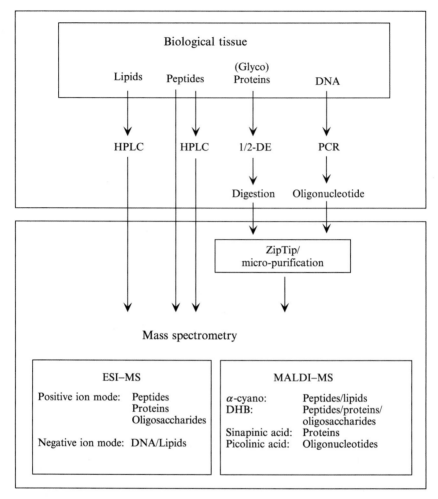

Fig. 1. Outline of main biomolecule classes and common purification and sample preparation strategies for MS. Biological tissue contains diverse biomolecules that require different purification strategies. Some common strategies are indicated. Once the compound of interest is isolated, it may be directly amendable to mass analysis (e.g., in case of an HPLC fraction), or it can be mass analyzed after a sample cleanup step employing microtip C18 columns or ZipTips (e.g., in case of samples extracted from acrylamide gels). ESI–MS and MALDI–MS are the most commonly used mass spectrometric techniques, and both techniques are suitable for analysis of the various biomolecule classes. HPLC, high-performance liquid chromatography; 1/2-DE, one- or two-dimensional gel electrophoresis; ESI, electrospray ionization; MALDI, matrix-assisted laser desorption/ionization; MS, mass spectrometry; Pos., positive; Neg., negative; α-cyano, α-cyano-4-hydroxy-*trans*-cinnamic acid; DHB, 2,5-dihydroxybenzoic acid; picolinic acid, 3-hydroxy-picolinic acid.

but rather by the presence of contaminants, such as detergents and salts, in the sample. The most powerful way to concentrate and desalt samples is by using microtip reversed-phase columns. In addition, several other strategies may be followed to remove contaminants (see upcoming paragraphs).

Different "on-target" sample preparation methods have been developed. The most frequently used is the "dried-droplet" method, in which the matrix is simply added to the sample on the sample stage. Alternatively, a thin layer of matrix material from a fast-evaporating solvent may be prepared first, onto which the sample is subsequently deposited. The latter sample preparation method has resulted in improved sensitivity and mass accuracy of peptides (Vorm and Roepstorff, 1994). Matrices that allow a degree of on-target washing, which is useful for removing contaminants, are α-cyano-4-hydroxy-*trans*-cinnamic acid and sinapinic acid. Moreover, the addition of nitrocellulose to the matrix solution has resulted in better peptide binding to the target, allowing for more extensive rinsing of the sample (Shevchenko *et al.*, 1996a,b). Alternatively, water-soluble 2,5-DHB acid may also be used because this matrix excludes impurities during crystallization. Especially when complex peptide mixtures are analyzed, the method of sample preparation as well as the matrix solvent composition can significantly affect the results due to differential mass discrimination/ion suppression effects (Cohen and Chait, 1996). Therefore, if a high protein coverage is required, spectra should be acquired from the sample prepared with different matrix solution conditions to identify as many components of the mixture as possible. Alternatively, a complex sample may be fractionated using nanoliquid chromatography (LC) and automatically spotted onto the sample stage, prior to MALDI–MS analysis.

Dried-Droplet Method

Materials

Matrices suitable for peptide analysis are α-cyano-4-hydroxy-*trans*-cinnamic acid (20 mg/ml in 0.1% trifluoroacetic acid (TFA)/50% acetonitrile or in acetone water [99:1, v/v]) and 2,5-DHB acid (20 mg/ml in 0.1% TFA/30% acetonitrile). DHB is also the preferred matrix for oligosaccharide analysis. The percentage of acetonitrile may vary depending on the hydrophobicity of the analyte analyzed. The matrix additive 2-methoxy-5-benzoic acid may be added (1 mg/ml) to the DHB matrix solution for better detection of larger (>3000 Da) peptides. For protein and glycopeptide analysis, sinapinic acid (20 mg/ml) in 0.1% TFA/50% acetonitrile or in acetone water (99:1, v/v) can be used. For oligonucleotide analysis,

4-methoxy cinnamic acid or 3-hydroxy-picolinic acid (50 mg/ml in 0.05 *M* ammonium citrate/30% acetonitrile) can be used.

Procedure

1. Clean the sample target with the appropriate solvents (usually water and methanol are sufficient). Use a pipette for a 0.5–1 μl sample to be placed onto the target, and add 0.5–1 μl matrix; mix by pipetting up and down. Let the sample air-dry. Keep in mind that samples may also be prepared by mixing the sample and matrix solution in a microfuge tube (recommended when α-cyano-4-hydroxy-*trans*-cinnamic acid is used), from which 0.5–1 μl is applied to the target.

2. Apply an appropriate standard mixture to the target in the vicinity of the sample or at appropriate locations for external calibration. For best calibration, the analyte should fall within the mass area covered by the standards. For highest mass accuracy, internally calibrate in-gel digests by using trypsin autolysis products (e.g., at mass-to-charge ratio 841.51 and 2211.10). Alternatively, include internal standards in the sample; however, do not add an excess of internal standard because this may result in ion suppression of the sample (Fig. 2) (Cohen and Chait, 1996).

3. Record the mass spectrum, which can be done manually in a few minutes or automatically on all newer instruments. To measure manually,

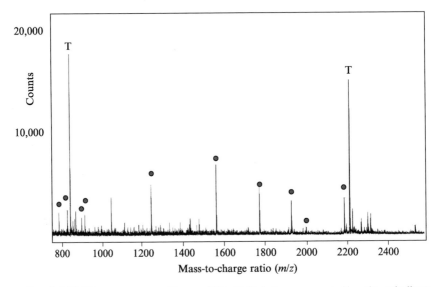

Fig. 2. MALDI mass spectrum (Voyager Elite, DeStr) of an unseparated in-gel tryptic digest of an excised two-dimensional gel spot. Ions marked with dots were used for database searching (www.prospector.ucsf.edu) and matched a heat-shock protein 40 homologue. Unmarked ions represent background peaks. *m/z*, mass-to-charge ratio; T, trypsin autolysis product.

insert sample target into the mass spectrometer and mass analyze as recommended by the manufacturer. Irradiate the sample just above the threshold for appearance of the ions. Generally, the ion signals of peptides correspond to $(M + H)^+$, whereas oligonucleotides and oligosaccharides generally yield $(M + Na)^+$. Larger peptides may yield doubly protonated ions that appear at half the peptide mass $(M/2)$.

4. Store the sample target with matrix-sample preparations in the dark at ambient temperature; this storage is appropriate for a number of days without significant loss of signal. For longer storage, keep the target at $-20°$ in a sealed container.

Critical Parameters

Contaminants such as salts and detergents should be less than 50 mM. When crystallization is impaired due to the presence of contaminants, the matrix-sample preparation (not suitable for DHB) should be rinsed one to three times with ice-cold 0.1% TFA or 10% formic acid and then blown off using pressurized air or carefully pipetted off after a few seconds. Very hydrophilic peptides, however, may be lost in this washing procedure. Alternatively, the sample can be diluted in 1% TFA/20–30% acetonitrile and applied again to the target, the rapid evaporation method can be used that allows for more extensive rinsing, or microscale solid phase extraction columns can be used to desalt the sample (see upcoming paragraphs).

Due to microheterogeneity of the matrix crystals (especially when DHB is used), spectra should be accumulated from several different crystals of the matrix-sample preparation to obtain a representative sum spectrum. With the inclusion of an internal standard in the matrix solution, the relative peak abundancies may even give a semiquantitative indication of the relative concentrations of peptides in a mixture (Jiménez et al., 1998).

Rapid Evaporation Method

Materials

Nitrocellulose (5 mg/ml) and α-cyano-4-hydroxy-*trans*-cinnamic matrix (20 mg/ml) in 0.5 ml acetone should first be dissolved (vortex if necessary for 1 to 2 min) before adding 0.5 ml isopropanol (Gobom et al., 1999).

Procedure

1. Deposit ~0.5 μl of the matrix/nitrocellulose solution on a clean sample target. Pipet quickly because of the high volatility of the solvent. The matrix/nitrocellulose solution will spread out rapidly, allowing fast evaporation of the solvent and the formation of a uniform white layer.

2. Apply ~0.5 μl of sample solution onto the matrix surface on the sample stage. Let the sample evaporate at ambient temperature. The sample solution should not completely dissolve the matrix layer. Therefore, keep the concentration of organic solvent in the sample at less than 30% and make sure that the pH is acidic. Attain these conditions by applying 0.5 μl water or 0.5 μl 10% formic acid, respectively, onto the matrix surface, into which the sample is subsequently pipetted.

3. Rinse the sample with ~10 μl cold 10% formic acid. Leave the solution on the sample for 10 sec, and then blow off using pressurized air, or carefully pipet off. Repeat the rinse step with water. Note: This rinsing step is strongly recommended for unseparated peptide digest mixtures but can be omitted for samples that are free of salts (e.g., high-performance liquid chromatography or HPLC fractions).

4. For best ion intensities, immediately analyze matrix-sample preparations made using the rapid evaporation method.

Direct Single-Cell Peptide Profiling Using MALDI–MS

Since the beginning of the twenty-first century, MALDI–MS has emerged as a powerful tool for the direct analysis of peptide profiles in very complex samples (i.e., single neuroendocrine cells and tissue biopsies of mollusks, toads, and rats) (Jimenez and Burlingame, 1998; Li et al., 2000). Neuropeptide profiling is based on the measurement of the molecular weights of neuropeptides that are abundantly (femtomole range) expressed in a given cell type. The molecular weight information together with information of prohormone sequences synthesized in the cell of interest is often sufficiently specific to identify the peptides. Otherwise, targeted MS/MS may be used to elucidate the peptide structure. This approach has been important for the analysis of cell-specific expression and processing of neuropeptides (Fujisawa et al., 1999; Jimenez et al., 1994; van Strien et al., 1996). Since neuropeptide profiling is an open screening approach, it can also identify novel (putative) peptides (i.e., molecular weights that do not match any predicted peptide encoded by known prohormones). To characterize these putative peptides, MS/MS and PSD sequencing have been successfully employed (Jimenez and Burlingame, 1998; Jimenez et al., 2004; Li et al., 2000). Finally, it should be noted that peptides are not the only abundant cellular components of peptidergic neurons. However, with the direct MALDI peptide profiling method described below, peptides are preferentially detected, a feature that may be dependent on the hydrophobicity of the matrix solvent.

Materials and Procedure

The preferred matrix for the direct analysis of the contents of single cells and tissue biopsies is 2,5-DHB (20 mg/ml in 0.1% TFA/30% acetonitrile). Samples for single-cell analysis are prepared using the dried-droplet method in DHB (see previous section). DHB is the matrix of choice because this matrix excludes impurities during crystallization. The procedure can be explained as follows: a single freshly dissected peptidergic neuron from the brain (snail) or a dissociated peptidergic neuron from a culture dish (rat) is sucked up into a fine glass pipette (tip diameter 2–50 μm, depending on the cell diameter) and is directly transferred to a droplet of DHB matrix solution (20 mg DHB in 1 ml 0.1% TFA/20% acetonitrile) on the sample stage, where the cell is ruptured by pipetting up and down.

An example of a MALDI spectrum (measured on a home-built reflectron instrument) of a single rat melanotrope cell (cell diameter ~10 μm) is shown in Fig. 3B (Jiménez *et al.*, unpublished data). The molecular ions identify most of the reported processing products of the rat proopiomelanocortin (POMC) prohormone (Fig. 3A). All the acetylation variants of the α-melanocortin stimulating hormone (αMSH) and β-endorphin (βEND) are detected as well as a phosphorylated and several truncated forms of the corticotropin intermediate peptide (CLIP). In addition,

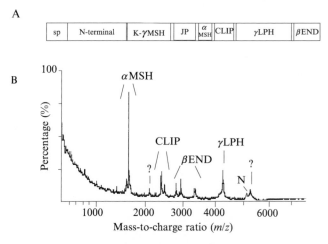

FIG. 3. Neuropeptide profile generated by direct MALDI–MS (home-built) of a single melanotrope cell of the rat pituitary gland. (A) Structural organization of the propriomelanocortin (POMC) prohormone from which multiple peptides are cleaved. (B) MALDI spectrum of a single melanotrope cell (cell diameter ~10 μm). Question marks indicate unknown biomolecule; *m/z*, mass-to-charge ratio.

two molecular ions at mass-to-charge ratio (m/z) 1458 and 5418 (mass accuracy 0.05–0.1%) are present and do not match any of the predicted POMC products. Therefore, these ions may represent novel processing/ posttranslationally modified products.

Critical Parameters

When isolating cells for MALDI–MS, care should be taken that the total volume of the cell in saline does not exceed ~100 nl; otherwise, salts impair the crystallization of the matrix.

Sample Preparation for NanoESI–MS

ESI–MS is a different but complementary method for global profiling and identification of peptides and proteins. A solution containing the sample is pumped through a metal capillary and dispersed at high voltage, resulting in rapidly evaporating peptide- or protein-containing droplets. After being desolved, the protein/peptide molecules remain charged by attachment of one or more protons and are drawn into the vacuum of a mass spectrometer. Two approaches have been developed to analyze peptide samples with ESI– MS. The first is an off-line approach, also called nanoelectrospray (nanoESI) (Wilm and Mann, 1996). The second approach utilizes liquid chromatography on-line coupled to mass spectrometry (LC–MS) and is not the focus of this chapter.

For direct infusion, nanoESI samples need to be free of salts and in a compatible solvent. To this end, samples such as in-gel digests are first micropurified using microtips with a small bed volume (<1 μl) of C18 or poros material (see following paragraphs) and subsequently eluted into a very fine-tipped spraying needle. As little as 1 μl of a sample can be measured during 20 to 40 min with detection limits in the femtomole range (Shevchenko *et al.*, 1996a,b; Wilm *et al.*, 1996). Figure 4 shows an example of the nanoESI–MS spectrum of an in-gel digested serum albumen after a ZipTip micropurification. Numerous tryptic peptides derived from serum albumen with different charge states can be detected. Spraying times lasting up to 40 min allow for several MS/MS analyses per sample.

Procedure

The spraying needle is a glass capillary with a fine tip (1–3 μm) that can be purchased or can be made in a fast and reproducible way from standard borosilicate capillaries with a commercial micropipette puller (e.g., Model P-87 Puller, Sutter Instrument Company, Novato, California, www.sutter. com). The capillaries may be gold-coated in a sputtering tray used in

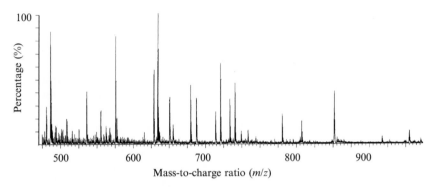

FIG. 4. NanoESI spectrum (acquired on a Micromass Q-Tof instrument) of in-gel digested serum albumen after sample preparation using Millipore ZipTips; m/z, mass-to-charge ratio.

sample preparation for electron microscopy. As an alternative to metallic sputter coating, an internal wire electrode may be used to make electrical contact between the source probe and the analyte solution (Alving *et al.*, 1999). For operation, the spraying needle is immobilized onto a metal holder that can be positioned in front of the orifice of the mass spectrometer under microscopic view. For an analysis, ~1 μl (or ~5 μl if an internal wire is used) of sample is directly loaded into the needle from the back using a gel loader pipette tip that can be inserted into the capillary. The sample is sprayed purely electrostatically in positive or negative ion mode, without the need of a nebulizer, sheath flow, or gas. For each sample, a new capillary is used; therefore, no cross-contamination between samples can occur. Altogether, this approach works very well although it is labor intensive because manual loading of each sample is required.

Desalting and Concentration of Samples Using Microtip Solid Phase Extraction Columns

The success of MS analysis, especially direct infusion using nanoESI, is critically dependent on a proper sample preparation. The presence of salts, detergents, or incompatible solvents may reduce the spectrum quality considerably.

Microscale solid phase extraction columns (bed volume in the submicroliter range) of C18 or poros R2 chromatography cast in pipette tips have been successfully used for purifying and concentrating femtomoles to picomoles of protein, peptide, or oligonucleotide samples prior to analysis (Erdjument-Bromage *et al.*, 1998; Gobom *et al.*, 1999; Shevchenko *et al.*, 1996b). These micropurification devices have become commercially available (ZipTips from Millipore, Billerica, MA, www.millipore.com) (Pluskal, 2000). ZipTips are easy to use (bidirectional flow is possible) and work well (highly improved peptide detection). For economical reasons, however, it is advantageous to make one's own microtip columns.

The use of these microtip columns for desalting and concentration of "dirty" samples such as in-gel digests is absolutely required for mass analysis using nanoESI (see Fig. 4) but is also recommended for MAL-DI–MS. The quality of the MALDI mass spectrum is significantly improved after microtip purification of an in-gel digest (see Fig. 5), and at low

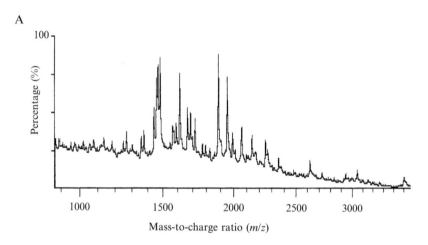

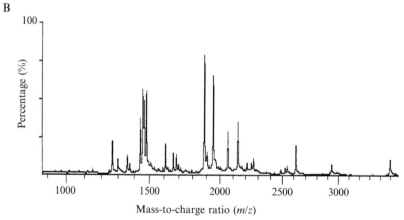

FIG. 5. Improved quality of MALDI mass spectrum after microtip purification of an in-gel digest. (A) Extracted peptides before sample clean-up. (B) Same extract as in (A) after purification using home-made poros microtips.

femtomole levels, it may even determine whether the spectrum displays signals at all. The following subsection describes how to use ZipTips and provides some guidelines for making/using homemade microtip columns.

Procedure Materials: ZipTips

- C18 (e.g., peptides, proteins, oligosaccharides) or C4 (e.g., larger proteins) ZipTips (from Millipore) with bed volume 0.6 or 0.2 μl.
- Wetting solution of 50% acetonitrile.
- Equilibration and washing solution of 0.1% TFA or 5% formic acid/ 5% MeOH.
- Elution solution of 50–75% ACN/0.1% TFA or 5% formic acid/60% MeOH. (For direct spotting in matrix, elute with desired matrix, such as with cyano-4-hydroxy-cinnamic acid at 10 mg/mL in 50% ACN/ 0.1% TFA.)

ZipTips Protocol

1. Prewet the ZipTip by slowly pipetting up and down, depressing the pipette plunger to a dead stop using the maximum volume setting (usually 10 μl), and aspirating the wetting solution into the tip and dispensing it to waste. Repeat this step.

2. Equilibrate ZipTip for binding by repeating step 1 with the equilibration solution.

3. Acidic (pH < 4) samples (usually containing 0.1–1 % TFA or 5% formic acid), ideally in a volume of 10 μl, are applied by slowly pipetting up and down about 10 times.

4. Wash ZipTip by aspirating the washing solution into the tip and dispensing it to waste (about three times).

5. Place 1–4 μl (or 0.5–2 μl for the 0.2-μl bed tips) eluant in a clean low-retention propylene tube (e.g., use 500-μl Eppendorf tubes from Eppendorf North America, www.eppendorfna.com). Elute the sample into vial by slowly pipetting the eluant up and down (about 5 to 10 times, without introducing air). The sample may also be dispensed (at the final dispense step) directly onto the MALDI target. Sample recovery can be improved (at the expense of concentration) by increasing elution volume to 10 μl. In case of lower eluant volumes, work quickly because acetonitrile is very volatile.

Procedure Materials: Poros Microtips

- Gel loader pipette tips (Eppendorf), Poros 10R2 (Perspective Biosystems, Framingham, MA, www.appliedbiosystems.com) (suspension of 5 mg/300 μl in ethanol/water, ratio 2:1), and glass beads (Emergo, Landsmeer, Netherlands, www.emergolab.com/info.htm).

- Preparation solution of 60% MeOH.
- Equilibration and washing solution of 5% formic acid/5% MeOH.
- Elution solution of 5% formic acid/60% MeOH (For direct spotting in matrix, elute with desired matrix, such as with cyano-4-hydroxy-cinnamic acid at 10 mg/mL in 50% ACN/0.1% TFA.)

Poros Microtips Protocol

1. Frit the gel loader pipette tips with a tiny glass bead (400–500 μm). Pipet the poros suspension (5–10 μl) into the tip; when flicking the tip, the suspension is pushed down. Apply all solvents for the top using a pipette, and push solvents through using a plastic syringe with a piece of tube as an adaptor to secure it to the gel loader tip.

2. Wash the microtip column with 100-μl preparation solution (this will also pack the poros suspension into a small (bed volume <1 μl) column.

3. Equilibrate with 100-μl equilibration/washing solution. Make sure not to push through any air.

4. To apply the sample (pH < 4), pipet the sample into the microtip column, and flick the tip to push it onto the column material without any air bubbles.

5. Wash with 100-μl equilibration/washing solution.

6. To elute, apply 1–10 μl elution solution (see step 4), and slowly push it through and collect eluant in an Eppendorf tube, electrospray needle, or onto a MALDI target.

Acknowledgment

I would like to thank Dr. K. W. Li at the Vrije Universiteit for reading the manuscript and his useful comments.

References

Aebersold, R., and Mann, M. (2003). Mass spectrometry-based proteomics. *Nature* **422,** 198–207.

Alving, K., Paulsen, H., and Peter-Katalinic, J. (1999). Characterization of O-glycosylation sites in MUC2 glycopeptides by nanoelectrospray QTOF mass spectrometry. *J. Mass Spectrom.* **34,** 395–407.

Cohen, S. L., and Chait, B. T. (1996). Influence of matrix solution conditions on the MALDI–MS analysis of peptides and proteins. *Anal. Chem.* **68,** 31–37.

Erdjument-Bromage, H., Lui, M., Lacomis, L., Grewal, A., Annan, R. S., McNulty, D. E., Carr, S. A., and Tempst, P. (1998). Examination of microtip reversed-phase liquid chromatographic extraction of peptide pools for mass spectrometric analysis. *J. Chromatogr. A* **826,** 167–181.

Fujisawa, Y., Furukawa, Y., Ohta, S., Ellis, T. A., Dembrow, N. C., Li, L., Floyd, P. D., Sweedler, J. V., Minakata, H., Nakamaru, K., et al. (1999). The *Aplysia mytilus* inhibitory peptide-related peptides: Identification, cloning, processing, distribution, and action. *J. Neurosci.* **19**, 9618–9634.

Gobom, J., Nordhoff, E., Mirgorodskaya, E., Ekman, R., and Roepstorff, P. (1999). Sample purification and preparation technique based on nanoscale reversed-phase columns for the sensitive analysis of complex peptide mixtures by matrix-assisted laser desorption/ ionization mass spectrometry. *J. Mass Spectrom.* **34**, 105–116.

Jimenez, C. R., and Burlingame, A. L. (1998). Ultramicroanalysis of peptide profiles in biological samples using MALDI mass spectrometry. *Exp. Nephrol.* **6**, 421–428.

Jimenez, C. R., ter Maat, A., Pieneman, A., Burlingame, A. L., Smit, A. B., and Li, K. W. (2004). Spatio-temporal dynamics of the egg-laying-inducing peptides during an egg-laying cycle: A semiquantitative matrix-assisted laser desorption/ionization mass spectrometry approach. *J. Neurochem.* **89**, 865–875.

Jimenez, C. R., van Veelen, P. A., Li, K. W., Wildering, W. C., Geraerts, W. P., Tjaden, U. R., and van der Greef, J. (1994). Neuropeptide expression and processing as revealed by direct matrix-assisted laser desorption ionization mass spectrometry of single neurons. *J. Neurochem.* **62**, 404–407.

Li, L., Garden, R. W., and Sweedler, J. V. (2000). Single-cell MALDI: A new tool for direct peptide profiling. *Trends Biotechnol.* **18**, 151–160.

Pandey, A., and Mann, M. (2000). Proteomics to study genes and genomes. *Nature* **405**, 837–846.

Pluskal, M. G. (2000). Microscale sample preparation. *Nat. Biotechnol.* **18**, 104–105.

Shevchenko, A., Jensen, O. N., Podtelejnikov, A. V., Sagliocco, F., Wilm, M., Vorm, O., Mortensen, P., Boucherie, H., and Mann, M. (1996a). Linking genome and proteome by mass spectrometry: Large-scale identification of yeast proteins from two dimensional gels. *Proc. Natl. Acad. Sci. USA* **93**, 14440–14445.

Shevchenko, A., Wilm, M., Vorm, O., and Mann, M. (1996b). Mass spectrometric sequencing of proteins silver-stained polyacrylamide gels. *Anal. Chem.* **68**, 850–858.

van Strien, F. J., Jespersen, S., van der Greef, J., Jenks, B. G., and Roubos, E. W. (1996). Identification of POMC processing products in single melanotrope cells by matrix-assisted laser desorption/ionization mass spectrometry. *FEBS Lett.* **379**, 165–170.

Villanueva, J., Philip, J., Entenberg, D., Chaparro, C. A., Tanwar, M. K., Holland, E. C., and Tempst, P. (2004). Serum peptide profiling by magnetic particle-assisted, automated sample processing and MALDI–TOF mass spectrometry. *Anal. Chem.* **76**, 1560–1570.

Vorm, O., and Roepstorff, P. (1994). Peptide sequence information derived by partial acid hydrolysis and matrix-assisted laser desorption/ionization mass spectrometry. *Biol. Mass Spectrom.* **23**, 734–740.

Wilm, M., and Mann, M. (1996). Analytical properties of the nanoelectrospray ion source. *Anal. Chem.* **68**, 1–8.

Wilm, M., Shevchenko, A., Houthaeve, T., Breit, S., Schweigerer, L., Fotsis, T., and Mann, M. (1996). Femtomole sequencing of proteins from polyacrylamide gels by nanoelectrospray mass spectrometry. *Nature* **379**, 466–469.

[3]   In-Solution Digestion of Proteins for Mass Spectrometry

By KATALIN F. MEDZIHRADSZKY

Abstract

Mass spectrometry (MS) has gradually replaced classical methods as a major tool in protein sequencing and characterization. However, the sample preparation repertoire has not changed very much; it has just been adjusted to the needs of the new analytical method. In this chapter frequently used in-solution enzymatic digestions and chemical cleavages are reviewed. In addition, some practical recommendations as well as the advantages and shortcomings of the methods are discussed.

Introduction

MALDI and electrospray ionization permit intact molecular weight determination of proteins. However, protein identification, the assignment of covalent modifications, the determination of sequence errors, as well as *de novo* sequencing are carried out usually at the peptide level. The digestion method and the conditions have to be carefully selected based on the protein sequence as well as on the desired information to be obtained. In-solution digestion is preferred over in-gel digestion for more control over the outcome of the process. The conditions (i.e., the pH, protein concentration, digestion buffers, additives, the proteolytic enzyme, and enzyme/substrate ratio) can be altered for in-solution digestions more easily according to the researchers' needs, and the recovery of the digestion products is more reliable. Recombinant proteins produced for therapeutical purposes are usually characterized by mass spectrometry (MS) after carefully controlled in-solution digestion (Bloom *et al.*, 1996; Guzzetta *et al.*, 1993; Ling *et al.*, 1991; Medzihradszky *et al.*, 1997; Rush *et al.*, 1995). Limited proteolysis is carried out frequently when the research is focused on the identification of structural and functional domains of proteins (Ostrelund *et al.*, 1999). Similarly, in-solution digestion is applied usually when protein/protein interactions are probed. For example, an antigen/antibody complex can be subjected to proteolysis, during which the "protected" portion of the antigen remains intact (Jemmerson and Paterson, 1986; Sheshberadaran and Payne, 1988) and can be separated from the antibody to be characterized by MS (van de Water *et al.*, 1997).

METHODS IN ENZYMOLOGY, VOL. 405   0076-6879/05 $35.00
DOI: 10.1016/S0076-6879(05)05003-2

The idea of "protected" surfaces has been utilized in identifying interactive surfaces of proteins as well as substrate binding pockets. Sequence stretches involved in protein/protein interactions or "covered" with the substrate are not available for H-D exchange (Mandell *et al.*, 1998, 2001) or chemical derivatization (Everett *et al.*, 1990) and thus can be identified by mass spectrometric characterization following the appropriate proteolysis. Obviously, in order to gain information on the tertiary structure of proteins, one has to retain the native conformation as much as possible. In-solution covalent labeling or intramolecular cross-linking of proteins may be then followed by in-solution digestion and MS characterization of the products (Tschirret-Guth *et al.*, 1999; Young *et al.*, 2000).

This chapter presents an overview of some frequently used in-solution cleavages, with practical hints, as well as the advantages and shortcomings of some methods listed.

Steps Prior to the Digestion

If the final protein purification step is gel-electrophoresis, electroelution can be used to transfer the protein into solution. However, successful elimination of the sodium dodecyl sulfate (SDS) by acetone precipitation (Konigsberg and Henderson, 1983) as well as by chromatography (Simpson *et al.*, 1987) may be possible although their use certainly leads to significant protein losses. Some researchers recommend a simple elution from the gel, using a formic acid/acetonitrile/isopropanol/water 50:25:15:10 mixture (v/v/v/v) (Feick and Shiozawa, 1990).

In general, the purification protocols have to be altered so that the resulting protein solution will not contain anything that is interfering with the following digestion and cannot be easily removed prior to the analysis steps if necessary. Traditional protocols frequently called for complex buffers with high salt concentrations, for additives to stabilize the enzymes or to slightly increase their activity, and for detergents to disrupt the tertiary structure of the protein to be digested. For digests prepared for analysis by MS, volatile buffers of lower salt concentrations [25–100 mM] are recommended; additives, unless absolutely necessary, are eliminated. To keep the protein soluble and to denature it, urea, guanidine hydrochloride, or organic solvents, such as acetonitrile, are preferable to detergents.

Denaturing proteins disrupts their tertiary structure and, thus, may provide access for the digesting enzymes to all sites. However, secreted and membrane proteins usually feature disulfide bridges that will not be disrupted by simple denaturing measures. Disulfide bonds have to be reduced, and to prevent reoxidation, the newly formed free sulfhydryls have to be blocked prior to the digestion. For best results, the protein should be dissolved in 6 M guanidine hydrochloride and in a buffer of pH ~8.0. (If solid guanidine hydrochloride is added to a solution to achieve a final concentration of 6 M, the volume of the solution will increase 1.75-fold!). Approximately 500-fold excess of dithiothreithol (DTT) at 60° for 1 h will complete the reduction. A 1100-fold excess of iodoacetic acid sodium salt is then added to the mixture and is incubated at room temperature for 1.5 h, under Ar, in the dark. Iodoacetic acid may be preferred to other derivatizing agents because most proteolytic enzymes are active at mildly basic pH, and the additional acidic residues may make the proteins more soluble under these conditions. Once the derivatization is complete, the reagent excess has to be removed either by high-performance liquid chromatography (HPLC) or by dialysis. "Leftover" alkylating agents at pH ~8 may derivatize the ε-amino groups of Lys residues, the side chains of Met and His residues, as well as the N-termini of the peptides formed during the digestion. Both modes of purification lead to some protein losses. In general, dialysis is more recommended than HPLC, since for stickier proteins, it can be performed with urea- or guanidine hydrochloride-containing buffers that may prevent protein precipitation. If this removal step is skipped, the reduction/alkylation should be performed in the presence of a denaturing agent tolerated by the endoprotease, and the solution should be diluted accordingly prior to the digestion. However, the recommended protein concentration for digestion is not lower than 25 μg/ml. Other alkylating agents, such as iodoacetamide, vinylpyridine, etc., also can be applied. Using volatile reducing and alkylating agents, such as triethylphosphyne and iodoethanol, respectively, may permit the elimination of the desalting step (Hale *et al.*, 2004). Glycosylation of the protein also may hinder its accessibility to proteolytic digestions. Thus, in certain instances, the removal of the N-linked carbohydrates (i.e., incubation with peptide N-glycosidase F [PNGase F]) has to be considered. In such cases, one has to keep in mind that PNGase F converts the previously glycosylated Asn residues into aspartic acids.

Enzymatic Digestions

A wide variety of endoproteases can be used for protein cleavages. Digestion protocols may have to be altered to eliminate some of the originally used components, salts, and detergents that are not compatible with mass spectrometry as discussed earlier. In addition, for mass spectrometric purposes, enzymes of high specificity are preferred because the masses of the expected products have to be predicted.

Each protein requires different conditions for optimal digestion. While bovine fetuin (341 residues, N- and O-glycosylated protein) is fully digested with 1% (w/w) trypsin in 1 h (Medzihradszky *et al.*, 1994), Factor VIII polypeptides (368, 372, 643, and 909 residues, variably glycosylated) require a much longer digestion time (16 – 20 h) with 4% enzyme (Medzihradszky *et al.*, 1997). The same protein may be readily accessible for certain enzymes, while it may also require the use of denaturing agents for others. The relatively small shark liver fatty-acid-binding protein (132 residues) required the presence of 2 M urea for endoprotease Lys-C or Glu-C digestions, while it was completely digested with trypsin without denaturation (Medzihradszky *et al.*, 1992).

Tables I and II show the cleavage specificity, pH optimum, and recommended digestion conditions for frequently used endoproteases (Allen, 1989; Riviere *et al.*, 1991). In addition, urea, guanidine hydrochloride, and acetonitrile concentrations are listed when the enzyme still retains a significant portion of its activity. However, long incubations in the presence of urea may result in carbamoylation of the available N-termini (43 Da mass increase) (Wen *et al.*, 1992) as well as the ε-amino groups of Lys-residues (+43 Da and prevented tryptic or Lys-C cleavage). It is important to remember that endoproteases are usually very poor exoproteases. When there are a series of potential cleavage sites in close proximity, once one has been cleaved, the others may be too close to the termini to be digested. Similarly, it has to be considered that most enzymes lose specificity during long incubation times. For example, endoprotease Lys-C will cleave at most tryptic sites. Trypsin will show more chymotryptic activity, but given enough time, "tryptic" cleavage has been observed at about any amino acid. In addition, one has to consider that these enzymes will digest every protein present, including themselves, in a given mixture.

Though endoproteases are added to digestion mixtures, usually in substoichiometric quantities, autolysis products are frequently detected. Side chain-protected trypsin manufactured by reductive alkylation of the ε-amino groups of Lys residues was developed to minimize autolysis by blocking some of the cleavage sites (Rice *et al.*, 1977). However, this attempt was not completely successful. Table III shows some CID-characterized autolysis products of the side-chain-modified porcine trypsin (Promega, Madison, WI). Our results, i.e., the observation of N-terminally and C-terminally modified peptides (Table III), suggest that some of the autolysis must have happened during the derivatization of the protease. Interestingly, the removal of Lys residues as cleavage sites promotes hydrolysis at the C-terminus of Asn residues. This phenomenon was also observed and reported for wild-type trypsin, though at a lesser extent (Vestling *et al.*, 1990). Enzymes can be tricked by chemical modifications: for example, *O*-glycosylated, *O*-phosphorylated, or sulfated Ser and Thr residues after β-elimination and Michael addition of 2-aminoethanethiol

TABLE I
HIGHLY SPECIFIC ENDOPROTEASES

Enzyme (IUBMB enzyme nomenclature)	Specificity[a]	pH	Tolerates[b]	Recommended (37°)
Trypsin[c](EC 3.4.21.4)	Arg↓, Lys↓	7–8.5	[d]Urea: 6 M / Gu.HCl: 1 M / MeCN: 40%	1–5%; 2–18 h
Endoprotease Arg-C[e] / Clostridium histolyticum (EC 3.4.22.8)	Arg↓	7.2–8	Urea: 4 M / MeCN: 10%	0.5–2%; 1–18 h / 1–5 mM DTT[e]
Endoprotease Glu-C / Staphylococcus aureus V8 (EC 3.4.21.19)	Glu↓, Asp↓	7.5–8.5	Urea: 2 M / Gu.HCl: 1 M / MeCN: 20%	1–5%; 4–18 h
Endoprotease Lys-C / Achromobacter lyticus (EC 3.4.21.50)	Lys↓	8–9.5	Urea: 8 M / Gu.HCl: 2 M / MeCN: 40%	0.5–2%; 2–6 h
Endoprotease Asp-N / Pseudomonas fragi mutant (EC 3.4.24.33)	↓Asp, ↓cysteic acid (↓Glu)	6–8.5	Urea: 1 M / Gu.HCl: 1 M / MeCN: 10%	0.5–5%; 2–18 h
Prolyl endopeptidase[f] / Flavobacterium meningosepticum (EC 3.4.21.26)	Pro↓	7–7.5		0.1–1%; 1–4 h

[a] Certain residues adjacent to the cleavage site may slow down the hydrolysis process. Pro residues usually have a prohibitive affect.
[b] This recommendation is based on observations in the authors' laboratory, on manufacturers' recommendations, and data published (Riviere et al., 1991). Gu.HCl stands for guanidine hydrochloride; MeCN stands for acetonitrile.
[c] Trypsin preparations always display some chymotryptic activity, and autolysis may yield peptides with C-terminal Asn residues, especially if the Lys-cleavage sites are blocked (see Table III).
[d] Long incubation in the presence of urea may result in the carbamoylation of the newly formed N-termini as well as the ε-amino groups of Lys residues, thus eliminating this residue as tryptic or Lys-C cleavage site.
[e] Requires the presence of thiol for full activity.
[f] This enzyme does not digest proteins but will cleave peptides approximately up to 20 to 30 residues.

TABLE II

LESS SPECIFIC ENDOPROTEASES

Enzyme (IUBMB enzyme nomenclature)	Preference[a]	pH	Tolerates[b]	Recommended
Chymotrypsin (EC 3.4.21.1)	Phe↓, Trp↓, Tyr↓, (Leu↓, Met↓) (Adjacent Pro prevents cleavage)	7–9	Urea: 2 M Gu.HCl; 2 M MeCN; 30%	1-5%; 2-18 h, 37°
Pepsin[c] (EC 3.4.23.1)	Phe↓, Met↓, Leu↓ (Adjacent hydrophobic residues preferred; prolyl peptide bonds are not cleaved)	2–4		1%1-2 h, RT
Proteinase K Tritirachium album (EC 3.4.21.64)	Aliphatic residue↓ Aromatic residue↓ Hydrophobic residue↓	6.5–9.5	Urea: 2 M	1-2°;1-8 h, 37°
Thermolysin (EC 3.4.24.27)	↓Leu, ↓Ile, ↓Phe, ↓Trp, ↓Met, ↓Val (Residue cannot have a Pro at its C terminus)	7–9	Up to 80° Urea: 8 M	0.5-5%; 45°, 2-6 h

[a] While chymotrypsin exhibits specificity for aromatic amino acids, the other enzymes cleave rather randomly, displaying some preferences.
[b] These recommendations are based on observations in the authors' laboratory, on manufacturers' recommendations, and data published (Allen, 1989; Riviere et al., 1991). Gu.HCl stands for guanidine hydrochloride; MeCN stands for acetonitrile.
[c] For quick digestions, immobilized pepsin may be used at a 1:1 ratio for a few minutes (Mandell et al., 1998).

TABLE III

FREQUENTLY OBSERVED AUTOLYSIS PRODUCTS OF SIDE CHAIN-PROTECTED PORCINE TRYPSIN (PROMEGA), CHARACTERIZED BY LOW ENERGY ESIMS–CID

$MH^+_{calc.}$	Position	Sequence
515.3306	[46–49]	IQVR
842.5100	[100–107]	(R)VATVSLPR(S)
856.5256	[100–107]	(R)V*ATVSLPR(S)
870.5412	[100–107]	(R)V**ATVSLPR(S)
1045.5642	[90–99]	(K)LSSPATLNSR(V)
1126.5645	[70–79]	(K)IITHPNFNGN(T)
1420.7225	[212–223]	(N)YVNWIQQTIAAN(<)
1531.8405	[84–97]	(N)DIM(O)LIKLSSPATLN(S)
1940.9354	[50–66]	(R)LGEHNIDVLEGNEQFIN(A)
2003.0734	[80–97]	(N)TLDNDIM(O)LIK**LSSPATLN(S)
2211.1046	[50–69]	(R)LGEHNIDVLEGNEQFINAAK(I)
2225.1202	[50–69]	(R)LGEHNIDVLEGNEQFINAAK*(I)
2239.1358	[50–69]	(R)LGEHNIDVLEGNEQFINAAK**(I)
2283.1807	[70–89]	(K)IITHPNFNGNTLDNDIMLIK(L)
2299.1756	[70–89]	(K)IITHPNFNGNTLDNDIM(O)LIK(L)
2678.3822	[76–99]	(N)FNGNTLDNDIM(O)LIK**LSSPATLNSR(V)
2807.3145	[20–45]	(N)SGSHFC*GGSLINSQWVVSAAHC*YK**SR(I)
2914.5062	[50–75]	(R)LGEHNIDVLEGNEQFINAAK**IITHPN(F)
3094.6246	[70–97]	(K)IITHPNFNGNTLDNDIMLIK**LSSPATLN(S)
3337.7577	[70–99]	(K)IITHPNFNGNTLDNDIMLIK**LSSPATLNSR(V)
3353.7526	[70–99]	(K)IITHPNFNGNTLDNDIM(O)LIK**LSSPATLNSR(V)

Note: N-terminal amino acids or Lys residues labeled with * or ** indicate the presence of 1 or 2 methylgroups at the terminus or on the side chain, respectively; C* = half cystine; Met(O) = Met sulfoxide. For a more complete list, visit http://prospector.ucsf.edu.

will become Lys analogues and, thus, cleavage sites for trypsin or endopro-tease Lys-C (Rusnak *et al.*, 2002). The specificity of recombinant enzymes may be altered by design: for example, a Tyr-specific trypsin mutant has been described (Pal *et al.*, 2004).

Immobilized Enzymes

A variation for in-solution digestion is the utilization of immobilized enzymes. Some enzymes that are immobilized on HPLC cartridges are commercially available; these include trypsin, endoprotease Glu-C, and pepsin (e.g., from Pierce, Rockford, IL). The cartridges are reusable. Small volume samples can be digested on-column with little dilution. The high enzyme excess that can be applied accelerates the digestion process: the reaction time can be controlled by the flow rate. Autolysis products can be eliminated this way. The digestion and the following fractionation readily can be automated (Hara *et al.*, 2000; Hsieh *et al.*, 1996). The determination of *in vivo* adducts of mitochondrial aldehyde dehydrogenase with disulfi-ram, a drug used in the aversion therapy treatment of alcoholics, is an excellent demonstration for the combination of on-line in-solution diges-tion and mass spectrometry (Shen *et al.*, 2001). Immobilized enzymes are preferred when studying H-D-exchanged proteins. To prevent the loss of the labels, the proteolysis as well as MS analysis has to be accelerated. These studies are carried out performing digestions at 0° with immobilized pepsin (1:1 w/w) that reduces the digestion time to approximately 10 min, and the samples are analyzed by MALDI mass spectrometry with "frozen" sample introduction (Mandell *et al.*, 1998). Trypsin and endoprotease Glu-C also have been used immobilized on paramagnetic beads (Krogh *et al.*, 1999). This approach may be preferable when a series of different enzymes have to be used to achieve the desired results. Microfluidic reac-tors containing immobilized enzymes also have been fabricated (Krenkova and Foret, 2004).

Chemical Cleavages

Only a handful of chemical methods yield relatively predictable cleav-age products; however, these chemicals may succeed when enzymes fail or provide more specific alternative methods for sequences that can be di-gested only by nonspecific endoproteases (Allen, 1989; Smith, 1997). Prior reduction of the disulfide bridges and alkylation of the sulfhydryls that helps enzymatic digestions also may lead to higher yields for chemical cleavages.

Cleavage of Asn-Gly Bonds

The cleavage of Asn-Gly bonds can be performed with 2-M hydroxylamine solution at pH 9 in the presence of 2-M guanidine hydrochloride at 45° for 4 h. Since the Gly residue does not have a side-chain, there is no steric hindrance, and the Asn residue may form a cyclic imide. Its β-amide reacts with the amide N of the peptide bond between the Asn and Gly residues:

$$-NH-CH-CO-N-CH_2-CO-$$
$$\begin{array}{c} | \quad \diagup \\ CH_2\text{-}CO \end{array}$$

This succinimide reacts with the hydroxylamine, yielding a free N-terminus at the Gly-side and an aspartyl hydroxamate as the C terminus of the other new peptide, which alters its elemental composition and the mass accordingly by an additional O atom (Blodgett *et al.*, 1985; Bornstein and Balian, 1977). Both the α-~NH-CH(CH$_2$-COOH)-CO-NHOH and β-aspartyl hydroxymate ~NH-CH(CH$_2$-CO-NHOH)-COOH may form. The reaction can be stopped by acidifying the mixture.

With extended reaction time, Asn-any residue bonds may be cleaved, and Asn and Gln residues may yield hydroxamate derivatives (Bornstein and Balian, 1977). Similarly, some posttranslational modifications may not survive cleavage conditions: for example, fatty acids attached to Cys residues may be released (Weimbs and Stoffel, 1992).

Cleavage of Asp-Xxx Bonds

This cleavage is carried out with diluted (~10 mM) hydrochloric acid in a sealed tube at 108° for 2 h (Smith, 1997). Under these conditions, a series of side reactions can be expected, such as the deamidation of Gln and Asn residues, their potential cyclization, the peptide bond cleavage (see previous paragraphs), and the decomposition of Trp residues. In case of glycoproteins, acid-sensitive neuraminic acid will be lost as well.

Cleavage of Cys-Xxx Bonds

The Cys residues are converted to S-cyanocysteine by reaction with 2-nitro-5-thiocyanobenzoate at pH 8. The cyanylated protein is cleaved at the Cys residues by incubation at pH 9 at 37° for 16 h or longer (Jacobson *et al.*, 1983). The new peptides will bear an iminothiazolidine-4-carboxyl residue:

$$NH=C \overset{\displaystyle \overset{NH-CH-CO-}{\diagup \quad |}}{\underset{\diagdown_{S}\diagup}{\quad \quad CH_2}}$$

at their N-termini, while the new C-termini will be free of carboxylic acids. Wu and Watson (1997) modified this protocol for disulfide-bridge assignment. Partial reduction was carried out under acidic conditions (pH ~3) to prevent disulfide shuffling, and newly formed sulfhydryls were immediately cyanylated by 1-cyano-4-dimethylamino-pyridinium tetrafluoroborate (Wakselman and Guibe-Jampel, 1976).

Extended incubation at high pH may lead to the cleavage of additional peptide bonds as well as to the β-elimination of O-linked carbohydrates, H_3PO_4 or H_2SO_4, from modified Ser and Thr residues (Medzihradszky *et al.*, 2004; Rusnak *et al.*, 2002).

Cleavage of Met-Xxx Bonds

The cleavage is performed with large excess of CNBr in acid at room temperature for 12 to 24 h. The new peptides have a homoserine: -NH-CH(CH$_2$-CH$_2$-OH)-COOH open (residue weight 101.0477) or forming a lacton ring at their C-termini (Gross and Witkop, 1961). Met-Thr, and at a lesser extent Met-Ser, bonds may be cleaved with a lower yield even when the Met is converted to homoserine (Schroeder *et al.*, 1969). Because of the low pH and long incubation times, Asp-Xxx, especially Asp-Pro bonds, may be cleaved, and other side reactions listed previously for acidic conditions may occur. Traditionally, the reaction was carried out in 70% formic acid. Incubation with formic acid causes the formylation of hydroxy-amino acids, the newly formed homoserines included (Beavis and Chait, 1990). This esterification is reversible by incubation with 0.1% trifluoroacetic acid (TFA) in water at room temperature for approximately 24 h. To prevent such side reactions, the formic acid can be replaced by 50 to 70% TFA, which does not produce esterified side chains.

Analysis of the Digests

When the analysis has to be accelerated, for example, in H-D exchange studies, MALDI–TOF analysis of the unseparated digest is the method of choice (Mandell *et al.*, 1998, 2001).

Whenever high sequence coverage is a prerequisite for the success in order to achieve this outcome successfully, the digest should be fractionated by reversed-phase HPLC and subjected to MS analysis on-line or off-line.

Since the presence of TFA adversely affects the detection sensitivity in electrospray ionization, formic acid is recommended as the ion pair-forming reagent of the mobile phase for on-line liquid chromatography (LC)–MS analysis. To achieve chromatographic resolution comparable to that with the TFA-containing solvents, the acetonitrile may be replaced by an ethanol/propanol 5:2 mixture (Medzihradszky et al., 1994). In addition, when formic acid is the ion-pairing agent negatively charged molecules, such as sialylated glycopeptides, phospho- and sulfo-peptides will feature longer retention times than their unmodified counterparts (Medzihradszky et al., 1994, 2004).

For comprehensive protein characterization, picomoles, if not nanomoles, of the protein should be available, even if we have instrumentation that is routinely capable of femtomole-level sample detection. One of the best-documented examples of comprehensive protein characterization is bovine fetuin. From a single LC-MS analysis of about 20 picomoles of a tryptic digest, reproducibly almost the entire sequence is covered, as shown here:

^{1}IPLDPVAGYK EPACDDPDTE QAALAAVDYI NKHLPRGYKH TLNQIDSVKV^{50}
^{51}WPRRPTGEVY DIEIDTLETT CHVLDPTPLA N*CSVRQQTQH AVEGDCDIHV^{100}
^{101}LKQDGQFSVL FTKCDSSPDS AEDVRKLCPD CPLLAPLN*DS RVVHAVEVAL^{150}
^{151}ATFNAESN*GS YLQLVEISRA QFVPLPVSVS VEFAVAATDC IAKEVVDPTK^{200}
^{201}CNLLAEKQYG FCKGSVIQKA LGGEDVRVTC TLFQTQPVIP QPQPDGAEAE^{250}
^{251}APSAVPDAAG PTPSAAGPPV ASVVVGPSVV AVPLPLHRAH YDLRHTFSGV^{300}
^{301}ASVESSSGEA FHVGKTPIVG QPSIPGGPVR LCPGRIRYFK I^{341}

In addition, from the masses observed, the carbohydrate-heterogeneity at the N-glycosylation sites (Asn-81, Asn-138, and Asn-158 [labeled with asterisks]) could be addressed. It can also be determined that one of these sites, Asn-158, is not 100% occupied. Similarly, the carbohydrate heterogeneity for the O-linked glycopeptide [228–288] can be addressed (Medzihradszky et al., 1994). However, the structure of carbohydrates and the sites of O-glycosylation cannot be determined from a single experiment. In addition, another O-linked glycopeptide, [316–330], which is present at much lower quantities, goes usually undetected without prior enrichment of the O-linked species by Jacaline-agarose chromatography (R. R. Townsend and K. F. Medzihradszky, unpublished results). Fetuin also contains phosphopeptides that were discovered only recently (Thompson et al., 2003). This is a common problem: covalent labels, xenobiotic, or posttranslational modifications are usually present in a digest in substoichiometric quantities. In addition, some of them, like small, highly glycosylated peptides, phosphopeptides, or sulfopeptides may be too hydrophilic to be retained on the column during fractionation. Very hydrophobic species, such as palmitoylated peptides or transmembrane regions, may never elute. The fractionation has to be tailored to the project. Thus, special tracking, such as monitoring a characteristic UV-absorbance or radioactivity, could be employed during fractionation if the study is aimed at the characterization of some specific modification.

Fragmentation induced in the ion source may permit the identification of modified compounds that produce unique fragment ions. For example, glycopeptide-containing fractions can be identified by monitoring the presence of a mass-to-charge ratio (m/z) 204 ion, an oxonium fragment for N-acetylhexosamines (Huddleston *et al.*, 1993), while phosphopeptides and sulfopeptides yield diagnostic negative ion fragments at m/z 79 and 80, respectively (Bean *et al.*, 1995). Precursor ion scanning also may be employed to identify peptides containing a certain residue or modification. Carbohydrate ions may be used for glycopeptide identification (Carr *et al.*, 1993), and immonium ions can be used to identify peptide ions barely above the noise level (Wilm *et al.*, 1996). Scanning for the precursors of the phosphorylated immonium ion of Tyr aids the identification of such modified peptides (Steen *et al.*, 2001). In general, phosphopeptides can be identified as precursors of the m/z 79 negative ion (Carr *et al.*, 1996). Similarly, isotope-coded affinity tag (ICAT)-modified peptides (Gygi *et al.*, 1999) can be identified as precursors of diagnostic fragments (Baldwin *et al.*, 2001). However, with proper LC-MS conditions, most components elute in narrow peaks (i.e., on-line precursor ion scanning may not be possible at the desired sensitivity level).

Thus, precursor ion scanning can be performed most efficiently on nanospray-introduced HPLC fractions of the digests. Frequently, a special enrichment method can be applied instead of or prior to reversed-phase chromatography. Such purification methods are, for example, the immobilized metal ion affinity chromatography (IMAC)-enrichment of phosphopeptides (Ficarro *et al.*, 2002; Neville *et al.*, 1997; Nuwaysir and Stults, 1993; Posewitz and Tempst, 1999; Zhou *et al.*, 2000), the lectin-based affinity chromatography of sugars (Hortin, 1990; Krogh *et al.*, 1999; Treuheit *et al.*, 1992), as well as the extraction of biotinylated peptides with avidin (Girault *et al.*, 1996). Even the enriched fractions are usually subjected to multiple MS analyses, first to determine the complexity of the mixture and the peptide masses and then to gain more detailed structural information by the MS/MS analysis of the selected components.

Summary

Endoproteases or chemical methods can be applied under controlled conditions to produce peptides with predictable results. A wide variety of software programs are available for such predictions, such as MS-digest in ProteinProspector (University of California, San Francisco, http:// prospector.ucsf.edu), PAWS (ProteoMetrics, New York, http://prowl. rockefeller.edu), or Sherpa (Biochemistry Department, University of Washington, http://hairyfatguy.com/Sherpa). In-solution digestions are recommended when the native conformation of the protein has to be retained as well as when complete sequence coverage is desirable. For the analysis of such digests on-line LC-MS with high chromatographic resolution is preferable. For comprehensive protein characterization, multiple digestions and analytical steps may be necessary. Thus, the sample requirement for such analyses may be much higher than the detection sensitivity of one's mass spectrometer.

Acknowledgment

This work was supported by NIH grants NCRR RR01614, RR01296, RR014606, and RR015804 to the UCSF Mass Spectrometry Facility Director A. L. Burlingame.

References

Allen, G. (1989). Specific cleavage of the protein. *In* "Sequencing of Proteins and Peptides," 2nd ed., pp.73–104. Elsevier, Amsterdam.

Baldwin, M. A., Medzihradszky, K. F., Lock, C. M., Fisher, B., Settineri, C. A., and Burlingame, A. L. (2001). Matrix-assisted laser desorption/ionization coupled with quadrupole/orthogonal acceleration time-of-flight mass spectrometry for protein discovery, identification, and structural analysis. *Anal. Chem.* **73**, 1707–1720.

Bean, M. F., Annan, R. S., Hemling, M. E., Mentzer, M., Huddleston, M. J., and Carr, S. A. (1995). LC–MS methods for the selective detection of post-translational modifications in proteins: Glycosylation, phosphorylation, sulfation, and acylation. *In* "Techniques in Protein Chemistry VI" (J. W. Crabb, ed.), pp. 107–116. Academic Press, San Diego, CA.

Beavis, R. C., and Chait, B. T. (1990). Rapid, sensitive analysis of protein mixtures by mass spectrometry. *Proc. Natl. Acad. Sci. USA* **87**, 6873–6877.

Blodgett, J. K., Londin, G. M., and Collins, K. D. (1985). Specific cleavage of peptides containing an aspartic-acid (beta-hydroxamic acid) residue. *J. Am. Chem. Soc.* **107**, 4305–4313.

Bloom, J. W., Madanat, M. S., and Ray, M. K. (1996). Cell line and site specific comparative analysis of the N-linked oligosaccharides on human ICAM-1des454–532 by electrospray ionization mass spectrometry. *Biochemistry* **35**, 1856–1864.

Bornstein, P., and Balian, G. (1977). Cleavage at Asn-Gly bonds with hydroxylamine. *Meth. Enzymol.* **47**, 132–145.

Carr, S. A., Huddleston, M. J., and Bean, M. F. (1993). Selective identification and differentiation of N- and O-linked oligosaccharides in glycoproteins by liquid chromatography–mass spectrometry. *Protein Sci.* **2**, 183–196.

Carr, S. A., Huddleston, M. J., and Annan, R. S. (1996). Selective detection and sequencing of phosphopeptides at the femtomole level by mass spectrometry. *Anal. Biochem.* **239**, 180–192.

Everett, E. A., Falick, A. M., and Reich, N. O. (1990). Identification of a critical cysteine in EcoRI DNA methyltransferase by mass spectrometry. *J. Biol. Chem.* **265**, 17713–17719.

Feick, R. G., and Shiozawa, J. A. (1990). A high-yield method for the isolation of hydrophobic proteins and peptides from polyacrylamide gels for protein sequencing. *Anal. Biochem.* **187,** 205–211.

Ficarro, S. B., McCleland, M. L., Stukenberg, P. T., Burke, D. J., Ross, M. M., Shabanowitz, J., Hunt, D. F., and White, F. M. (2002). Phosphoproteome analysis by mass spectrometry and its application to *Saccharomyces cerevisiae. Nat. Biotechnol.* **20,** 301–305.

Girault, S., Chassaing, G., Blais, J. C., Brunot, A., and Bolbach, G. (1996). Coupling of MALDI–TOF mass analysis to the separation of biotinylated peptides by magnetic streptavidin beads. *Anal. Chem.* **68,** 2122–2126.

Gross, E., and Witkop, B. (1961). Selective cleavage of the methionyl peptide bonds in ribonuclease with cyanogen bromide. *J. Am. Chem. Soc.* **83,** 1510–1511.

Guzzetta, A. W., Basa, L. J., Hancock, W. S., Keyt, B. A., and Bennett, W. F. (1993). Identification of carbohydrate structures in glycoprotein peptide maps by the use of LC–MS with selected ion extraction with special reference to tissue plasminogen activator and a glycosylation variant produced by site directed mutagenesis. *Anal. Chem.* **65,** 2953–2962.

Gygi, S. P., Rist, B., Gerber, S. A., Turecek, F., Gelb, M. H., and Aebersold, R. (1999). Quantitative analysis of complex protein mixtures using isotope-coded affinity tags. *Nat. Biotechnol.* **17,** 994–999.

Hale, J. E., Butler, J. P., Gelfanova, V., You, J. S., and Knierman, M. D. (2004). A simplified procedure for the reduction and alkylation of cysteine residues in proteins prior to proteolytic digestion and mass spectral analysis. *Anal. Biochem.* **333,** 174–181.

Hara, S., Katta, V., and Lu, H. S. (2000). Peptide map procedure using immobilized protease cartridges in tandem for disulfide linkage identification of *neu* differentiation factor epidermal growth factor domain. *J. Chromatogr. A.* **867,** 151–160.

Hortin, G. L. (1990). Isolation of glycopeptides containing O-linked oligosaccharides by lectin affinity chromatography on jacalin-agarose. *Anal. Biochem.* **191,** 262–267.

Hsieh, Y. L. F., Wang, H., Elicone, C., Mark, J., Martin, S. A., and Regnier, F. (1996). Automated analytical system for the examination of protein primary structure. *Anal. Chem.* **68,** 455–462.

Huddleston, M. J., Bean, M. F., and Carr, S. A. (1993). Collisional fragmentation of glycopeptides by electrospray ionization LC–MS and LC–MS/MS: Methods for selective detection of glycopeptides in protein digests. *Anal. Chem.* **65,** 877–884.

Jacobson, G. R., Schaffer, M. H., Stark, G. R., and Vanaman, T. C. (1983). Specific chemical cleavage in high yield at the amino peptide bonds of cysteine and cystine residues. *J. Biol. Chem.* **248,** 6583–6591.

Jemmerson, R., and Paterson, Y. (1986). Mapping epitopes on a protein antigen by the proteolysis of antigen/antibody complexes. *Science* **232,** 1001–1004.

Konigsberg, W., and Henderson, H. L. (1983). Removal of sodium dodecyl sulfate from proteins by ion-pair extraction. *Meth. Enzymol.* **91,** 254–259.

Krenkova, J., and Foret, F. (2004). Immobilized microfluidic enzymatic reactors. *Electrophoresis* **25,** 3550–3563.

Krogh, T. N., Berg, T., and Hojrup, P. (1999). Protein analysis using enzymes immobilized to paramagnetic beads. *Anal. Biochem.* **274,** 153–162.

Ling, V. A., Guzzetta, W., Canova-Davis, E., Stults, J. T., Hancock, W. S., Covey, T. R., and Shushan, B. I. (1991). Characterization of the tryptic map of recombinant DNA-derived tissue plasminogen activator by high-performance liquid chromatography–electrospray ionization mass spectrometry. *Anal. Chem.* **63,** 2909–2915.

Mandell, J. G., Baerga-Ortiz, A., Akashi, S., Takio, K., and Komives, E. A. (2001). Solvent accessibility of the thrombin/thrombomodulin interface. *J. Mol. Biol.* **306,** 575–589.

Mandell, J. G., Falick, A. M., and Komives, E. A. (1998). Identification of protein/protein interfaces by decreased amide proton solvent accessibility. *Proc. Natl. Acad. Sci. USA* **95,** 14705–14710.

Medzihradszky, K. F., Maltby, D. A., Hall, S. C., Settineri, C. A., and Burlingame, A. L. (1994). Characterization of protein N-glycosylation by reversed-phase microbore liquid chromatography/electrospray mass spectrometry, complementary mobile phases, and sequential exoglycosidase digestion. *J. Am. Soc. Mass Spectrom.* **5,** 350–358.

Medzihradszky, K. F., Darula, Z., Perlson, E., Fainzilber, M., Chalkley, R. J., Ball, H., Greenbaum, D., Bogyo, M., Tyson, D. R., Bradshaw, R. A., and Burlingame, A. L. (2004). O-sulfonation of serine and threonine: Mass spectrometric detection and characterization of a new post-translational modification in diverse proteins throughout the eukaryotes. *Mol. Cell. Proteomics* **3,** 429–440.

Medzihradszky, K. F., Besman, M. J., and Burlingame, A. L. (1997). Structural characterization of site-specific N-glycosylation of recombinant human factor VIII by reversed-phase high-performance liquid chromatography-electrospray ionization mass spectrometry. *Anal. Chem.* **69,** 3986–3994.

Medzihradszky, K. F., Gibson, B. W., Kaur, S., Yu, Z., Medzihradszky, D., Burlingame, A. L., and Bass, N. M. (1992). The primary structure of fatty-acid-binding protein from nurse shark liver. Structural and evolutionary relationship to the mammalian fatty-acid-binding family. *Eur. J. Biochem.* **203,** 327–339.

Neville, D. C. A., Rozanas, C. R., Price, E. M., Grius, D. B., Verkman, A. S., and Townsend, R. R. (1997). Evidence for phosphorylation of serine 753 in CFTR using a novel metal-ion affinity resin and matrix-assisted laser desorption mass spectrometry. *Protein Sci.* **6,** 2436–2445.

Nuwaysir, L. M., and Stults, J. T. (1993). Electrospray ionization mass spectrometry of phosphopeptides isolated by on-line immobilized metal-ion affinity chromatography. *J. Am. Soc. Mass Spectrom.* **4,** 662–669.

Ostrelund, T., Beussman, D. J., Julenius, K., Poon, P. H., Linse, S., Shabanowitz, J., Hunt, D. F., Schotz, M. C., Derewenda, Z. S., and Holm, C. (1999). Domain identification of hormone-sensitive lipase by circular dichroism and fluorescence spectroscopy, limited proteolysis, and mass spectrometry. *J. Biol. Chem.* **274,** 15382–15388.

Pal, G., Patthy, A., Antal, J., and Graf, L. (2004). Mutant rat trypsin selectively cleaves tyrosyl peptide bonds. *Anal. Biochem.* **326,** 190–199.

Posewitz, M. C., and Tempst, P. (1999). Immobilized gallium(III) affinity chromatography of phosphopeptides. *Anal. Chem.* **71,** 2883–2892.

Rice, R. H., Means, G. E., and Brown, W. D. (1977). Stabilization of bovine trypsin by reductive methylation. *Biochem. Biophys. Acta.* **492,** 316–321.

Riviere, L. R., Fleming, M., Elicone, C., and Tempst, P. (1991). Study and applications of the effects of detergents and chaotropes on enzymatic proteolysis. *In* "Techniques in Protein Chemistry II" (J. J. Villafranca, ed.), pp. 171–179. Academic Press, San Diego, CA.

Rush, R. S., Derby, P. L., Smith, D. M., Merry, C., Rogers, G., Rohde, M. F., and Katta, V. (1995). Microheterogeneity of erythropoietin carbohydrate structure. *Anal. Chem.* **67,** 1442–1452.

Rusnak, F., Zhou, J., and Hathaway, G. M. (2002). Identification of phosphorylated and glycosylated sites in peptides by chemically targeted proteolysis. *J. Biomol. Tech.* **13,** 228–237.

Schroeder, W. A., Shelton, J. B., and Shelton, J. R. (1969). An examination of conditions for the cleavage of polypeptide chains with cyanogen bromide: Application to catalase. *Arch. Biochem. Biophys.* **130,** 551–556.

Shen, M. L., Johnson, K. L., Mays, D. C., Lipsky, J. L., and Naylor, S. (2001). Determination of *in vivo* adducts of disulfiram with mitochondrial aldehyde dehydrogenase. *Biochem. Pharmacol.* **61**, 537–545.

Sheshberadaran, H., and Payne, L. G. (1988). Protein antigen-monoclonal antibody contact sites investigated by limited proteolysis of monoclonal antibody-bound antigen: Protein "footprinting." *Proc. Natl. Acad. Sci. USA* **85**, 1–5.

Simpson, R. J., Moritz, R. L., Nice, E. E., and Grego, B. (1987). A high-performance liquid chromatography procedure for recovering subnanomole amounts of protein from SDS-gel electroeluates for gas-phase sequence analysis. *Eur. J. Biochem.* **165**, 21–29.

Smith, B. (1997). Chemical cleavage of polypeptides. In "Protein Sequencing Protocols. Methods in Molecular Biology" (B. J. Smith, ed.), pp. 55–72. Humana Press, Totowa, NJ.

Steen, H., Küster, B., Fernandez, M., Pandey, A., and Mann, M. (2001). Detection of tyrosine phosphorylated peptides by precursor ion scanning quadrupole TOF mass spectrometry in positive ion mode. *Anal. Chem.* **73**, 1440–1448.

Thompson, J. S., Hart, R., Franz, C., Barnouin, K., Ridley, A., and Cramer, A. (2003). Characterization of protein phosphorylation by mass spectrometry using immobilized metal ion affinity chromatography with on-resin beta-elimination and Michael addition. *Anal. Chem.* **75**, 3232–3243.

Treuheit, M. J., Costello, C. E., and Halsall, H. B. (1992). Analysis of the five glycosylation sites of human alpha 1-acid glycoprotein. *Biochem. J.* **283**, 105–112.

Tschirret-Guth, R. A., Medzihradszky, K. F., and deMontellano, P. R. O. (1999). Trifluoromethyldiazirinylphenyldiazenes: New hemoprotein active site probes. *J. Am. Chem. Soc.* **121**, 4731–4737.

van de Water, J., Deininger, S. O., Macht, M., Przybylski, M., and Gerschwin, M. E. (1997). Detection of molecular determinants and epitope mapping using MALDI–TOF mass spectrometry. *Clin. Immunol. Immunpathol.* **85**, 229–235.

Vestling, M. M., Murphy, C. M., and Fenselau, C. (1990). Recognition of trypsin autolysis products by high-performance liquid chromatography and mass spectrometry. *Anal. Chem.* **62**, 2391–2394.

Wakselman, M., and Guibe-Jampel, E. (1976). 1-cyano-4-dimethylamino-pyridinium salts— New water-soluble reagents for cyanylation of protein sulfhydryl groups. *J. Chem. Soc. Chem. Commun.* 21–22.

Wen, D. X., Livingston, B. D., Medzihradszky, K. F., Kelm, S., Burlingame, A. L., and Paulson, J. C. (1992). Primary structure of Galβ1,3(4)GlcNAc α2,3-sialyltransferase reveals a conserved region in the sialyltransferase family. *J. Biol. Chem.* **267**, 21011–21019.

Weimbs, T., and Stoffel, W. (1992). Proteolipid protein (PLP) of CNS myelin: Positions of free, disulfide-bonded, and fatty acid thioester-linked cysteine residues and implications for the membrane topology of PLP. *Biochemistry* **31**, 12289–12296.

Wilm, M., Neubauer, G., and Mann, M. (1996). Parent ion scans of unseparated peptide mixtures. *Anal. Chem.* **68**, 527–533.

Wu, J., and Watson, J. T. (1997). A novel methodology for assignment of disulfide bond pairings in proteins. *Protein Sci.* **6**, 391–398.

Young, M. M., Tang, N., Hempel, J. C., Oshiro, C. M., Taylor, E. W., Kuntz, I. D., Gibson, B. W., and Dollinger, G. (2000). High throughput protein fold identification by using experimental constraints derived from intramolecular cross-links and mass spectrometry. *Proc. Natl. Acad. Sci. USA* **97**, 5802–5806.

Zhou, W., Merrick, B. A., Khaledi, M. G., and Tomer, K. B. (2000). Detection and sequencing of phosphopeptides affinity bound to immobilized metal ion beads by matrix-assisted laser desorption/ionization mass spectrometry. *J. Am. Soc. Mass Spectrom.* **11**, 273–282.

[4] Identification of Phosphorylation Sites Using Microimmobilized Metal Affinity Chromatography

By GARRY L. CORTHALS, REUDI AEBERSOLD, AND DAVID R. GOODLETT

Abstract

One of the most important roles that mass spectrometry (MS) has played in the late twentieth and early twenty-first centuries has been to assist in the growth of knowledge of dynamic phosphorylation events. Not only has MS allowed researches to pinpoint the site of phosphorylation, but it has also enabled them to identify the kinase/phosphatase pairs responsible for regulation of a specific modification as well as to follow the functional consequences of the observed phosphorylation events on the biology of the system. For phosphorylation analysis, the important contribution of MS has been critical but not definitive. There are numerous methods that have been applied with success, yet none are generally applicable to all analyses. So, for the time being, researchers in the field must select from a panel of methods to find (de)phosphorylation events. In the work described in this chapter, a collection of integrated methods are presented. A detailed account is provided for phosphorylation capture via on- and off-line immobilized metal affinity chromatography (IMAC). This is followed by a suite of useful strategies for discovery of phosphorylation positioning through sequence determination by phosphate-specific diagnostic ion scans, including precursor and product ion scans, neutral loss scans, and in-source dissociation and post-source decay.

Introduction

More than two hundred different types of posttranslational protein modifications have been described (Krishna and Wold, 1993), but to date, only a few have been shown to be reversible and of regulatory importance in biological systems. Protein phosphorylation has received the most attention because protein phosphorylation and dephosphorylation events result in functional consequences (Hunter, 1995, 2000). Two counteracting enzyme systems, kinases and phosphatases, regulate protein phosphorylation and dephosphorylation, respectively. The structures, specificities, and regulation of the most common of these have been extensively studied and are under continued investigation (Hunter, 1995, 2000). There are assumed to be hundreds of protein kinases/phosphatases differing in their substrate

METHODS IN ENZYMOLOGY, VOL. 405
0076-6879/05 $35.00
DOI: 10.1016/S0076-6879(05)05004-4

specificities, kinetic properties, tissue distribution, and association with regulatory pathways. For yeast (*Saccharomyces cerevisiae*) alone, analysis of the complete genomic sequence for motifs that are thought to be indicative of protein kinases and phosphatases predicts 123 different protein kinases and 40 protein phosphatases. Thus, approximately 2% of expressed yeast proteins are involved in performing protein (de)phosphorylation.

The most common types of protein (de)phosphorylation encountered are in serine, threonine, and tyrosine. The phosphoramidates of arginine, histidine, and lysine also occur as do acyl derivatives of aspartic and glutamic acid although they are less abundant. Some of these modifications are not typically observed unless specific precautions are taken to prevent their elimination during protein isolation (Krishna and Wold, 1993). Since all mass spectrometry (MS) methods rely on analysis of specific mass alterations to amino acids to discriminate the various types of phosphorylation, it is important to view the structural similarities and differences between the most commonly encountered amino acids (see Table I).

There are three principal goals in protein phosphorylation. First, a researcher should determine the amino acid residues that are phosphorylated *in vivo* for a protein present in a cell in a given biological state. Second, it is important for a researcher to identify the kinase/phosphatase pair responsible for regulation of a specific modification. Third, a researcher must also understand the functional consequences of the observed phosphorylation events on the biology of the system. Among these aims, the first is where the use of MS can make a significant impact. Isolation of the phosphoprotein and then isolation of the phosphopeptide are critical to the success of any protein phosphorylation study. This task is typically difficult because many phosphorylated proteins are present in cells in extremely small amounts. It is often the case that even when a phosphoprotein is expressed at amounts amenable to MS analysis, phosphopeptide analysis is frequently complicated by the low stoichiometry of phosphorylation (i.e., only a small fraction of a given protein may be phosphorylated and be beyond the limits of current biochemical analysis). Multiple differentially phosphorylated protein isoforms also exist, which further complicates site-specific analysis. Thus, researchers are presented with a formidable challenge, which is to isolate (*in vivo*) quantities of phosphorylated proteins that are sufficiently available for analysis.

Separation Methods

Characterization of the site(s) of protein phosphorylation usually follows a general scheme in which first the phosphoprotein is purified. Second, enzymatic or chemical cleavage of the phosphoprotein into peptides is performed. Third, isolation and separation of the phosphopeptides

TABLE I
PHOSPHOAMINO ACID AND PHOSPHATE ELIMINATION

Phosphotheonine	Peptide loss of 98 or 80 amu	79, 97

Phosphothreonine	Peptide loss of 98 or 80 amu	79, 97

Phosphotyrosine	Peptide loss of 80 amu	79, 97

Note: The table shows elimination of the phosphate groups for serine, threonine, and tyrosine phosphorylated residues during CID or in-source fragmentation. Typical immonium ion products for the phosphates are listed on the right of each diagram.

from nonphosphorylated peptides through enrichment and concentration procedures is achieved. Finally, structural characterization of the phosphopeptides by mass spectrometry is attempted.

Among the separation techniques available, two-dimensional phosphopeptide mapping (2-DPP) (Boyle *et al.*, 1991), reversed-phase high-performance liquid chromatography (RP-HPLC) (Becker *et al.*, 1998), high-resolution two-dimensional gel electrophoresis (Storm and Khawaja, 1999), and immobilized metal affinity chromatography (IMAC) (Andersson and Porath, 1986; Porath *et al.*, 1975) have all been successfully used for the separation of phosphopeptides. Peptide separation techniques help to concentrate and enrich phosphopeptides and therefore increase the signal-to-noise ratio. Radiolabelled phosphopeptides can be used to quantitatively determine changes in the phosphorylation state of a protein as a function of time or cellular state. Peptide separation methods also effectively remove nonpeptidic contaminants, thus facilitating the detection and analysis of low-abundance phosphopeptides by decomplexing the total number of analytes. The following sections highlight some micro-IMAC methods that have successfully been applied in the past to assist the site-specific characterization of phosphorylation.

Metal Ion Affinity Chromatography

Immobilized metal ion affinity chromatography (IMAC) was first introduced by Porath *et al.* (1975). The IMAC interaction relies on specific binding between an analyte and an immobilized metal ion. Initially, immobilized metal ions (e.g., Ni^{2+}, Co^{2+}, Zn^{2+}, Mn^{2+}) were shown to bind strongly to proteins with a high density of histidines. The immobilized metal ions of Fe^{3+}, Ga^{3+}, and Al^{3+}, however, have shown strong binding characteristics with phosphopeptides, thus opening roads for development of methods that can selectively enrich for phosphopeptides from complex protein and peptide mixtures. The use of IMAC is now a commonly used process of reversibly capturing phosphopeptides; allowing for preconcentration and selective retention of phosphopeptides; and removing salts, detergents, and nonspecific contaminants not compatible with MS analysis.

Phosphopeptides are acidic by virtue of the phosphate group and bind preferentially to chelated metal ions. However, other peptides, particularly those containing strings of acidic amino acids, are also coenriched. In general, the strength of binding is dependent on numerous factors, such as the degree of phosphorylation, the pH of solutions, salt and peptide concentrations, chelated metal ions, temperature, and the degree of exposure of chelated ions interacting with the peptide side chains.

Because the enrichment of phosphopeptides prior to MS reduces ion suppression effects that would otherwise occur with untreated complex mixtures, enrichment allows for a higher success rate in the assignment of site-specific phosphorylation. The use of IMAC either on-line and off-line followed by MS analysis directly or coupled to other capillary electrophoresis (CE) or liquid chromatography (LC) systems is discussed below.

Off-Line Analysis

Optimal conditions for the elution of phosphopeptides from the IMAC column directly into an electrospray/ionization–mass spectrometry (ESI–MS) analysis do not exist for low-level analysis; therefore, integrated methods are required. In-line IMAC–HPLC–ESI–MS has previously been used successfully (Becker *et al.*, 1998), and while this approach is still useful, optimal low-level analysis cannot be achieved due to the relatively large bore of the columns described. We have adapted this approach for its use off-line with a microcapillary setup. With this new configuration a wide range of sample volumes can be loaded (1–100 μL), and samples can rapidly be loaded with a pressure vessel (Corthals *et al.*, 1999; Karlsson and Novotny, 1988; Kennedy and Jorgenson, 1989). Construction of the column is basically as follows: to each end of a 10-cm long piece of Teflon tubing (1/16 in outer diameter (O.D.) × 0.0001 in inner diameter (I.D.)), a piece of polyimide-coated fused silica capillary (e.g., Polymicro Technologies, Tucson, AZ) is inserted and held in place by a union (e.g., Valco, Houston, TX). Prior to fixing the second of the two polyimide capillaries in place, the open Teflon end is placed in a slurry of POROS-MC (Applied Biosystems, Foster City, CA) inside a vessel pressurized by helium; the IMAC column is packed to a length of approximately 5 cm under 500-psi pressure. The second piece of fused silica capillary is then fixed in place with a second union (see Fig. 1). For operation, one of the two polyimide capillaries is placed in a helium pressure vessel, and the other serves as an outlet.

Using this approach, one can elute (radiolabelled) phosphopeptides from the IMAC column. Further analysis can be performed by either RP microcapillary LC–MS/MS (Becker *et al.*, 1998) or SPE–CE–MS/MS (Figeys *et al.*, 1999). A successful elution procedure that the authors of this discussion have employed in the past is provided in Table II. For direct LC–MS/MS analysis, samples are first concentrated by evaporation and subsequently reconstituted in the starting LC buffer. We have successfully reconstituted samples in starting buffer (0.4% acetic acid, 0.005% heptafluorbutyric acid) and then injected onto the HPLC column. Phosphopeptides were then further separated using a 50-μm I.D. RP column and eluted using a linear gradient from 0–60% acetonitrile containing 0.4% acetic acid and n-Heptafluorobutyric acid (0.005% HFBA). Phosphopeptide fractions could be further chosen for fractionation by solid-phase extraction (SPE–CE) prior to MS/MS analysis using an eloquently employed estimation of phosphorylation by detecting radiolabelled peptides with Cerenkov counting, as shown in Fig. 3.

In addition, the identification of eNOS phosphopeptides of low abundance was maximized by increasing the time available for peptide analysis;

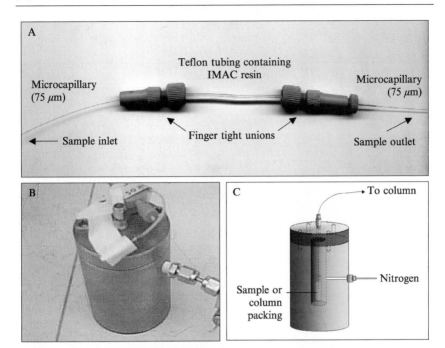

FIG. 1. (A) Detailed IMAC column setup including 6 cm of Teflon tubing packed with NTA-sepharose. (B) Photo of IMAC column in use. Setup was used for identification of eNOS phosphorylation sites[16,17]. (C) Cartoon of pressure vessel and sample flow path. Table II lists the protocol for enrichment of phosphopeptides using this setup. (See color insert.)

a procedure known as *peak parking* is used (Fig. 4): the flow rate of eluting peptides is reduced by inducing a voltage drop to allow the MS to include many more peptides for collision-induced dissociation (CID) otherwise omitted from fragmentation due to the duty cycle of the mass spectrometer. The utility of this method is highlighted in Fig. 4.

Posewitz and Tempst (1999) were able to show a distinct advantage of Ga (III)-bound ions (see Fig. 2), packed in microtips, as developed by Erdjument-Bromage (Erdjument-Bromage *et al.*, 1998) in terms of affinity, selectivity, and efficiency. Here, metal affinity microtips were packed with 10 μl (50 mg/ml) slurry of POROS MC resin (Perseptive Biosystems) using a gel-loading tip plugged with a small trifluoroacetic acid (TFA)-treated glass fiber disk. This column was activated with metal ions by slowly passing 75 μl of 100 mM aqueous metal ion solution over it followed by a 0.1% acetic acid wash to remove unbound ions; a similar approach is detailed in Table III.

TABLE II
MICRO-IMAC COLUMN PACKING AND PREPARATION FOR THE ANALYSIS OF PHOSPHOPEPTIDES
FOR OFF-LINE AND ON-LINE MS ANALYSIS

Off-line IMAC column	
Column specifications	Capillaries: 360 μm O.D./50 μm I.D.
	Teflon (containing IMAC resin): 250 μm I.D.
	Teflon tube (6 cm)
Packing material	100 μl of 50% chelating sepharose slurry
	in 20% ethanol; fill tube 3 cm, and insert
	5 cm of capillary column

Protocol for phosphopeptide purification[a]	
Wash column	H$_2$O, 5 min @ 500 psi
Wash column	EDTA 0.1 M, 2.5 min (until ΔpH) @ 500 psi
Wash column	H$_2$O, 5 min @ 500 psi
Wash column	HOAc 0.1 M, 5 min @ 500 psi
Activate column[b]	FeCl$_3$ 0.1 M, in HOAc 0.1 M, 5 min @ 500 psi
Wash column	HOAc 0.1 M, 10 min @ 500 psi
Load sample	@ 200 psi (preferably from small volume <5 μl)
Wash column	HOAc 0.1 M, 2.5 min @ 200 psi
Wash column	H$_2$O, 2.5 min @ 200 psi
Elute sample	0.1% ammonium-acetate pH8, in 50 mM
	Na$_2$HPO$_4$ @ 200 psi
Regenerate column	Flush with H$_2$O, 2.5 min @ 500 psi
	Flush with 0.1 M EDTA, 2.5 min @ 500 psi
	Flush with H$_2$O, 2.5 min @ 500 psi
	Flush with HOAc 0.1 M, 2.5 min @ 500 psi
Column storage	In HOAc

[a] Monitor pH after each step except sample.
[b] Other metals can also be used: Cupric sulphate, nickel chloride, or gallium nitrate in Milli-Q grade water.

On-Line Analysis

On-line analysis of phosphopeptides is desirable to minimize sample handling and loss. Even though procedures described previously are successful, there are losses associated with each handling step. Moreover, in many cases, there is a need to provide the ability to characterize minor species (low abundance) in peptide mixtures generated by digestion of phosphoproteins. The advantages of applying IMAC for the selective retention of phosphopeptides is clear; however, direct coupling of IMAC columns to an MS source is not practically possible because it leads to lower detection limits. Adaptation to a CE instrument overcomes the problems associated with MS detection following IMAC while allowing a

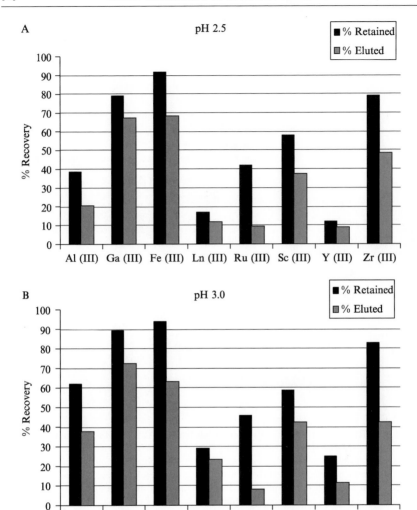

FIG. 2. Metal-dependent recovery of phosphopeptides from IMAC microtips21. Percentage of 32P-radiolabeled phosphopeptides retained (black bars) and eluted (gray bars) are shown for selected metal affinity resins. (A) Aliquots (500 fmol) of phosphopeptide in 0.1 M ammonium bicarbonate containing 1.0% Zwittergent 3–16 adjusted with 20% acetic acid to pH 2.5. (B) Aliquots in A adjusted to pH 3.0. Aliquots were applied to IMAC microcolumns (10–12 μl bed volume of POROS MC beads with metal). Columns were washed with 0.1% acetic acid and 0.1% acetic acid/30% acetonitrile, 0.1% acetic acid; they were then eluted in 15–20 μl of 0.2 M sodium phosphate (pH 8.4). Data were averaged from six experiments and for three different peptides in duplicate. The percentage retained (% retained) is defined as the ratio of radiation levels remaining on the column following washing to those loaded as determined by Cerenkov counting; the percentage eluted (% eluted) indicates the ratio of counts eluted from the column to those loaded.

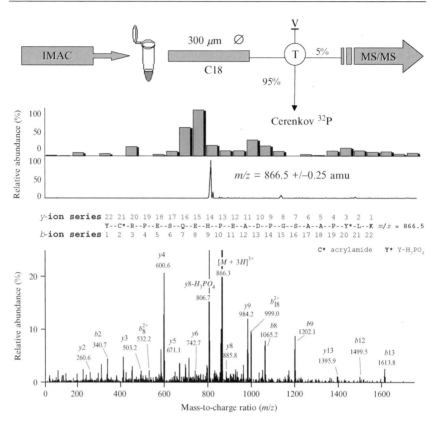

FIG. 3. Representative flow diagram of off-line IMAC RP–MS/MS identification of phosphopeptide. The tryptic digest of the complex phosphoprotein sample is separated by IMAC for enrichment of the phosphopeptide. Pooled phosphopeptides undergo further fractionation by microbore RP-HPLC. Flow-split contributes 5% of this sample for sequence identification of the phosphoamino acid, and the remainder of the sample is checked for the relative abundance of phosphopeptide elution by Cerenkov counting. The histogram shows the Cerenkov counts in the fractions that were collected post-HPLC separation by flow-splitting and indicates phospho-containing peptides. The lower diagram of the reconstructed ion chromatogram shows the elution of ions with an m/z ratio of 866.5 amu, the triply charged mass of a tyrosine-containing peptide from Stat3βtc[2]. The tandem mass spectrum shows the fragmentation pattern of this ion and sequence determination.

concentration limit of detection in the nanomole range. An integrated peptide enrichment and separation system consisting of an IMAC–CE–ESI–MS/MS was developed by Cao and Stults (1999) for the analysis of a trypticly digested solution of alpha and beta casein. Here, the IMAC column (5 cm × 150 μm I.D., 360 μm O.D.) was fitted over the smaller

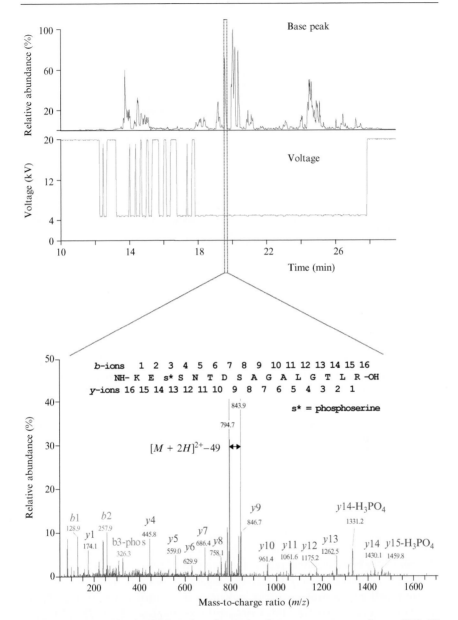

FIG. 4. Peak parking in which each peak entering the mass spectrometer from a SPE–CE fractionation triggered a voltage drop, providing sufficient time for the acquisition of most abundant peptides in each fraction. The base peak trace shows the complexity of the sample despite two stages of chromatographic fractionation. The phosphopeptide identified was from the eNOS protein17. (See color insert.)

TABLE III

MICRO-IMAC COLUMN PACKING AND PREPARATION FOR THE ANALYSIS OF PHOSPHOPEPTIDES FOR
OFF-LINE AND ON-LINE MS ANALYSIS*

Gel-loader tip IMACα for off-line analysis[a]	
Column specifics	Constricted gel-loader tip
Packing material	Ni-NTA resin in 30% ethanol (7 μl), wash with 0.1 M EDTA
Flow rate	Air pressure with a 10 mL syringe
Activation[b]	Ratio of 0.1 M acetic acid : 0.1 M FeCl$_3$ (10 μl), 1:1
Wash	
Sample	Sample solubilized in 20 μl of 0.1 M CH$_3$CN and 0.1 M acetic acid (10 μl)
Wash	0.1 M acetic acid (10 μl)
	25% acetonitrile (10 μl)
	H$_2$O (10 μl)
Elute	NH$_4$OH pH 10.5 (4 μl)
	NH$_4$OH pH 10.5 / 30% acetonitrile (4 μl)

*Table based on Larsen et al. (2001).
[a] Packed IMAC tips can also be obtained from Millipore, Billerica, MA, www.millipore.com.
[b] Other metals can also be used: cupric sulphate, nickel chloride, or gallium nitrate in Milli-Q grade water.

CE column (75 cm × 75 μm I.D., 150 μm O.D.) containing a low binding polyvinylidene fluoride (PVDF) frit. A 1-cm bed of activated Fe (III) POROS beads were pulled into the IMAC capillary by applying vacuum and sealing with a terminating frit. An ESI-ion trap was used to scan a range of mass-to-charge ration (m/z) 450–1500 and to automatically isolate and fragment target ions as retained by IMAC and separated by CE at the low picomole level.

Sequence Determination

There are a number of different MS methods for determining which amino acid residues in a peptide are phosphorylated, and these fall into two general themes. The first method relies on the chemical ability of the phosphoester bonds to phosphoserine, phosphothreonine, and phosphotyrosine. These phosphoester bonds can easily be induced to fragment in a collision cell or ion source of ESI instruments or during post-source decay (PSD) in a matrix-assisted laser desorption/ionization (MALDI)–MS, resulting in loss of phosphate from the peptide. Phosphopeptides that lose phosphate can then be identified by implementing one of several possible phosphate-specific diagnostic ion scans, which include precursor ion scans, neutral loss scans, and in-source dissociation. The phosphate-specific diagnostic ions (i.e., $H_2PO_4^-$ [97 u], PO_3^- [79 u], and PO_3^- [63 u]) generated by ESI in negative ion mode during in-source CID (Katta *et al.*, 1991) can be monitored to identify phosphopeptides (Huddleston *et al.*, 1993). In general, methods that produce some sort of phosphate-specific ion (i.e., a diagnostic ion) are useful when incorporation of ^{32}P is not possible (Meyer *et al.*, 1993) or when the radiolabel has decayed past the point of detection.

The second method relies on detection of the mass added to a peptide by the phosphate group. Typically, in protein phosphorylation studies, the amino acid sequence of the protein investigated is known. Therefore, phosphopeptides derived from the protein can, in principle, be detected by a net mass differential of 80 u that occurs when phosphate is added to serine, threonine, or tyrosine. Thus, a peptide mass map of the proteolytically fragmented phosphoprotein can potentially identify the phosphorylated peptide by comparison to the theoretical peptide map. Neither method, however, identifies the phosphorylated amino acid residue(s) within the peptide directly, except cases in which the peptide sequence contains only a single possible phosphorylation site, and the phosphorylated residue is effectively located by default. The information obtained by the specific scanning methods is often designed to differentiate phosphopeptides from nonphosphopeptides rather than to provide sequence information. These methods can successfully identify a phosphorylated residue only when the peptide sequence is known and when it can be shown that only one of the three types of hydroxyl amino acids is present in the peptide. If this is not the case, then tandem MS (MS/MS) is necessary for locating the phosphorylated amino acid residue.

The following subsections present several types of MS-based approaches to phosphopeptide analysis following IMAC enrichment.

In-Source CID

When in-source CID is combined on-line with HPLC, a chromatographic trace is established that identifies the elution time of a phosphopeptide. If carried out as done by Carr *et al.* (1996), then both the chromatographic marker and the phosphopeptide molecular weight are determined in the same scan. The generation of fragment ions is accomplished by use of a high-orifice potential across the two skimmers prior to Q1, while the low *m/z* range is scanned for the low-mass diagnostic ions. The orifice potential is returned to a normal voltage that does not induce fragmentation, and a scan of high *m/z* is done. A similar experiment can be done on instruments for which a heated capillary replaces the first skimmer (Aebersold *et al.*, 1998). Here, an alternating scan approach is used—selected ion monitoring of appropriate diagnostic ions at a high octapole offset voltage is followed by two full scans. The first full scan is conducted at the same high offset voltage as the selected ion monitoring (SIM) experiment, providing signals for the deprotonated phosphopeptide molecular ion and the phosphopeptide molecular ion minus phosphate. The second full scan is done at a normal octapole offset voltage to provide a reference to the full scan at high octapole offset. This series of MS scans are repeated continuously throughout the LC separation. Such an experiment provides the same information as the methods of Carr *et al.* (1996), but because SIM is used rather than scanning to detect diagnostic ions, one cycle of scans is faster and potentially more sensitive. The first full scan at high octapole offset is compared to the full scan at low octapole offset to provide a clue as to which peptide ion is phosphorylated when, as often occurs with peptide mixtures on microcapillary columns, peptides "co-elute." Such techniques are generally good down to a few femtomoles of phosphopeptide standards loaded on column, but with real *in vitro* or *in vivo* samples sensitivities are often in the picomole range.

Of course, it would be advantageous in one microcapillary separation on-line with ESI–MS to detect phosphopeptides in negative ion mode and then switch to positive ion mode for a CID experiment because negative ion CID spectra generally produce insufficient fragment ions for sequence elucidation. To date, this result has been technically difficult to achieve in the same analytical run where diagnostic ion scanning is carried out, but it may be possible in the future using nonscanning mass spectrometers. The difficulty in doing such an experiment on a scanning mass spectrometer, such as a quadrupole, lies in the time required to switch between positive and negative ion mode in real time.

Neutral Loss

Neutral loss scanning for phosphopeptide detection and analysis was first described by Covey *et al.* (1991) and further developed by Huddelston *et al.* (1993). It is carried out in positive ion mode with ESI in a triple quadrupole MS. Instead of using Q1 to select specific ions for fragmentation in Q2, Q1 and Q3 are scanned over a different m/z range to Q1. Thus, the two quadrupoles are simultaneously scanned over two different m/z ranges, the difference of which corresponds to the m/z value for the neutral molecule that is being lost. For a neutral loss of phosphate from a $[M + 2H]^{2+}$ phosphopeptide ion, the offset value is 49 m/z. As shown in Table I, only phosphoserine and phosphothreonine may undergo a neutral loss of 98 by ß-elimination (Gibson and Cohen, 1990) and not phosphotyrosine because the ß-carbon proton that needs to be abstracted by a lone pair of electrons from the phosphate moiety is now too far removed for facile loss. The method has not been as popular as the aforementioned in-source CID methods because of false positives and the need to know the charge state of the ion losing the phosphate. An advantage of the method is that it is carried out in positive ion mode and can be used with data-dependent scanning to acquire CID in the same experiment using the detection of a neutral loss of phosphate as a trigger to initiate CID.

Precursor Ion

In the precursor ion method, negative ion ESI is carried out with continuous scanning of Q1. All ions are fragmented in Q2, and Q3 passes only one ion, which for phosphopeptides is usually m/z 79 (i.e., loss of PO_3^-). Consequently, the resultant mass spectrum shows only ions that lost m/z 79 (Neubauer and Mann, 1999; Wilm *et al.*, 1996). This greatly simplifies mixture analysis and is best done during direct infusion with a nanoelectrospray source. Again, as described for the phosphate diagnostic ion scans, there is a problem associated with sequencing in positive ion mode immediately after detecting the loss of phosphate in negative ion mode.

Product Ion

As a general trend during low energy CID of phosphopeptides, it has been observed that phosphate tends to be lost from shorter phosphopeptides more readily than longer phosphopeptides. This loss occurs because roughly the same amount of energy for collision is dispersed across fewer bonds, such as from phosphoserine more readily than phosphothreonine and from phosphothreonine more readily than phosphotyrosine. Interestingly, it is rare to observe the immonium ions for phosphoamino acids that form as a result of dehydroalanine and dehydroamino-2-butyric acid breaking down after loss of phosphate. However, using an ion trap mass spectrometer (DeGnore and Qin, 1998) and monitoring the CID of a phosphopeptide ion, dehydroamno-2-butyric acid has been observed in place of threonine in the peptide fragment ion.

PSD

Metastable decay of phosphopeptides has been typically observed during PSD–MALDI–time-of-flight (TOF) and provides a method to sequence peptides in a single stage instrument. While not popular for the aforementioned reasons, PSD has been successfully applied to sequence analysis of phosphopeptides (Larsen *et al.*, 2001). Phosphopeptide enrichment with this method has predominantly been achieved in combination with Gel-loader tips of which the protocol is provided in Table III.

Enzymatic Dephosphorylation

A method that does not receive much attention but can provide the location of a phosphorylated amino acid in a phosphopeptide involves use of phosphatases (Zhang *et al.*, 1998). Using a MALDI–TOF instrument, the masses of the peptides resulting from proteolytic digestion of the phosphoprotein are acquired. Then the same sample is treated with phosphatase to remove phosphate and the masses acquired again. Now any mass that decreases by 80 u will be a clue regarding which peptide is phosphorylated. An advantage to conducting such an experiment by MALDI is that peptide ions produced tend to be singly rather than multiply protonated, and this makes interpretation easier than with ESI.

References

Aebersold, R., Figeys, D., Gygi, S., Corthals, G., Haynes, P., Rist, B., Sherman, J., Zhang, Y., and Goodlett, D. (1998). Towards an integrated analytical technology for the generation of multidimensional protein expression maps. *J. Protein Chem.* **17,** 533–555.

Andersson, L., and Porath, J. (1986). Isolation of phosphoproteins by immobilized metal (Fe3$^+$) affinity chromatography. *Anal. Biochem.* **154,** 250–254.

Becker, S., Corthals, G. L., Aebersold, R., Groner, B., and Muller, C. W. (1998). Expression of a tyrosine phosphorylated, DNA binding Stat3beta dimer in bacteria. *FEBS Lett.* **441,** 141–147.

Boyle, W. J., Smeal, T., Defize, L. H., Angel, P., Woodgett, J. R., Karin, M., and Hunter, T. (1991). Activation of protein kinase C decreases phosphorylation of *c-Jun* at sites that negatively regulate its DNA-binding activity. *Cell* **64,** 573–584.

Cao, P., and Stults, J. T. (1999). Phosphopeptide analysis by on-line immobilized metal-ion affinity chromatography–capillary electrophoresis–electrospray ionization mass spectrometry. *J. Chromatogr. A.* **853,** 225–235.

Carr, S. A., Huddleston, M. J., and Annan, R. S. (1996). Selective detection and sequencing of phosphopeptides at the femtomole level by mass spectrometry. *Anal. Biochem.* **239,** 180–192.

Corthals, G. L., Gygi, S. P., Aebersold, R., and Patterson, S. D. (1999). Identification of proteins by mass spectrometry. *In* "Proteome research: 2-D Gel Electrophoresis and Detection Methods" (T. Rabilloud, ed.). Springer, New York.

Covey, T. R., Huang, E. C., and Henion, J. D. (1991). Structural characterization of protein tryptic peptides via liquid chromatography/mass spectrometry and collision-induced dissociation of their doubly charged molecular ions. *Anal. Chem.* **63,** 1193–1200.

DeGnore, J. P., and Qin, J. (1998). Fragmentation of phosphopeptides in an ion trap mass spectrometer. *J. Am. Soc. Mass Spectrom.* **9,** 1175–1188.

Erdjument-Bromage, H., Lui, M., Lacomis, L., Grewal, A., Annan, R. S., McNulty, D. E., Carr, S. A., and Tempst, P. (1998). Examination of microtip reversed-phase liquid chromatographic extraction of peptide pools for mass spectrometric analysis. *J. Chromatogr. A.* **826,** 167–181.

Figeys, D., Corthals, G. L., Gallis, B., Goodlett, D. R., Ducret, A., Corson, M. A., and Aebersold, R. (1999). Data-dependent modulation of solid-phase extraction capillary electrophoresis for the analysis of complex peptide and phosphopeptide mixtures by tandem mass spectrometry: Application to endothelial nitric oxide synthase. *Anal. Chem.* **71**, 2279–2287.

Gibson, B. W., and Cohen, P. (1990). Liquid secondary ion mass spectrometry of phosphorylated and sulfated peptides and proteins. *Methods Enzymol.* **193**, 480–501.

Huddleston, M. J., Bean, M. F., and Carr, S. A. (1993). Collisional fragmentation of glycopeptides by electrospray ionization LC–MS and LC–MS/MS: Methods for selective detection of glycopeptides in protein digests. *Anal. Chem.* **65**, 877–884.

Hunter, T. (1995). Protein kinases and phosphatases: The yin and yang of protein phosphorylation and signaling. *Cell* **80**, 225–236.

Hunter, T. (2000). Signaling—2000 and beyond. *Cell* **100**, 113–127.

Karlsson, K. E., and Novotny, M. (1988). Separation efficiency of slurry-packed liquid chromatography microcolumns with very small inner diameters. *Anal. Chem.* **60**, 1662–1665.

Katta, V., Chowdhury, S. K., and Chait, B. T. (1991). Use of a single-quadrupole mass spectrometer for collision-induced dissociation studies of multiply charged peptide ions produced by electrospray ionization. *Anal. Chem.* **63**, 174–178.

Kennedy, R. T., and Jorgenson, J. W. (1989). Preparation and evaluation of packed capillary liquid chromatography columns with inner diameter from 20 to 50 μm. *Anal. Chem.* **61**, 1128–1135.

Krishna, R. G., and Wold, F. (1993). Post-translational modification of proteins. *Adv. Enzymol. Relat. Areas Mol. Biol.* **67**, 265–298.

Larsen, M. R., Sorensen, G. L., Fey, S. J., Larsen, P. M., and Roepstorff, P. (2001). Phospho-proteomics: Evaluation of the use of enzymatic dephosphorylation and differential mass spectrometric peptide mass mapping for site-specific phosphorylation assignment in proteins separated by gel electrophoresis. *Proteomics* **1**, 223–238.

Meyer, H. E., Eisermann, B., Heber, M., Hoffmann-Posorske, E., Korte, H., Weigt, C., Wegner, A., Hutton, T., Donella-Deana, A., and Perich, J. W. (1993). Strategies for nonradioactive methods in the localization of phosphorylated amino acids in proteins. *FASEB. J.* **7**, 776–782.

Neubauer, G., and Mann, M. (1999). Mapping of phosphorylation sites of gel-isolated proteins by nanoelectrospray tandem mass spectrometry: Potentials and limitations. *Anal. Chem.* **71**, 235–242.

Porath, J., Carlsson, J., Olsson, I., and Belfrage, G. (1975). Metal chelate affinity chromatography, a new approach to protein fractionation. *Nature* **258**, 598–599.

Posewitz, M. C., and Tempst, P. (1999). Immobilized gallium(III) affinity chromatography of phosphopeptides. *Anal. Chem.* **71**, 2883–2892.

Storm, S. M., and Khawaja, X. Z. (1999). Probing for drug-induced multiplex signal transduction pathways using high-resolution two-dimensional gel electrophoresis: Application to beta-adrenoceptor stimulation in the rat C6 glioma cell. *Brain Res. Mol. Brain Res.* **71**, 50–60.

Wilm, M., Neubauer, G., and Mann, M. (1996). Parent ion scans of unseparated peptide mixtures. *Anal. Chem.* **68**, 527–533.

Zhang, X., Herring, C. J., Romano, P. R., Szczepanowska, J., Brzeska, H., Hinnebusch, A. G., and Qin, J. (1998). Identification of phosphorylation sites in proteins separated by polyacrylamide gel electrophoresis. *Anal. Chem.* **70**, 2050–2059.

[5]   Mapping Posttranslational Modifications of Proteins by MS-Based Selective Detection: Application to Phosphoproteomics

By STEVEN A. CARR, ROLAND S. ANNAN, AND MICHAEL J. HUDDLESTON

Abstract

This chapter outlines general principals that apply to the analysis of posttranslational modifications of proteins, with an emphasis on phospho-proteins. Mass spectrometry (MS)-based approaches for selective detection and site-specific analysis of posttranslationally modified peptides are de-scribed, and an MS-based method that relies on production and detection of fragment ions specific for the modification(s) of interest and that was developed in the authors' laboratory is described in detail. The method is applicable to selective detection of *N*- and *O*-linked carbohydrates in glycoproteins, *O*-linked sulfate, and *N*- and *O*-linked lipids. Detailed pro-cedures for application of this strategy to phosphorylation-site mapping are presented here.

Introduction

Proteins that have covalent modifications are the rule rather than the exception in nature. Nearly 200 structurally distinct covalent modifications have been identified thus far, ranging in size and complexity; these mod-ifications result from a variety of factors, from the conversion of amides to carboxylic acids (delta mass +0.9840) to the attachment of multiple com-plex oligosaccharides each with molecular masses up to several thousand daltons (Graves *et al.*, 1994; Krishna and Wold, 1993; Wold, 1981). Most post-translational modifications are introduced by enzymes. The presence of the modification is often required for normal biological function or tissue disposition of the protein, although, in many cases, the role of the modification is as of yet unknown.

Phosphorylation and glycosylation are two of the most biologically relevant and ubiquitous posttranslational modifications of proteins. In both cases, an organism's commitment to these modifications, as measured by the estimated numbers of genes involved, is substantial. Protein kinases may constitute as much as 3% of the entire eukaryotic genome (Cohen, 1992; Hubbard and Cohen, 1993; Hunter, 1991), while gene products involved in oligosaccharide biosynthesis may represent as much as 1% of

METHODS IN ENZYMOLOGY, VOL. 405   0076-6879/05 $35.00
DOI: 10.1016/S0076-6879(05)05005-6

the genome (Varki and Marth, 1995). It is estimated that as many as one-third of proteins present in typical mammalian cells are phosphorylated (Hubbard et al., 1993), and up to half of the proteins are glycosylated (Apweiler et al., 1999).

Phosphorylation is the most common and physiologically important reversible regulatory modification. An exquisitely complex, integrated network of protein kinases and phosphatases controls many aspects of cell growth, metabolism, division, motility, and differentiation through selective phosphorylation (often at multiple sites) and dephosphorylation of cellular proteins (Cohen, 1992; Hubbard and Cohen, 1993; Hunter, 1991). Phosphorylation of intracellular proteins plays an essential role in signal transduction. This is the process by which extracellular signals are communicated to the cell's nucleus by the binding of components such as cytokines, neurotransmitters, and hormones to cell-surface receptors, thereby regulating an array of intracellular physiological processes (Daum et al., 1994; Eck, 1995; Pawson, 1995; Schlessinger, 1994).

Posttranslational modifications frequently complicate or even prevent the use of classical tools for protein sequence analysis, such as automated Edman degradation. In addition, the presence of lipid or carbohydrate on proteins can dramatically decrease the accuracy of molecular weight estimates obtained by sedimentation velocity, gel permeation, or SDS–PAGE measurements. Unlike classical biochemical techniques, mass spectrometry (MS) relies on entirely different principles to accomplish structure analysis. Because MS measures the mass of a molecule, it is uniquely suited for detection and structural characterization of covalent posttranslational modifications of proteins that involve either a mass change to an individual amino acid in the sequence or a removal of a portion of the N and/or C terminus of the protein. From the authors' viewpoint, the potential for MS in the study of posttranslational modifications is virtually unlimited.

This chapter outlines general principals that apply to the analysis of posttranslationally modified proteins, with an emphasis on phosphoproteins. The chapter then describes an MS-based approach for selective detection and site-specific analysis of posttranslationally modified peptides. It is important to note that the method does not require or depend on incorporation of a tag or label (e.g., a radiolabel like ^{32}P or ^{33}P or an isotopic label like 18$_O$) but rather on production and detection of fragment ions specific for the modification of interest. Detailed procedures for application of this strategy to phosphorylation-site mapping are presented as well (Annan et al., 1997, 2001; Bean et al., 1995; Carr et al., 1996; Hill et al., 1994; Huddleston et al., 1993a; Hunter et al., 1994; Neubauer et al., 1997; Verma et al., 1997); other related strategies are referenced in Table I.

TABLE I
MARKER-IONS DERIVED FROM COMMON POSTTRANSLATIONAL MODIFICATIONS

Fragment ions[a]	m/z	Origin	Indication	References
PO_2^-/PO_3^-	63/79	O-linked phosphate	Phosphopeptides containing pThr, pSer, and pTyr; phosphocarbohydrates	Allen et al., 1997; Annan and Carr, 1997; Annan et al., 2001; Azzam et al., 2004; Bean et al., 1995; Carr et al., 1996; Chen et al., 2002; Crabb, in press; Huddleston et al., 1993a; Hunter and Games, 1994; Jedrzejewski and Lehmann, 1997; Le Blanc et al., 2003; Neubauer and Mann, 1997; Sulivan et al., 2004; Till et al., R 1994; Verma et al., 1997; Wilm et al., 1996; Watty et al., 2000; Zappacosta, 2002
$C_8H_{10}NO_4P^+$	216.043[b]	Immonium ion of pTyr	pTyr	Steen et al., 2001, 2003
SO_3^-	80	O-linked sulfate	Sulfopeptides; sulfocarbohydrates	Bean et al., 1995

	204	N- or O-linked HexNAc (e.g., GlcNAc, GalNAc)[c]	Glycopeptide or carbohydrate containing any N- or O-linked sugar	Carr et al., 1993; Greis et al., 1996; Hayes and Aebersold, 2000; Huddleston et al., 1993b; Kragten et al., 1995; Mazsaroff et al., 1997; Medzihradszky et al., 1997, 1998; Roberts et al., 1995; Rush et al., 1995; Schindler et al., 1995; Sullivan et al., 2004
	292	N-acetyl neuraminic acid	Glycopeptides or carbohydrates containing complex-type N- or O-linked sugars	Carr et al., 1993; Greis et al., 1996; Hayes and Aebersold, 2000; Huddleston et al., 1993b; Kragten et al., 1995; Mazsaroff et al., 1995; Medzihradszky et al., 1997, 1998; Roberts et al., 1995; Schindler et al., 1995

(continued)

TABLE I (continued)

Fragment ions[a]	m/z	Origin	Indication	References
	366	N- or O-linked Hex- HexNAc (e.g. Gal-GlcNAc, Gal-GalNAc)[c]	Glycopeptides or carbohydrates containing any N-linked and most O-linked sugars	Carr et al., 1993; Greis et al., 1996; Hayes and Aebersold, 2000; Huddleston et al., 1993b; Kragten et al., 1995; Mazzaroff et al., 1997; Medzihradszky et al., 1997, 1998; Roberts et al., 1995; Schindler et al., 1995
($C_{15}H_{31}CONH_2+$ and $C_{15}H_{31}CO+$)	256 239	N-linked palmitic acid[d]	Lipopeptides or lipoproteins	Bean et al., 1995; Gu et al., 1997; Rush et al., 1995; Sullivan et al., 2004

[a] Charge on fragment indicates required analysis mode (positive or negative ion).

[b] Numerous potential interferences exist at nominal mass 216, which requires use of instruments capable of accurate mass and high resolution.

[c] Fragment ions shown will form regardless of sequence position of indicated sugar in the carbohydrate or if it is directly linked to the peptide; ions due to consecutive losses of water (18 Da) are also commonly observed.

[d] Other lipids (saturated and unsaturated) will produce analogous marker-ions at the expected masses.

Similar methods for the analysis of *N*- and *O*-linked carbohydrates in glycoproteins (Carr *et al.*, 1993b; Greis *et al.*, 1996; Gu *et al.*, 1997; Hayes *et al.*, 2000; Huddleston *et al.*, 1993b; Kragten *et al.*, 1995; Mazsaroff *et al.*, 1997; Medzihradszky *et al.*, 1997; Roberts *et al.*, 1995; Schindler *et al.*, 1995), *O*-linked sulfate (Bean *et al.*, 1995), and *N*- and *O*-linked lipids (Bean *et al.*, 1995; Gu *et al.*, 1997) have also been developed (see Table I). Detailed procedures for analysis of these posttranslational modifications may be found in the references cited at the end of this chapter.

Since this chapter was written in 2000, there has been a dramatic increase in the phosphoproteomics literature. The expansion in phosphoproteomics has been aided in large part by significant improvements in the ability of mass spectrometers to carry out data-dependent experiments, such as the ability to automatically select and further fragment ions that have a mass-to-charge ratio (m/z) corresponding to neutral loss of phosphoric acid (Bateman *et al.*, 2002; Covey *et al.*, 1991; Schroeder *et al.*, 2004). In addition, selective enrichment strategies for phosphoproteins and phosphopeptides based on the use of immunoprecipitation-capable antibodies (Rush *et al.*, 2005) and immobilized metal affinity chromatography (Ficarro *et al.*, 2005) have seen great improvement through 2005. For a recent review of the phosphoproteomics literature, which emphasizes strategies that have been proven useful for identification of previously unknown phosphorylation sites, we recommend the review by Loyet *et al.*, (2005).

Special Issues with Respect to the Analysis of Posttranslationally Modified Proteins: Focus on Phosphorylation-Site Mapping

The goals of any posttranslational modification analysis are to provide as complete a map as possible of the modification sites in a protein and to define the structure(s) of the modifications at each specific attachment site. This requires detection and analysis of peptides covering *all* of the potential modification sites. It is important to note that this is a much more stringent analytical requirement than for protein identification by MS and database searching (a subset of "proteomics"), which can often be accomplished with molecular weight and partial sequence for *any* peptide (preferably more than one) derived from the protein (Larsen *et al.*, 2000).

A reasonable starting strategy is to obtain coverage of the known consensus sites for a given modification. However, it must always be kept in mind that consensus sites serve only as a guide. An exclusive focus on predicted sites will work against finding sites of attachment that violate the "known" consensus rules (a fairly common occurrence in the case of phosphorylation). The goal in analyzing posttranslationally modified proteins should be to obtain as complete a sequence coverage as possible. This goal is often difficult to achieve in practice for single phosphoproteins, and it is presently unachievable for the analysis of large numbers of phosphoproteins as complex mixtures (phosphoproteomics).

Site-specific analysis of posttranslational modifications by either conventional approaches or MS requires that the modified protein first be cleaved enzymatically or chemically into peptides of a size suitable for sequence analysis. In the case of MS, this size is ideally between 500 Da and 3000 Da. Trypsin is usually the first choice because the rules for fragmentation of peptides having Lys or Arg at their C termini are well understood and predictable. It is possible to directly analyze unfractionated protein digests by techniques such as nanoelectrospray mass spectrometry (nanoESMS) and matrix-assisted laser desorption/ionization (MALDI). Methods employing direct analysis of unfractionated peptide digests minimize losses that invariably occur in sample handling at low levels. However, direct analysis of complex mixtures suffers from well-known problems of suppression effects (particularly for highly charged peptides like phosphopeptides), charge-state overlap, limited dynamic range for peptide signal detection, and limited sampling frequency of the mass spectrometer.

To minimize these problems and to maximize sequence coverage of posttranslationally modified proteins, it is necessary to employ chromatographic or electrophoretic separation to fractionate the mixture prior to or during MS analysis. Several studies of more highly phosphorylated proteins illustrate the advantages of employing a separation stage prior to analysis (Watty *et al.*, 2000; Wu *et al.*, 2000). Use of high-performance liquid chromatography (HPLC) during on-line liquid chromatography (LC)–ESMS (as employed in the strategy detailed below) is particularly effective as it accomplishes both desalting and peptide separation in one step. A separation step is also required after digestion and prior to analysis by MS if the protein has been reduced and alkylated (to maximize sequence coverage) or if the protein was derived from digestion of a gel band. In both instances, the sample will contain excess reagents, by-products, or impurities that would severely compromise or defeat analysis by either MALDI–MS or ES–MS if not removed.

In addition to the general concerns previously noted for analysis of posttranslationally modified proteins, phosphoproteins present some unique challenges. Many phosphoproteins of interest are present in cells at only very low concentrations, and often only femtomole to low picomole amounts of phosphopeptides may be recovered for analysis following enzymatic digestion and chromatographic isolation (Bean *et al.*, 1995; Boyle

et al., 1991; Luo *et al.*, 2005). Phosphoproteins are also usually phosphorylated on a number of different sites throughout the protein, with individual
sites being phosphorylated to varying degrees (e.g., 1 to 100%). Upon
digestion with trypsin (or another enzyme), it is very common to generate
peptides containing more than one potential phosphorylation site (Ser,
Thr, and Tyr). Furthermore, an observed molecular mass for a phosphopeptide can sometimes be assigned to more than one reasonable peptide
sequence from the protein. Thus, it is necessary to sequence phosphopeptides by MS/MS to assign the part of the protein sequence from which that
the peptide derives and to establish which residues in that peptide are
phosphorylated. To estimate the stoichiometry of modification at each site,
both the modified and unmodified forms of each phosphopeptide must be
detected so that the ratio of the ion abundances can be determined.

The serious constraints of low phosphopeptide yield and the mixture
complexity noted earlier make it desirable, if not essential, to have analytical methods capable of preferentially detecting and analyzing phosphopeptides. By far, the most commonly employed techniques for the isolation of
phosphopeptides from phosphoproteins have employed cells metabolically
labeled *in vivo* or *in vitro* with [32P]-phosphate or proteins that are labeled -
in vitro with [32P]-phosphate by reaction with purified kinase (Boyle
et al., 1991; Kuiper *et al.*, 1995; Luo *et al.*, 2005; Wettenhall *et al.*, 1995;
Winz *et al.*, 1994). Efforts to identify and/or isolate phosphopeptides that
circumvent the requirement for radiolabeling have centered mainly on two
approaches: (i) use of immobilized metal-ion affinity techniques both off-
line (Posewitz and Tempst, 1999) and coupled on-line with an ES mass
spectrometer (Posewitz and Tempst, 1999; Watts *et al.*, 1994) and (ii)
selective detection in MS-based methods on the unique fragmentation
behavior of phosphopeptides (Carr *et al.*, 1996; Huddleston *et al.*, 1993a).
The latter MS-based methods are described in detail in the upcoming
paragraphs.

As already noted, since this chapter was written in 2000, a number of
advances have been introduced for phosphopeptide detection and sequencing. A recent survey of the literature that is current to 2004 can be found in
Loyet *et al.* (2005).

Selective Detection of Posttranslational Modifications by MS

Post-translational modifications on proteins are often more susceptible
to cleavage by collision-induced fragmentation in the mass spectrometer
than the peptide backbone. The methods the authors of this chapter and
others have developed for selective detection of such modifications take
advantage of this fact using three approaches: (i) detection of the loss of the

modification from the peptide precursor, (ii) detection of the low mass "signature" or "marker" ions from the modification itself, or (iii) selective detection of the precursor ions that give rise to the low mass signature peaks (referred to as precursor-ion scanning). Neutral loss monitoring has long been used in ESMS for detection of phosphorylation (Covey et al., 1991). This approach has been automated on orthogonal time-of-flight mass spectrometers (Bateman et al., 2002) and on ion-trap instruments (Schroeder et al., 2004). The approach relies on the detection of the loss of 98, 98/2, 98/3, and so on from singly, doubly, triply, and other multiply charged precursors of phosphopeptides. Detection of the loss can be used to trigger acquisition of MS/MS spectra for sequencing of the peptide. Neutral loss of phosphoric acid from pSer and pThr is a facile process on any mass spectrometer, but it is especially dominant on ion trap instruments that induce fragmentation by resonance excitation. To compensate for the lack of sequence informative backbone cleavage in the ion trap MS/MS spectra, automated procedures have been developed that carried out MS^3 on the neutral loss peak(s).

The strategies developed in the authors' laboratory utilize marker-ion production and precursor-ion scanning in specific combination to selectively identify peptides containing a particular covalent modification and to selectively fragment modified peptides to define the attachment site(s) of the modification (Annan et al., 2001; Zappacosta et al., 2002). Briefly, the experiments involve first a modification-specific marker-ion scan that enables selective detection and fractionation of modified peptides in a protein digest during LC–MS analysis. The marker-ions are produced by low energy collisions with gas molecules in the premass analysis region of the mass spectrometer (see Table I). Second, the experiments use a precursor-ion scan during LC–MS analysis or of fractions collected during a marker-ion analysis. This experiment identifies the molecular weights of only those peptides carrying the modification of interest. Third, the experiments perform sequence and structure analyses of the identified precursor ions by MS/MS. The first two steps just described detect the presence of a modification and establish the molecular masses of modified peptides in the presence of an excess of unmodified peptides, even in cases where the signal from the modified peptide is indistinguishable from background in the normal MS scan. Knowledge of the molecular masses of the modified peptides enables targeted use of MS/MS for sequencing. These experiments and how they interrelate are described in greater detail in the following paragraphs for phosphoprotein analysis.

*Selective Detection and Preparative Fractionation of Modified Peptides
Using Marker Ions*

Modified peptides may be selectively detected in mixtures of unmodified peptides by formation and detection of low-mass fragment ions that serve as specific markers for the modifications of interest. Abundant low-mass marker ions, however, are not usually observed in ESMS spectra under normal operating conditions. All mass spectrometers, by design or default, have one or more defined regions of reduced but still relatively high pressure between the atmospheric pressure ion source and the true high vacuum regions where the mass separation and detection elements of the mass spectrometer reside (Fig. 1A). In conventional LC–ESMS analysis, the velocity of the ions in these relatively high-pressure zones is kept low to minimize fragmentation that occurs as a result of collision with gas molecules and maximize transmission of intact parent ions. To produce abundant marker ions, the velocity of the ions in this region is increased to induce fragmentation via collision-induced dissociation (CID) of all ions *prior* to mass separation.

The highest sensitivity for marker-ion detection is obtained by selected ion monitoring (SIM) of just the marker ions of interest under conditions of continuous, high-level CID in the ion-sampling/collision region. Alternatively, CID may be "turned on" while scanning the lower mass range and then "turned off" during the remainder of each scan. In this stepped collision-energy scanning mode, peptide molecular weight information and modification-selective, low-mass marker ions may be detected in the *same* scan, albeit at lower sensitivity for detection of the marker ions. While these experiments can be carried out on any type of mass spectrometer, they are ideally suited to a single quadrupole mass spectrometer (or one mass analyzer of a triple quadrupole) because of the ability to accumulate ion current from one or a few masses/markers so as to increase sensitivity of detection.

The SIM traces for the marker ions indicate which regions of the chromatogram contain the modified peptides (Fig. 1B). Marker-ion traces are readily compared with the total-ion current (TIC) and/or the UV chromatogram traces to give an indication of peak complexity. In the case of phosphopeptides, the marker ion profile is analogous to the output from an HPLC radioactivity detector or the autoradiogram from a two-dimensional phosphopeptide map but without the requirement for ^{32}P or ^{33}P labeling of the protein. This trace also serves as a fingerprint for the phosphorylation profile of a protein. Changes in the phosphorylation state of a protein would be reflected by a change in the phosphorylation profile. Only those components of the profile that change would need further analysis. By splitting

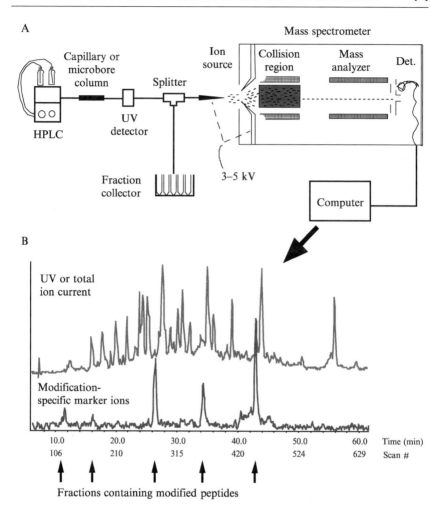

FIG. 1. Selective detection and preparative fractionation of modified peptides using marker ions during LC–ESMS. (A) Typical experimental arrangement for formation and detection of modification-specific marker ions during LC–MS analysis of proteins. (B) Representative data showing how fractions containing modified peptides are identified and collected. (See color insert.)

the column effluent, greater than or equal to 80% of the modified peptides are fraction-collected during the on-line LC–ESMS analysis (Fig. 1A). Columns with internal dimensions of 180 μm or larger may be effectively used in this first dimension step (see Zappacosta *et al.*, 2002 and detailed procedures described later in this chapter).

Table I contains a list of useful marker ions. The general approach used for identification of suitable marker ions for modifications of interest is to analyze model compounds, such as suitably modified peptides, proteins, and carbohydrates, in product-ion scanning, MS/MS mode to ascertain if there are appropriate diagnostic fragments produced and to determine the optimum collision energy for their formation. The model compounds are then added to defined mixtures of peptides or into a digest of a model protein or mixtures of proteins to test for specificity and sensitivity of detection. The m/z 63 and 79 marker ions appear to be highly specific for phosphopeptides. Other markers, such as m/z 204 used for carbohydrate analysis (see Table I), may occasionally be formed from CID of unmodified peptides. However, nonspecific fragment-ion formation is usually readily distinguished from modification-specific responses by the observation of synchronous changes of several (or all) of the marker ions for a given modification. For example, in the case of glycopeptides or carbohydrates, one would commonly monitor m/z 204, 292, and 366 (see Table I) in a single LC–ESMS experiment. Coincident increases in response for two out of three of these ion traces would indicate a specific rather than a nonspecific response. Ions due to loss of water from the ions just listed can also be monitored for an added degree of assurance.

The use of marker-ion formation and detection is illustrated for phosphorylation analysis of bovine α_s-casein in Fig. 2. Phosphopeptide content was assessed by SIM of m/z 63 and 79 during negative-ion LC–ESMS. Superposition of the SIM trace on the UV trace identified eight HPLC fractions eluting between 13 to 22 min as containing phosphopeptides. These collected fractions were further analyzed in the second and third dimension steps of the analysis described in the following paragraphs.

Defining the Molecular Masses of Modified Peptides by Precursor-Ion Scanning and Sequence Analysis by MS/MS

The fractions containing the modified peptides that were isolated in the marker-ion experiment almost always contain additional, unmodified peptides that can make recognition of the masses of modified peptides difficult or impossible. Furthermore, the modified peptides may represent only a few percent of the total peptide in the fraction and may not even be discernable in a conventional MS scan of the fraction. To selectively identify the masses of the modified peptides, a precursor-ion scan is used that only produces signals for those ions that fragment to yield the marker-ion monitored.

Figure 3 illustrates the precursor scan experiment in a triple quadrupole mass spectrometer for detection of phosphopeptide precursor ions. Triple

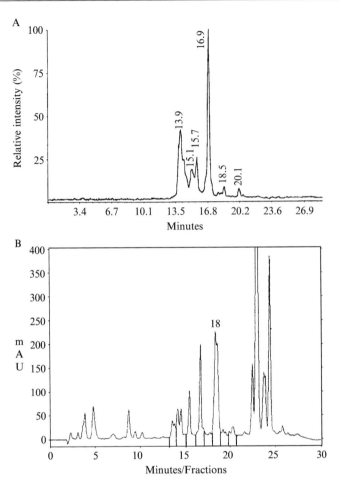

FIG. 2. Selective detection of phosphopeptides of bovine α_s-casein. (A) LC–ESMS SIM trace for m/z 63 and 79. Phosphopeptide content was assessed by selected ion monitoring (SIM) of the marker ions m/z 63 and 79 during negative-ion LC–ESMS. (B) LC–UV trace for tryptic digest of bovine α-casein. Superposition of the SIM trace on the UV trace identified eight HPLC fractions eluting between 13 and 22 min as containing phosphopeptides. These collected fractions were further analyzed in the second and third dimension steps of the analysis as described in the text.

quadrupole mass analyzers equipped with ion-counting detectors provide the highest sensitivity of any mass analyzer for precursor-ion scanning due to their capability to signal-time average very weak data and reject noise. In this simplified scheme, three precursor ions (M1, M2, and M3) are emerging from the electrospray needle at the start of a 2s-scan of the mass

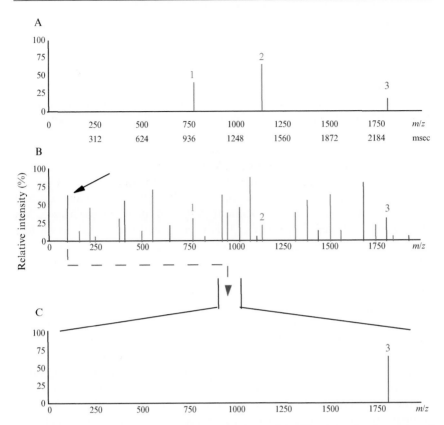

FIG. 3. Scheme illustrating precursor-ion scanning to determine the molecular masses of phosphopeptides for subsequent MS/MS analysis. (A) Q1: Normal scan (2s scan). Observed are (M − H) of all peptides. (B) Q2: Collison-induced decomposition of all ions. All peptides fragment; only phosphopeptides yield *m/z* 79 marker ion, which is observed only when peptide 3 fragments. (C) Q3: Selective detection of *m/z* 79. Observed intensity of *m/z* 79 was recorded at the *m/z* of the precursor that produced it in Q1. See text for detailed discussion. (See color insert.)

spectrometer's first mass analyzer (Fig. 3A). As the *m/z* scan of Q1 progresses, the three precursor ions are transmitted sequentially and individually into the true collision cell, Q2, where each fragments (Fig. 3B). The fragment patterns for all three are shown superimposed, color coded to the parent from which they derived. The second mass analyzer (Q3) is set to pass one marker ion (e.g., *m/z* 79 for phosphopeptides, Fig. 3C). Only precursor M3 fragments to produce a negative ion of *m/z* 79. The precursor-ion spectrum is obtained by recording the observed ion abundance of

m/z 79 at the m/z of the precursor that produced it, which in this case is M3. For simplicity, the authors show only a single peak for each precursor in Fig. 3A. However, ESMS spectra often exhibit multiple charge states for any given parent/precursor. In practice, each member of a charge series for a given phosphopeptide precursor will fragment to produce m/z 79, and so each will be recorded in the resulting precursor-ion scan. Precursor-ion scans may also be used to selectively detect modified forms of an intact protein analyzed by ESMS (Neubauer *et al.*, 1997).

Precursor-ion scanning can be accomplished in an on-line LCMS mode as was first demonstrated for glycopeptide analysis (Carr *et al.*, 1993) or in an off-line mode following collection of fractions identified to contain the modified peptide by marker-ion scanning. At the time, we, the authors, had developed this strategy, data-dependent control of the triple quadrupole MS instrument used was very limited. This necessitated the use of an off-line approach in which phosphopeptide-containing fractions were subsequently analyzed off-line by precursor-ion scanning using nanoelectrospray (Carr *et al.*, 1996; Wilm *et al.*, 1996). Improvements in control of scan functions and data-dependent experiments have enabled MS systems, such as the AB 4000 triple quadrupole, to automatically switch from negative-ion precursor-ion mode to positive-ion mode full-scan MS/MS mode following detection of the m/z 79 precursor ion, which permits the analysis to be carried out in the course of a single LC–MS experiment (Le Blanc *et al.*, 2003).

The selectivity of this step for phosphopeptide detection is illustrated in Fig. 4 for analysis of phosphopeptide-containing fraction 18 collected during the LC–ESMS analysis of bovine α_s-casein (see Fig. 2). Panels A and B compare the full-scan, negative-ion nanoelectrospray mass spectrum with the precursor ion spectrum for m/z 79. These data were acquired from circa 0.5 μL of the collected fraction after adjusting the pH to greater than 10 to improve the sensitivity for detection of the phosphopeptides (see following subsection entitled *Procedures*). The phosphopeptide of determined $M_r = 1832.8$ is clearly a very minor component and, in the absence of the precursor-ion data, may otherwise have gone unnoticed in this fraction. In contrast, only this phosphopeptide is evident in the precursor-ion spectrum (Fig. 2B).

Once the molecular weights of the phosphopeptide "needles in the haystack" are determined, it is straightforward to switch analysis modes and select the masses for $(M + 2H)^{2+}$, $(M + 3H)^{3+}$, and additional forms of the phosphopeptides in the positive ion data for sequencing by nanoESMS/MS (Fig. 4C). Generally, no additional sample is required because most of the 0.5–1 μl aliquot of sample that provided the precursor-ion scan data is usually remaining. The nanoESMS/MS product-ion spectrum of the $(M + 2H)^{2+}$ parent ion of m/z 917.4 is shown in Fig. 5. The series of ions marked

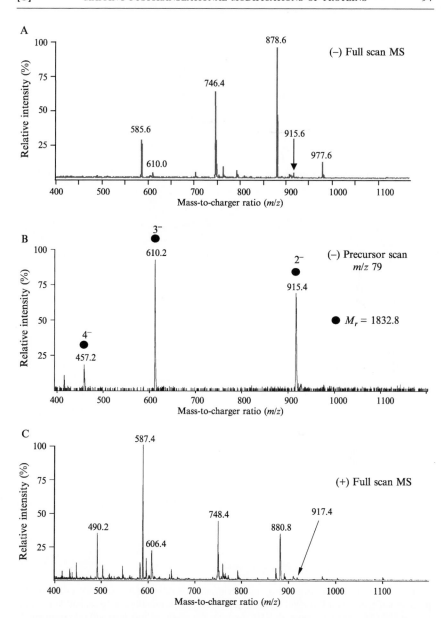

FIG. 4. Selective phosphopeptide analysis of fraction 18 collected during the LC–ESMS analysis of bovine α_s-casein (see Fig. 2). (A) Full scan negative-ion nanoelectrospray mass spectrum. (B) Precursor-ion mass spectrum for m/z 79. (C) Full scan positive-ion mass spectrum. The precursor-ion scan detects a single phosphopeptide with an apparent average mass of 1833.6. Arrows point to the $(M - 2H)^{2-}$ and $(M + 2H)^{2+}$ ions in the respective full-scan, negative-ion data (shown in A) and positive-ion data (shown in C) that are dominated by signals from nonphosphorylated peptides coeluting in this fraction.

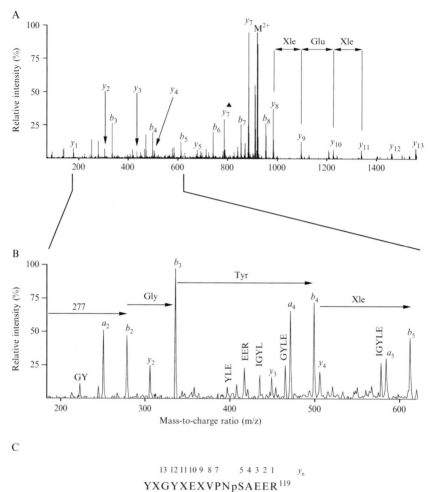

FIG. 5. Sequence localization of modification by MS/MS. (A) NanoES (+) ion CID product ion spectrum of m/z 917.2, the doubly charged ion for the phosphopeptide found in fraction 18. The partial sequence XEX (where X = Leu or Ile) is readily determined by inspection. This partial sequence matches part of a known phosphorylated sequence from αs1-casein, but the mass does not fit this phosphopeptide suggesting further modification. (B) Expansion of the low mass region of the spectrum shown in panel A. Internal fragment ions produced by cleavage of two amide bonds are represented by the single letter amino acid code for the residues included in the internal fragment. (C) Final sequence determined from the product ion spectrum. The amino acid sequence coverage provided by b_n and y_n fragment ions is shown. Sequence corresponds to a previously unknown variant of αs1-casein.

$y_8 - y_{11}$ in the spectrum clearly indicate the partial sequence XEX (X = Leu or Ile), which matches residues 109–111 from the αsl-casein tryptic peptide, VPQLEIVPNSAEER (residues 106–119, calc. M_r avg. 1660.8), which was identified in an earlier eluting fraction. Interpretation of the lower m/z region of the spectrum allowed us, the authors of this chapter, to extend the determined sequence toward the C-terminus using y_n ions, yielding the partial sequence XEXVPNpSAEER. Analysis of the low m/z region (Fig. 5B) led to the identification of a series of b-ions (and some a-ions) that overlapped with the sequence already defined and extended toward the N terminus of the peptide to yield (E,F or Y,X) GYLEIVPNpSAEER. The choice and order of the first two residues are defined by a series of internal fragment ions that included XGYL and XGYLE. The full sequence of the peptide is therefore YXGYLEIVPNpSAEER, which corresponds to a previously unreported variant of bovine αs1 casein.

There are several reasons to include sequencing in any modification mapping strategy. The most obvious is the case in which a given peptide contains more than one potential modification site. It is also relatively common to observe peptide signals for which no reasonable sequence assignment can be made based on the determined molecular weight or where the assignment is ambiguous. The quality of the MS/MS data required to confirm peptide identity from a protein of known sequence and to localize the modifications need not be as high as for *de novo* sequencing. Even weak, incomplete MS sequence data is often sufficient to answer the specific questions noted earlier. Nanoelectrospray provides long data acquisition times and very good sensitivity of analyte detection from very small sample volumes (Wilm and Mann, 1994a,b). This permits optimization of all experimental parameters and sensitivity improvement by accumulating ion counts for weak signals. This is especially important for large peptides and whenever a more complete sequence is required (e.g., in cases where the protein sequence is not known and is not in a database). Furthermore, several peptides can be sequenced without resorting to the use of another aliquot of sample. By making the pH of the spray solution basic, we demonstrated selective detection and sequencing of phosphopeptides in complex mixtures at the low (<10) femtomole range (Zappcosta *et al.*, 2002). Targeted on-line LC–MS/MS may also be used for sequencing of phosphopeptides with high sensitivity whose molecular masses have been established.

Stoichiometry of Modification: Phosphosite Occupancy

The extent to which each modification site is utilized often relates to its physiological relevance. For example, phosphoacceptor sites that are more heavily utilized are the first to be targeted for mutational analysis; this is a common way to assess the functional relevance of a phosphorylation site. A semiquantitative estimate of the extent to which each modification site is utilized can be obtained from the ratio of the ion abundances for the modified and unmodified forms of the peptides. For example, the ratio of MS response for phosphorylated to nonphosphorylated peptides in the full-scan, positive-ion ESMS data obtained at acidic pH may be used to determine phosphorylation stoichiometry with an accuracy of circa ±30% (Carr *et al.*, 1996). The ability to provide site-specific information about the extent of phosphate incorporation is a key advantage of the present approach relative to techniques that employ affinity-based methods like immobilized metal-ion affinity chromatography for selective fractionation of phosphopeptides. Finally, if low picomole amounts of intact phosphoprotein can be obtained in solution, the molecular weight of the intact protein is obtained by ESMS (or MALDI, especially in cases where ESMS is unsuccessful) so as to provide an overall check on the total number of moles of phosphate added to the protein and the relative distribution of protein molecules with different numbers of phosphate (Annan *et al.*, 2001; Verma *et al.*, 1997).

Sensitivity

As noted earlier, the sensitivity for precursor-ion scanning for selective detection and MS/MS for sequencing phosphopeptides by nanoESMS and nanoESMS/MS is in the low femtomole range (Carr *et al.*, 1996; Zappacosta *et al.*, 2002). However, the overall sensitivity of the method is currently limited by the first-dimension LC–ESMS analysis. ESMS (at flow rates above circa 50 nl/min) behaves as if it were a concentration-sensitive detector. Thus, smaller internal diameter HPLC columns will provide higher sensitivity analyses as the effective concentration for a given amount injected will be higher. These factors militate for use of as small an internal diameter (I.D.) column as possible. However, the off-line precursor-ion scanning method requires collection of fractions during LC–ESMS. Fraction collection, even by hand, requires a minimum of 1 to 2 μl per minute of collection flow in practice. This collection flow rate, together with the flow rate required by the electrospray source to provide a stable spray, sets the minimum flow rate necessary through the HPLC column. This flow rate, in turn, limits the I.D. of the HPLC column that can be practically employed. Microionspray sources (the part downstream of splitter and fraction collection in Fig. 1) are now available and produce stable sprays at flow rates of 100 to 200 nl per minute. These flow rates permit use of either 180 μm I. D. or 320 μm I.D. HPLC columns, with column flows of 2 to 4 μl per minute, which leaves sufficient flow for both MS detection and fraction collection. Using such a microionspray source flowing at circa 200 nl per minute and a 180-μm I.D. HPLC column flowing at rates of 3 to 4 μl per minute, the practical limit for phosphopeptide analysis is greater or equal to 50 fmoles on a column (Zappacosta *et al.*, 2002). Conditions are described in the next section named *Procedures*.

Procedures

This section presents the experimental procedures involved in optimizing the chromatographic conditions, electrospray source, and mass analyzer for each step of the multidimensional analysis of phosphopeptides derived from phosphopeptides. Wherever possible, we, the authors of this chapter, have provided benchmark data using readily available standards so that interested readers can work through optimization of the methods on their specific instruments.

Reagents. Phosphopeptides and their nonphosphorylated analogs (Table II) were obtained from the Protein and Carbohydrate Structure Facility, University of Michigan Medical School (Ann Arbor, MI, www. brcf.med.umich.edu). Other useful phosphopeptide standards are available from Bachem Bioscience (Philadelphia, PA, www.bachem.com). The phosphopeptide standards are dissolved in water, and 500 fmoles aliquots are dispensed into polypropylene microvials (Eppendorf, 0.5 ml size) and used fresh. Table II shows a test mixture of peptides and phosphopeptides that is used routinely to test system readiness for "real" samples. Peptides 9–13, Table II, were obtained as a mixture (Peptide Retention Standard, S1–S5) from Pierce Biotechnology (Rockford, IL). All other peptides were obtained from Peninsula Laboratories (Belmont, CA, www.penlabs.com) or Bachem Bioscience.

Evaluation of Mass Spectrometer Performance in Negative-Ion Mode. Conditions described here are specific to the acquisition of ES mass spectra on Sciex/Applied Biosystems quadrupole mass spectrometers (Concord, Ontario, Canada, www.appliedbiosystems.com). The tuning and calibration solution consists of a mixture of polypropylene glycol (PPG) 425, 1000, and 2000 ($3 \times 10^{-5} M$, $1 \times 10^{-4} M$, and $2 \times 10^{-4} M$, respectively) in 50/50/0.1 water/methanol/formic acid (v/v/v), 1 mM NH$_4$OAc. This mixture is used for tuning and calibrating the instrument in both positive- and negative-ion modes. The m/z range 10 to 2400 is calibrated in the negative-ion mode by multiple-ion monitoring of the isotope clusters of six PPG ion signals and two trifluoroacetic acid (TFA)-related ions, m/z 69 (CF$_3$) and m/z 113 (CF$_3$CO$_2$). The TFA-derived ions are present as background in any mass

TABLE II
PHOSPHORYLATED AND NON-PHOSPHORYLATED PEPTIDES USED FOR OPTIMIZATION AND
EVALUATION OF THE MULTIDIMENSIONAL PHOSPHOPEPTIDE ANALYSIS STRATEGY

Peptide number	Sequence	M_r (monoisotopic)
1	KRT(PO$_3$H$_2$)IRR (UOM 11)	908.5
2	KRTIRR (UOM 10)	828.5
3	KRPS(PO$_3$H$_2$)QRHGSKY (UOM 9)	1422.7
4	KRPSQRHGSKY (UOM 8)	1342.7
5	Ac-RRLIEDAEY(PO$_3$H$_2$)AARG-NH$_2$ (UOM 7)	1639.8
6	Ac-RRLIEDAEYAARG-NH$_2$ (UOM 6)	1559.8
7	TYSK	497.2
8	LGG	245.1
9	RGAGGLGLGK-NH$_2$	883.5
10	Ac-RGGGGLGLGK-NH$_2$	911.5
11	Ac-RGAGGLGLGK-NH$_2$	925.5
12	Ac-RGVGGLGLGK-NH$_2$	953.6
13	Ac-RGVVGLGLGK-NH$_2$	995.6
14	ISRPPGFSPFR	1259.7
15	MLF	409.2
16	DRVYIHPFHLLVYS	1757.9

spectrometer that is regularly exposed to TFA-containing HPLC mobile phases (see following paragraphs). Mass spectra are recorded at instrument conditions sufficient to resolve the first two isotopes of anion m/z 991.7 (PPG + HCO$_2$)$^-$ so that the valley between them is 55% of the height of the second isotope for the singly charged ion. At this resolution, singly charged ions can be distinguished from ions having two or more charges provided there are good ion statistics. In positive-ion mode, resolution is adjusted such that the first two isotopes of cation m/z 906.7 (PPG/NH$_4^+$) are resolved with a valley between them of 40% of the height of the second isotope for the singly charged ion. At this resolution, it is possible to assign charge states of 1+ or 2+ and to distinguish these from more highly charged ions. Mass-to-charge ratio assignments for the measured peak tops can be closer either to the monoisotopic or to the average M_r depending on the charge state and the isotopic distribution. Calibration and instrument resolution is checked prior to LC–ESMS using the background TFA anions at m/z 69 and 113. Alternatively, if these ions are weak or absent, one can use TFA-containing mobile phases and a high enough declustering potential to keep the m/z 69 and 113 ions of TFA from saturating the detector.

Prior to LC–ESMS, the declustering potential (source and collision region prior to first mass analyzer, Fig. 1) used to produce the marker ions is determined by infusion of a 5 pmol/μl solution of phosphopeptide standard 3 (Table I) in 30% HPLC mobile phase B containing 0.02% TFA (see following paragraphs). These test conditions are also used to routinely check for best performance by adjusting spray position, sample flow rates, gas flows, voltage settings, and other such factors. The S/N for m/z 63 and 79 are simultaneously monitored in real time by SIM (200 ms dwell per ion) and the signal for each maximized by adjusting the declustering potential for each. On the Sciex API-III, the optimal declustering voltages are circa -350 V for m/z 63 and -300 V for m/z 79 using TFA-containing mobile phases; these voltages are -250 V and -200 V, respectively, when 0.2% formic acid without TFA is used. These numbers may exceed the highest voltage provided on the particular MS system used. In this case, an external programmable power supply is added as previously described (Huddleston *et al.*, 1997). The appearance of the low-mass region containing the marker ions is shown in comparison to the solvent background in Fig. 6. The signal-to-background ratio for m/z 63 and 79 is typically circa 20:1 under these conditions.

HPLC Systems for Separation and Analysis of Phosphopeptides

The overall sensitivity of the method is currently limited by the flow rate and sample concentration requirements of the first dimension LC–ESMS analysis. HPLC columns with internal diameters of 0.5 mm work well for phosphopeptides in the range of 5 pmoles and higher. Columns with internal dimensions of 180-μm I.D. work well with phosphopeptide amounts in the 50 fmole to 5 pmole range. Procedures for these two levels of chromatography are given in the upcoming sections. While conventional 0.1% TFA-containing mobile phases may be employed (for examples, see Fig. 7), sensitivity for phosphopeptide detection is increased approximately 2.5-fold (as measured by signal-to-noise ratio) using a combination of 0.02% TFA and 0.1% formic acid. The degree to which the quality of the chromatographic separation is compromised can range from minimal to significant depending on the specific type of C18 employed. Therefore, columns should be tested with a mixture of standard phosphorylated and nonphosphorylated peptides (for example, see Table II and following text) prior to committing real samples for analysis. For maximum sensitivity of peptide analysis by the LC–ESMS positive ion, TFA may be eliminated altogether and replaced with 0.2% formic acid. However, no apparent benefit is seen for phosphate marker-ion sensitivity when doing LC–ESMS SIM with 0.2% formic acid. Some chromatographic peak

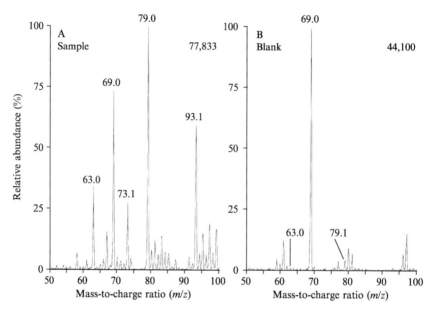

FIG. 6. (A) Typical appearance of the marker ion region for infusion of a 5-pmol/μl solution of a phosphopeptide standard (peptide 3, Table I). (B) TFA-containing mobile phases alone at a declustering potential of −350 V. On other instruments, one or more voltages may need to be adjusted to affect the extent of "in-source" CID so as to obtain a similar spectrum. The S/N for m/z 63 and 79 are simultaneously monitored in real time by SIM (200 ms dwell per ion), and the signal for each is maximized. On the Applied Biosystems Sciex API-III, the optimal declustering voltages are circa −350 V for m/z 63 and −300 V for m/z 79 using TFA-containing mobile phases; the voltages become −250 V and −200 V, respectively, when 0.2% formic acid without TFA is used. The signal-to-background ratios for m/z 63 and 79 are typically circa 20:1 under these conditions.

integrity is sacrificed under these conditions although certain column packings (such as the PepMap C18 phase, LC-Packings) can perform very well.

LC–ESMS Using 0.5 mm I.D. HPLC Columns. Any gradient HPLC system capable of forming reproducible gradients at flow rates of 20 μl per minute (either directly from the pumps or using an appropriate preinjector split flow system) with TFA-containing mobile phases may be employed. Desalting and preconcentration of dilute samples are accomplished as part of the sample injection step using a microprecolumn of C_{18} (PepMap phase, 1 mm × 5 mm LC-Packings, San Francisco, CA) as the sample loop of the injector. The HPLC column is a 0.5 mm × 150 mm C_{18} ("Magic" phase, 5 μm, 200-Å pore-size particles, Michrom BioResources, Auburn,

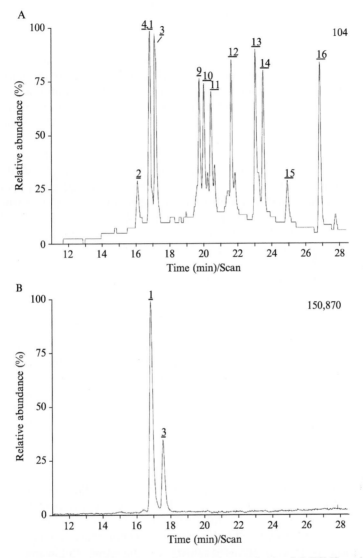

FIG. 7. Evaluation of overall system performance in negative-ion LC–ESMS mode. (A) UV chromatographic detection at 214 nm. (B) Selected-ion monitoring for phosphopeptide marker ions. A mixture of model peptides (for peptide identities, see Table II) at 20 pmoles was injected onto a 0.5-mm I.D. HPLC column using 0.1% TFA in the mobile phases (i.e., the least sensitive conditions; see text). Only the two phosphopeptides (Table II, lines 1 and 3) in the mixture are observed in the marker-ion trace (as shown in B). Marker ions at m/z 63 and 79 were monitored using a 0.2-Da window, 0.04-Da step, and a 100-msec dwell, making a 1.1 sec scan rate. The ion-spray capillary voltage was operated at −3.7 kV with the interface plate at −700 V. The nebulization gas consisted of zero-grade compressed air at 48 psi. Nitrogen (99.999%) at a flow rate of 0.7 L/min was used as the "curtain" gas. Peptides 5 and 6 (Table II) were not included in the mixture analyzed.

CA, www.michrom.com). The flow rate to the MS is reduced to 4 to 5 μl per minute after the column and UV detector (see Fig. 1) by splitting the flow using a Valco (Houston, TX) tee with an appropriate length of fused silica acting as a restrictor on the fraction collection leg. A length of 50-μm-I.D.-fused silica goes from the tee to the tip of the ionspray capillary. The mobile phases used for gradient elution consist of (A) water/acetonitrile 98:2 (v/v) and (B) acetonitrile/water 90:10 (v/v), with both A and B containing either 0.1% TFA or a combination of 0.02% TFA and 0.1% formic acid by volume (see earlier note). Typical gradient conditions are 5% B to 50% B linearly in 30 min, then linearly to 95% B in 5 min, and then a hold for 10 min. Prior to injecting samples, the guard column is conditioned using water/acetonitrile 98:2 (v/v) containing 0.1% TFA regardless of acid type or concentration used in mobile phases. The samples are made acidic with solvent A and then loaded onto the guard column, which is washed with solvent A. On this HPLC system, it is necessary to ground the metal tee to prevent charge flow to the UV electronics that would cause unacceptable UV baseline noise during LCMS.

LC–ESMS Using 180-μm I.D. HPLC Columns. Peptide mixtures are preconcentrated onto a PepMap C18 trap cartridge (300 μm × 5 mm; LC Packings) used in place of the injector loop. The trap is conditioned with 0.1% TFA, and the sample is then loaded. Samples are dissolved in or diluted with the 0.1% TFA (with or without 2% CH_3CN) prior to loading on the trap. Mobile phases are as noted previously, except that the concentration of TFA is reduced to 0.02%, and 0.1% formic acid is added to increase the sensitivity and improve electrospray stability without sacrificing chromatographic resolution. The sample is back-flushed off the cartridge onto the column at 4 μl per minute with a gradient from 0% to 50% B in 30 min and then from 50% to 95% B in 5 min with a 10-min hold. The flow from HPLC pumps (530 μl per minute) is reduced to 4 μl per minute using a microflow splitter (Accurate Splitter, LC-Packings). The 180-μm I.D. HPLC column (LC-Packings PepMap C18 capillary column, 15 cm long, 3 μm particles) is fitted directly into the injector. The column outlet, a 25-μm I.D. and 280-μm-O.D.-fused silica transfer line, is connected to the nanoflow electrospray block/splitter using a Teflon sleeve connector and a short piece (~15 cm) of fused silica with the same dimensions. A UV detector can be plumbed in-line prior to MS if desired. Flow is split for simultaneous fraction collection and MS using the microvolume Valco tee insert of the block (0.15-mm I.D. through-hole, Valco Instruments, Houston, TX, www.vici-store.com) to direct 0.6 μl per minute to the 20-μm-I.D.-fused silica (also tapered) ES tip (New Objective, Waltham, MA) and 3.4 μl per minute to the fraction collection line for manual collection into polypropylene microtubes. The fraction collection line is composed of two pieces of tubing connected in series through a stainless steel union (63-μm I.D. PEEK tubing from splitter and 50-μm I.D.-fused silica from union to fraction collector). The union is electrically grounded to prevent electrospray conditions at the fraction collector.

Evaluating the Overall Performance of the LC–ESMS System. The overall performance of the LC–ESMS system is evaluated using both UV detection at 214 nm and SIM of the respective marker ions. For a 0.5 mm scale chromatography, a mixture of model peptides (Table II) at 20 pmoles, each is injected. Representative UV and selected-ion current traces for this mixture using 0.1% TFA in the mobile phases (i.e., the least sensitive conditions) are shown Fig. 7 (mixture of all peptides except 5 and 6, Table II). All peptides in the mixture (with the frequent exception of Table II lines 7 and 8, which have weak UV response and elute at or near the injection void volume) produce peaks in the UV trace (Fig. 7A), whereas only the two phosphopeptides (Table II, lines 1 and 2) in the mixture are observed in the marker-ion trace (Fig. 7B).

Optimizing Nanoliter Flow Electrospray for Precursor-Ion Scanning and Peptide Sequencing by MS/MS

NanoES mass spectra are obtained using an articulated nanoelectrospray interface (Wilm and Mann, 1994a) and the "medium"-sized metal-coated capillary spray tips available commercially (Protana, Inc., Odense M, Denmark). Samples are loaded from the back of the electrospray capillary using an electrophoresis gel-loading pipette tip (Eppendorf Geloader tip, Brinkmann Instruments, Westbury, NY, www.brinkmann.com). Prior to sample loading, the gel-loading tips are washed with the solution in which the samples are dissolved. The dissolution phase (50 μl) is loaded from the top of the pipette and then pushed out through the tip using pressure from a 1-mL syringe. All plastic surfaces are washed, and freshly prepared solutions are used to remove compounds that may cause chemical noise in the mass spectra. The nanoelectrospray capillary tip is positioned in the ion source with the aid of the two cameras supplied with the setup from Protana. The optimal position in terms of S/N on the Sciex electrospray sources is approximately 1 to 2 mm directly in front of the orifice. To generate a stable signal and maintain a minimum flow rate (20–40 nl per minute), a positive air pressure on the sample in the capillary, of slightly more than atmospheric pressure, is usually required. The capillaries that work the best for the authors' research require touching the capillary to the gate valve plate (voltages off and applying air pressure) to initiate sample flow. With sample flow, the electronics are then turned on before the final

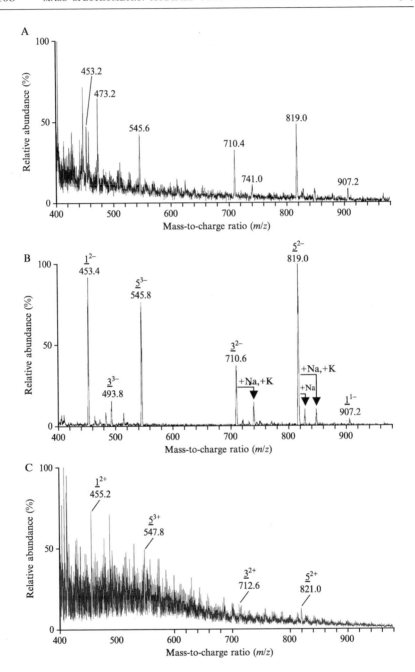

adjustment of moving the capillary toward the orifice. This last adjustment is made using the cameras and by monitoring the changes in the mass spectra from scan to scan in real time.

Tuning and calibration for full-scan data acquisition in both positive and negative ion nanoelectrospray modes are carried out as described earlier. A phosphopeptide standard is used to tune and optimize production and detection of m/z 79 (PO_3^-) marker ion in precursor-ion scanning mode. The resolution of Q3 is reduced to pass a 4 to 5 Da window around the precursor ion of interest to enhance precursor scan sensitivity. A further small gain in sensitivity is obtained by decreasing the resolution of Q1 to yield circa a 75% valley between singly charged ions. Depending on the instrument, all other important tuning parameters are checked, such as the quadrupole offsets (related to collision energy), lens settings, curtain gas (generally need to be reduced), and the collision gas pressure to maximize transmission of m/z 79. An argon:nitrogen (85:15) gas mix is used as the collision gas in Q2.

Tuning and calibration for peptide sequencing by MS/MS is carried out by nanoelectrospray of glu-fibrinopeptide B (Bachem Bioscience) at a concentration of 1 pmole/μl. On the Sciex triple quadrupole mass spectrometers, spectra are acquired over the desired mass range using a mass step of 1 Da with a dwell of 20 ms per mass step and a mass defect of 50 mmu per 100 Da. This approach produces analytically useful MS/MS data rapidly, allowing many product-ion spectra to be acquired using a single 1-μl loading of sample. The resolution of Q3 is adjusted to unresolved fragment-ion isotope clusters, scattered across the mass range, so as to maximize signal. The instrument is calibrated so that for the product ions, the most abundant peak in a cluster consisting of three 1-Da steps is the "monoisotopic peak." This method is used routinely when acquiring product-ion data at the femtomole level on precursor ions with 10^5 counts per second per scan or less. When sufficient amounts of peptide are available, or when isotope resolution is required, tuning and calibration in the product-ion scan mode is performed using a mass step of 0.2 Da with isotopes clearly resolved.

Optimizing Phosphopeptide Selective Detection by Precursor-Ion Scanning and Sequencing by Nanoelectrospray MS/MS. Phosphopeptide selective detection by precursor-ion scanning and sequencing by

FIG. 8. Optimizing phosphopeptide selective detection by precursor-ion scanning. (A) Negative-ion nanoelectrospray MS of a 100 fmole/μL solution of phosphopeptides 1, 3, and 5 (Table II) in basic water/methanol. (B) Precursor-ion mass spectrum of m/z 79. (C) Full scan positive-ion mass spectrum. See full text for details.

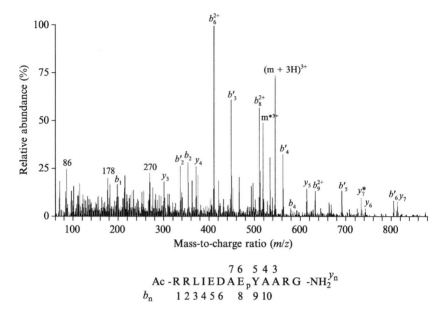

FIG. 9. (A) Representative MS/MS data for phosphopeptide 5 (Table II and panel B) obtained by switching to positive-ion mode during nanoelectrospray of the phosphopeptide-containing fraction in Fig. 8. The $(M + 3H)^{3+}$ of peptide 5 at m/z 547.8 was mass-selected for sequencing by CID MS/MS. Spectra were acquired over the mass range 50–1850 using a mass step of 1 Da, with a dwell of 20 ms per mass step and a mass defect of 50 mmu per 100 Da. (B) Sequence of peptide 5 (Table II) indicating y- and b-ions observed in MS/MS spectrum.

nanoelectrospray MS/MS are optimized using freshly prepared stock solutions of standard phosphopeptides (Figs. 8 and 9). The conditions are used to test detection limits and mass accuracy as well as to simulate conditions used to prepare a real sample for the same analyses. A dried aliquot containing 500 fmoles each of phosphopeptides 1, 3, and 5 (Table II) is diluted just prior to analysis to a final concentration of 100 fmole/μl using 5 μl of 50:40:10 methanol/water/ammonium hydroxide (v/v/v). The basic solution is made up fresh by diluting a 30% ammonium hydroxide solution (Instra-Analyzed reagent grade, J. T. Baker, Phillipsburg, NJ) with H_2O to make a 20% NH_4OH solution, which is then mixed 1:1 with methanol in a 1.5-ml Eppendorf tube. Then 1.5 μl of the standard solution is loaded into the nanoelectrospray capillary as described previously using prewashed gel-loader pipette tips. The experimental sequence is as follows:

1. Acquire full-scan, negative-ion data in a multichannel analyzer (MCA) mode (Fig. 8A). Monitor the data visually in real time, and terminate the acquisition when the overall ion statistics of the spectrum are satisfactory. To help prevent plugging of the capillary and, hence, deterioration of spray stability (which can happen during negative ion nanoelectrospray of basic solutions), use a slightly higher spray air pressure compared to that needed for positive ions. If plugging occurs, remove by increasing the air pressure to the capillary and by touching the tip to the entrance plate forcefully (with voltages off). When a droplet is observed, reduce the air pressure to normal operating levels.

2. Acquire m/z 79 precursor-ion spectra in the negative-ion, MCA mode to determine the molecular weights of the phosphopeptides (Fig. 8B). As previously noted, terminate the acquisition when the overall ion statistics of the spectrum are satisfactory (usually 3 to 6 min of data).

3. Acquire full-scan, positive-ion spectra in the MCA mode to identify species to select for CID tandem MS (Fig. 8C). Identify the ions corresponding to the $(M + 2H)^{2+}$, $(M + 3H)^{3+}$, and additional ions of the phosphopeptides of interest, bearing in mind that when comparing positive-ion to negative-ion data, there can be significant differences in signal intensities and in charge state distributions. If the parent ions of the phosphopeptides are not apparent above the background in the full-scan, positive-ion data, select for positive-ion MS/MS the most abundant charge state observed in the m/z 79 precursor-ion scan for each phosphopeptide as well as the next higher negative ion charge state (e.g., $M - 2H^{2-}$ observed, select $(M + 2H)^{2+}$ and $(M + 2H)^{3+}$ for MS/MS).

4. Acquire product-ion mass spectra of selected phosphopeptides and/ or nonphosphorylated peptides. If at anytime the overall signal appears unstable or weak, optimize the system before going on to the next experiment or before acquiring MCA data.

"Real" samples or HPLC fractions containing less than or equal to 200 fmoles of peptide are split, and the half that is dried down is taken through the latter protocol. The sample is redissolved in 2–3 μl basic solution, vortexed, and centrifuged; then 1–1.5 μl is loaded into the nanoelectrospray capillary. Sample dissolved in basic solution should be used the same day. If the positive-ion precursor(s) needed to acquire MS/MS data are weak but give a good response in negative-ion mode, an aliquot of the sample may be taken to dryness and brought up in 35:65:5 water/methanol/formic acid (v/v/v) and reanalyzed.

When acquiring data on samples that approach the limit of detection (less than 50 fmoles total), there may be little or no obvious peptide signals by negative ion. This makes it difficult to know if the flow rate and spray conditions are optimal. In such cases, phosphopeptide standard 3 can be added (as shown in Table II) as an internal standard to the sample that is dried down (50 fmole/μl when redissolved). A narrow m/z window around m/z 710.4, the $(M - 2H)^{2-}$ for this peptide, can be used to rapidly assess signal-to-noise and spray stability while optimizing air pressure (flow rate) and sprayer position. In precursor scan mode, it is also helpful to observe a phosphopeptide-related ion that can be monitored over time to detect a no-flow condition.

References

Allen, M., Anacleto, J., Bonner, R., Bonnici, P., Shushan, B., and Nuwaysir, L. (1997). Characterization of protein digest using novel mixed-mode scanning with a single quadrupole instrument. *Rapid Commun. Mass Spectrom.* **11**, 325–329.

Annan, R. S., and Carr, S. A. (1997). The essential role of mass spectrometry in characterizing protein structure: Mapping post-translational modifications. *J. Protein Chem.* **16**, 391.

Annan, R. S., Huddleston, M. J., Verma, R., Deshaies, R. J., and Carr, S. A. (2001). A multidimensional electrospray MS-based approach to phosphopeptide mapping. *Anal. Chem.* **73**, 393–404.

Apweiler, R., Hermjakob, H., and Sharon, N. (1999). On the frequency of protein glycosylation as deduced from analysis of the Swiss–Prot Database. *BBA* **1473**, 4–8.

Azzam, R., Chen, S. L., Shou, W., Mah, A. S., Alexandru, G., Nasmyth, K., Annan, R. S., Carr, S. A., and Deshaies, R. J. (2004). Phosphorylation by cyclin B-Cdk underlines releases of mitotic exit activator Cdc14 from the nucleolus. *Science* **305**, 516–519.

Bateman, R. H., Carruthers, R., Hoyes, J. B., Jones, C., Langridge, J. I., Millar, A., and Vissers, J. P. (2002). A novel precursor in discovery method on a hybrid quadrupole othrogonal acceleration time of flight (Q-TOF) mass spectrometer for studying protein phosphorylation. *J. Am. Soc. Mass Spectrom.* **13**, 792–803.

Bean, M. F., Annan, R. S., Hemling, M. E., Mentzer, M., Huddleston, M. J., and Carr, S. A. (1995). "Techniques in Protein Chemistry" (J. W. Crabb, ed.) Vol. VI. Academic Press, San Diego, CA.

Boyle, W. J., van der Geer, P., and Hunter, T. (1991). Phosphopeptide mapping and phosphoamino acid analysis by two-dimensional separation on thin-layer cellulose plates. *Meth. Enzym.* **201**, 110.

Carr, S. A., Huddleston, M. J., and Annan, R. S. (1996). Selective detection and sequencing of phosphopeptides at the femtomole level by mass spectrometry. *Anal. Biochem.* **239**, 180–192.

Carr, S. A., Huddleston, M. J., and Bean, M. F. (1993). Selective identification and differentiation of *N*-and *O*-linked oligosaccharides in glycoproteins by liquid chromatography–mass spectrometry. *Prot. Sci.* **2**, 183–196.

Chen, S. L., Huddleston, M. J., Wenying, S., Deshaies, R. J., Annan, R. S., and Carr, S. A. (2002). Mass spectrometry-based methods for phosphorylation site mapping of hyperphosphorylated proteins applied to *Net*1, a regulator of exit from mitosis in yeast. *Mol. Cell. Proteomics* **1**, 186–196.

Cohen, P. (1992). Signal integration at the level of protein kinases, protein phosphatases and their substrates. *Trends Biochem. Sci.* **17**, 408–413.

Covey, T., Shushan, B., Bonner, R., Schröder, W., and Hucho, F. (1991). Methods in protein sequence analysis. *In* "LC/MS and LC/MS/MS Screening for the Sites of Posttranslational Modification in Proteins." (H. Jörnvall, J. O. Höög, and A. M. Gustavsson, eds.). Birkhäuser Press, Basel, Switzerland.

Crabb, J. W. (ed.) (1994). "Techniques in protein chemistry," Vol. V. San Diego, CA: Academic Press.

Daum, G., Eisenmann-Tappe, I., Fries, H.-W., Troppmair, J., and Rapp, U. R. (1994). The ins and outs of *Raf* kinases. *Trends Biochem. Sci.* **19**, 474–480.

Eck, M. J. (1995). A new flavor in phosphotyrosine recognition. *Curr. Biol.* **3**, 421–429.

Ficarro, S. B., Salomon, A. R., Brill, L. M., Mason, D. E., Stettler-Gill, M., Brock, A., and Peters, E. C. (2005). Automated immobilized metal-affinity chromatography/nano-liquid chromatography/electrospray ionization mass spectrometry platform for profiling protein phosphylation sites. *Rapid Commun. Mass Spectrom.* **19**, 57–71.

Greis, K. D., Hayes, B. K., Comer, F. I., Kirk, M., Barnes, S., Lowary, T. L., and Hart, G. D. (1996). Selective detection and site-analysis of O-GlcNac-modified glycopeptides by beta-elimination and tandem electrospray mass spectrometry. *Anal. Biochem.* **234**, 38–49.

Gu, M., Kerwin, J. L., Watts, J. D., and Aebersold, R. (1997). Geramide profiling of complex lipid mixtures by electrospray ionization mass spectrometry. *Anal. Biochem.* **244**, 347–356.

Hayes, P. A., and Aebersold, R. (2000). Simultaneous detection and identification of GlcNAc-modified glycoproteins using liquid chromatography-tandem mass spectrometry. *Anal. Chem.* **72**, 5402–5410.

Hubbard, M. J., and Cohen, P. (1993). On target with a new mechanism for the regulation of protein phosphorylation. *Trends Biochem. Sci.* **18**, 172–177.

Huddleston, M. J., Annan, R. S., Bean, M. F., and Carr, S. A. (1993a). Selective detection of phosphopeptides in complex mixtures by electrospray liquid chromatography/mass spectrometry. *J. Am. Soc. Mass Spectrom.* **4**, 710.

Huddleston, M. J., Bean, M. F., and Carr, S. A. (1993b). Collisional fragmentation of glycopeptides by electrospray inonization LC–MS and LC–MS/MS: Methods for selective detection of glycopeptides in protein digests. *Anal. Chem.* **65**, 877–884.

Hunter, T. (1991). Protein kinase clarification. *Meth. Enzymol.* **200**, 3.

Hunter, A. P., and Games, D. E. (1994). Chromatographic and mass spectrometric methods for the identification of phosphorylation sites in phosphoproteins. *Rapid Commun. Mass Spectrom.* **8**, 559–570.

Jedrzejewski, P., and Lehmann, W. D. (1997). Detection of modified peptides in enzymatic digest by capillary liquid chromatography–electrospray mass spectrometry and a programmable skimmer CID acquisition routine. *Anal. Chem.* **69**, 294–301.

Kragten, E. A., Bergwerff, A. A., van Oostrum, J., Muller, D. R., and Richter, W. J. (1995). Site-specific analysis of the *N*-glycans on murine polymeric immunoglobin a using liquid chromatography–electrospray mass spectrometry. *J. Mass Spectrom.* **30**, 1679–1686.

Krishna, R. G., and Wold, F. (1993). Post-translational modification of proteins. *Adv. Enzymol. Relat. Areas Mol. Biol.* **67**, 265–298.

Larsen, M. R., Roepstorff, P., and Fresenius, J. (2000). *Anal. Chem.* **366**, 677.

Le Blanc, J. C. Y., Hager, J. W., Ilisiu, A. M. P., Hunter, C., Zhong, F., and Chu, I. (2003). Unique scanning capabilities of a new hybrid linear ion trap mass spectrometer (Qtrap) used for high-sensitivity proteomics applications. *Proteomics* **3**, 859–869.

Loyet, K. M., Stults, J. T., and Arnott, D. (2005). Mass spectrometric contributions to the practice of phosphorylation site mapping through 003: A literature review. *Mol. Cell. Proteomics* **4**, 235–245.

Luo, K., Hurley, T. R., and Sefton, B. M. (2005). Cyanogen bromide cleavage and proteolytic peptide mapping of proteins immobilized to membranes. *Methods in Enzymology* **201**, 149–152.

Mazsaroff, I., Yu, W., Kelley, B. D., and Vath, J. E. (1997). Quantitative comparison of global carbohydrate structures of glycoproteins using LC–MS and in-source fragmentation. *Anal. Chem.* **69**, 2517–2524.

Medzihradszky, K., Besman, M. J., and Burlingame, A. L. (1997). Structural characterization of site-specific *N*-glycosylation of recombinant human factor VIII by reversed-phase high-performance liquid chromatography–electrospray ionization mass spectrometry. *Anal. Chem.* **69**, 3986–3994.

Medzihradszky, K., Besman, M. J., and Burlingame, A. L. (1998). Reverse-phase capillary high-performance liquid chromatography–high-performance electrospray ionization mass spectrometry; an essential tool for the characterization of complex glycoprotein digests. *Rapid Commun. Mass Spectrom.* **12**, 472–478.

Neubauer, G., and Mann, M. (1997). Parent ion scans of large molecules. *J. Mass Spectrom.* **32**, 94–98.

Pawson, T. (1995). Protein modules and signaling networks. *Nature* **373**, 573–580.

Posewitz, M. C., and Tempst, P. (1999). Immobilized gallium(III) affinity chromatography of phosphopeptides. *Anal. Chem.* **71**, 2883–2892.

Roberts, G. D., Johnson, W. P., Burman, S., Anumula, K. R., and Carr, S. A. (1995). An integrated strategy for structural characterization of the protein and carbohydrate components of monoclonal antibodies: Application to anti-respiratory syncytial virus. *Mab. Anal. Chem.* **67**, 3613.

Rush, J., Moritz, A., Lee, K. A., Guo, A., Goss, V. L., Spek, E. J., Zhang, H., Zha, X. M., Polakiewicz, R. D., and Comb, M. J. (2005). Immunoaffinity profiling of tyrosine phosphorylation in cancer cells. *Nat. Biotechnol.* **23**, 94–101.

Rush, J., Moritz, A., Lee, K. A., Guo, A., Goss, V. L., Spek, E. J., Zhang, H., Zha, X. M., Schindler, P. A., Settineri, C. A., Collet, X., Fielding, C. J., and Burlingame, A. L. (1995). Site-specific detection and structural characterization of the glycosylation of human plasma proteins lecithin: Cholestrol acyltransferase and apolipoprotein D using HPLC–electrospray mass spectrometry and sequential glycosidase digestion. *Prot. Sci.* **4**, 791–803.

Schlessinger, J. (1994). SH2/SH3 signaling proteins. *Curr. Opin. Genet. Dev.* **4**, 25–30.

Schroeder, M. J., Shabanowitz, J., Schwartz, J. C., Hunt, D. F., and Coon, J. J. (2004). A neutral loss activation Method for improved phosphopeptide sequence analysis by quadrupole ion trap mass spectrometry. *Anal. Chem.* **76**, 3590–3598.

Steen, H., Fernandez, M., Ghaffari, S., Pandey, A., and Mann, M. (2003). Phosphotyrosine mapping in bCR/Abl oncoprotein using phosphotyrosine-specific immonium ion scanning. *Mol. Cell. Proteomics* **2**, 138–145.

Steen, H., Kuster, B., and Mann, M. (2001). Quadrupole time-of-flight versus triple-quadrupole mass spectrometry for the determination of phosphopeptides by precursor ion scanning. *J. Mass Spectrom.* **36**, 782–790.

Sullivan, B., Addona, T. A., and Carr, S. A. (2004). Selective detection of glycopeptides on ion trap mass spectrometers. *Anal. Chem.* **76**, 3112–3118.

Till, J. H., Annan, R. S., Carr, S. A., and Miller, W. T. (1994). Use of synthetic peptide libraries and phosphopeptide-selective mass spectrometry to probe protein kinase substrate specificity. *J. Biol. Chem.* **269,** 7423–7428.

Varki, A., and Marth, J. (1995). Oligosaccharides in vertebrate development. *Seminars Dev. Biol.* **6,** 127–138.

Verma, R., Annan, R. S., Huddleston, M. J., Carr, S. A., Reynard, G., and Deshaies, R. J. (1997). Phosphorylation of sic1p by G, Cdk required for it[s] degradation and entry into S phase. *Science* **278,** 455–460.

Watts, J. D., Affolter, M., Krebs, D. L., Wange, R. L., Samelson, L. E., and Aebersold, R. (1994). Identification by electrospray ionization mass spectrometry of the sites of tyrosine phosphorylation induced in activated Jurkat T cells on the protein tyrosine kinase ZAP-70. *J. Biol. Chem.* **269,** 29520–29529.

Watty, A., Neubauer, G., Dreger, M., Zimmer, M., Wilm, M., and Burden, S. J. (2000). The *in vitro* and *in vivo* phosphotoyrosine map of activated MuSk. *Proc. Nat. Acad. Sci.* **97,** 4585–4590.

Wold, F. (1981). *In vivo* chemical of proteins (post-translational modification). *Ann. Rev. Biochem.* **50,** 783–814.

Wilm, M., and Mann, M. (1994a). Electrospray and Taylor-Cone theory, sole's beam of macromolecules at last? *Int. J. Mass Spectrom. Ion Proc.* **136,** 167.

Wilm, M., and Mann, M. (1994b). Error-tolerant identification of peptides in sequences databases by peptide sequence tags. *Anal. Chem.* **66,** 4390–4399.

Wilm, M., Neubauer, G., and Mann, M. (1996). Analytical properties of the nanoelectrospray ion source. *Anal. Chem.* **68,** 1–8.

Winz, R., Hess, D., Aebersold, R., and Brownsey, R. W. (1994). Unique structural features and differential phosphorylation of the 280-kDa component (isozyme) of rat liver acetyl-CoA carboxylase. *J. Biol. Chem.* **269,** 14438–14445.

Wu, X., Ranganahan, V., Weisman, D. S., Heine, W. F., Ciccone, D. N., O'Neil, T. B., Crick, K. E., Pierce, K. A., Lane, W. A., Rathburn, G., Livingston, D. M., and Weaver, D. T. (2000). ATM phosphorylation of Nijmegen breakage syndrome gene production. *Nature* **405,** 477.

Zappacosta, F., Huddleston, M. J., Karcher, R. L., Gelfand, V. I., Carr, S. A., and Annan, R. S. (2002). Improved sensitivity for phosphopeptide mapping using capillary column HPLC and microion spray mass spectrometry: Comparative phosphorylation site mapping from gel-derived proteins. *Anal. Chem.* **74,** 3221–3231.

[6]   Characterization of Protein N-Glycosylation

By KATALIN F. MEDZIHRADSZKY

Abstract

Although mass spectrometry (MS)-based protein identification is a straightforward task, the characterization of most posttranslational modifications still represents a challenge. N-glycosylation with its well known consensus sequence, common core structure, and "universally" active endoglycosidase seems to belong to the easier category. In this chapter, MS methods for the analysis of N-glycosylated proteins are reviewed. In particular, LC–MS analysis of glycoprotein digests is discussed in detail. The examples included in this chapter illustrate the improved detection sensitivities achieved during the last decade. The characterization of site heterogeneity and of site occupancy is addressed. Low-energy collision-induced dissociation (CID) fragmentation of N-linked glycopeptides and their sodium-adducts is also described.

Introduction

Proteins are modified with carbohydrates at different amino acid side chains as follows. O-glycosylated proteins bear oligosaccharides at Thr and Ser residues (Varki, 1999; http://glycores.ncifcrf.gov). C-glycosylation has been reported at the side chain of Trp residues (Doucey et al., 1999; Hartmann and Hofsteenge, 2000; Hofsteenge et al., 1994). N-glycosylated species are modified at Asn residues and display a series of common features, including a consensus sequence, a common core structure, and an enzyme that removes most of the N-linked structures. The consensus sequence for N-glycosylation has been established as AsnXxxSer/Thr, where Xxx cannot be Pro (Pless and Lennarz, 1977). However, the presence of such a sequence only indicates the possibility of glycosylation because not all potential sites are modified. All N-linked oligosaccharides feature a common core structure of $Man_3GlcNAc_2$. Originally a $Man_9GlcNAc_2$ structure terminated with three Glcs at one antenna is transferred by an oligosaccharyltransferase to an appropriate Asn residue of the nascent protein. While the protein is traveling in the endoplasmic reticulum (ER) and the Golgi, this structure then undergoes enzymatic degradation and rebuilding (Varki, 1999; http://glycores. ncifcrf.gov). Depending on what kind of sugar residues are used for the

METHODS IN ENZYMOLOGY, VOL. 405
0076-6879/05 $35.00
DOI: 10.1016/S0076-6879(05)05006-8

elongation of the antennae oligomannose-type, hybrid and complex carbo-hydrates are formed, and the latter ones may feature repetitive lactosamine units (polylactosamine). Most common *N*-linked carbohydrates can be removed using an endoglycosidase enzyme, peptide *N*-glycosidase F (PNGase F), that leaves the oligosaccharides intact while hydrolyzing the originally glycosylated Asn residue to Asp.

Glycosylation is one of the most "mysterious" posttranslational mod-ifications. Glycoproteins display remarkable heterogeneity: *N*- and *O*-glycosylation sites may feature a wide array of different oligosaccharides, while some protein molecules bear no sugar at the same position. Proteins isolated from different species, or just from different tissues, may display completely different arrays of carbohydrates, and the site heterogeneity and occupancy may also be affected (Bloom *et al.*, 1996; Hironaka *et al.*, 1993; Suzuki *et al.*, 2001; Zamze *et al.*, 1998). Similarly, the glycosylation pattern may be altered by physiological changes or disease (Hakomori, 2002; Landberg *et al.*, 2000; Nemansky *et al.*, 1998; Wada *et al.*, 1992; Yamashita *et al.*, 1989). Despite extensive studies on protein glycosylation, we still know very little about the oligosaccharide function(s) and the reason (s) for the heterogeneity, both in structures and site occupation. Actually, this heterogeneity is mainly responsible for the lack of understanding: no existing method permits the isolation of homogenous glycoprotein populations.

Studying the enzymatically (O'Neill, 1996) or chemically (Patel and Parekh, 1994) released carbohydrates has been the path followed for decades in glycoprotein analysis (Rudd and Dwek, 1997). A wide variety of methods have been utilized for characterizing the free carbohydrates, such as affinity chromatography with immobilized lectins (Harada *et al.*, 1987), high pH anion exchange chromatography (Townsend and Hardy, 1991), capillary electrophoresis (Guttman, 1997; Guttman and Ulfelder, 1997), fluorophore-assisted carbohydrate electrophoresis (FACE) (Kumar *et al.*, 1996; Stack and Sullivan, 1992), sequential exoglycosidase digestions (Holmes *et al.*, 1996; Kawasaki *et al.*, 2003; Tyagarajan *et al.*, 1996; Watzlawick *et al.*, 1992), mass spectrometry (MS) (Duffin *et al.*, 1992; Fu *et al.*, 1994; Gallego *et al.*, 2001; Harvey, 2001; Papac *et al.*, 1997; Stephens *et al.*, 2004), and, NMR (Fu *et al.*, 1994; Gallego *et al.*, 2001; Staudacher *et al.*, 1991). In most cases, only the combination of different separation techniques and analytical methods delivers comprehensive structural infor-mation (Stroop *et al.*, 2000). Oligosaccharides of glycoproteins separated by two-dimensional electrophoresis have been released by in-gel or on-the-blot endoglycosidase digestion; the glycan mixtures were then derivatized with 3-acetamido-6-aminoacridine and were subjected to high-performance liquid chromatography (HPLC) and matrix-assisted laser desorption/ionization (MALDI) analyses (Charlwood et al., 2000).

Some recent proteomics papers focused on the selective identification of N-glycosylated proteins (Bunkenborg et al., 2004; Hagglund et al., 2004; Zhang et al., 2003). Glycoproteins and peptides were isolated and enriched either utilizing chemical modification (Zhang et al., 2003) or lectin chromatography (Bunkenborg et al., 2004) or hydrophilic interaction (Hagglund et al., 2004). Prior to the mass spectrometry analysis, the carbohydrates were removed by endoglycosidase treatment. Thus, these studies provided information only about the sites occupied and no information on the carbohydrate structures or on the degree of site occupancy.

This chapter will focus on methods utilized when the carbohydrate is still attached to the protein.

Methods for N-Linked Glycopeptide Characterization

The potential glycosylation sites have to be separated to address site occupancy and site-specific heterogeneity. The ultimate goal is to generate peptides featuring only a single consensus site, so the enzyme(s) or chemical cleavage method(s) are selected accordingly. The digestion can be performed in solution as well as in-gel (see Kuster et al., 2001, for a review on the analysis of in-gel digested glycoproteins). For analyzing the resulting mixtures, MS has become the method of choice. Analysis of unseparated enzymatic digests, commonly used for protein identification, cannot be utilized reliably since the glycopeptides may be discriminated against, both because of their size and charge—sialic acid, phosphorylation, sulfation may introduce partial negative charges. In addition, glycopeptides are usually present in substoichiometric quantities because of the carbohydrate heterogeneity at any given site. Thus, fractionation of the digests is required.

On-Line Electrospray Mass Spectrometry

Reversed phase HPLC directly coupled with electrospray ionization mass spectrometry (LC–ESIMS) provides comprehensive information on glycoproteins (Bloom et al., 1996; Carr et al., 1993; Dage et al., 1998; Huddleston et al., 1993; Kapron et al., 1997; Ling et al., 1991; Medzihradszky et al., 1994, 1997; Rush et al., 1995; Schindler et al., 1995; Wada et al., 1992). This approach offers multiple advantages. First of all, direct coupling of the HPLC to the mass spectrometer ensures that every component that is eluting is subjected to MS analysis. Minor components that would be "lost" in the "shadow" of abundant peptide ions may yield good quality mass spectra when analyzed separately. In addition, a series of different experiments can be performed on-line on the different components. For example, in-source fragmentation can be induced, and the fragments will be informative of the molecules eluting in that time frame. Tandem instruments also permit MS/MS experiments, some with both fragment and precursor ion scanning on-line. Oligosaccharides readily undergo fragmentation, yielding diagnostic carbohydrate ions, such as mass-to-charge ratio (m/z) 204, 292, 366, and 657 (i.e., the oxonium ions of N-acetylhexosamine, neuraminic (sialic) acid, hexosyl-N-acetylhexosamine, and sialyl-hexosyl-N-acetylhexosamine, respectively). Thus, glycopeptide-containing fractions can be selectively identified by either inducing in-source fragmentation or using precursor ion scanning in tandem instruments (Carr et al., 1993; Huddleston et al., 1993). The more sensitive way is the first approach that will identify only the time-window when the glycopeptides elute (Medzihradszky et al., 1997; Schindler et al., 1995). On-line glycopeptide detection is usually accomplished by "ping-pong" data acquisition (i.e., alternating two distinct sets of operating parameters): the intact molecules are detected under the first parameter set, while in-source fragmentation is induced by applying the required voltages in every second scan. This approach is also very useful for the localization of specific structures, such as the sialyl Lewis[X] antigen that is characterized by its m/z 803 oxonium ion (Dage et al., 1998). Figure 1 illustrates the utilization of in-source fragmentation and selected ion monitoring (SIM) on a tryptic digest of a mixture of human lecithin: cholesterol acyltransferase (LCAT) and apolipoprotein D (Schindler et al., 1995). Though the data presented here were acquired from 300-pmol injected digest, improved HPLC techniques and increased MS detection sensitivity would allow this experiment to be performed at much lower sample levels. Precursor ion scanning that may be performed on different tandem instruments will identify not only the glycopeptide-containing fractions but the ions that produce the diagnostic carbohydrate fragment (Carr et al., 1993; Huddleston et al., 1993). Note that in some cases, 204-positive fractions do not contain glycopeptides because tryptic peptides with a GlyLys C-terminus may yield a y_2 fragment (Biemann, 1990)—in practice a protonated Gly-Lys dipeptide—also at m/z 204 (Medzihradszky, 2002).

Glycopeptide Identification by Mass

Glycopeptides are frequently represented by a single ion of relatively high charge state. Thus, even when the glycopeptide-containing fraction or the glycopeptide ions themselves were identified in order to determine the

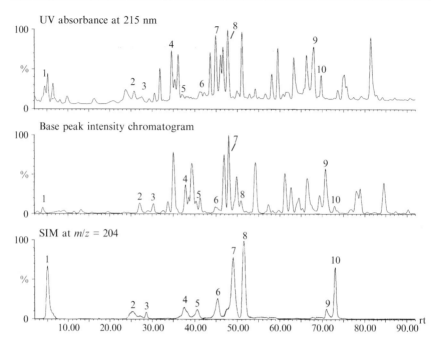

FIG. 1. UV absorbance (215 nm), base peak intensity, and selected ion monitoring (SIM) (m/z 204) chromatograms obtained from LC–MS analysis of a tryptic digest of human lecithin: cholesterol acyltransferase and apolipoprotein D. This experiment was carried out on a VG Bio-Q triple quadrupole mass spectrometer. Approximately 300 pmole of the digest was separated by gradient elution RP–HPLC using a microbore C18 column. 0.1% TFA was used as ion-pairing agent in both water and acetonitrile. To enhance the MS detection sensitivity, a make-up solvent of 2-methoxyethanol/2-propanol (1:1) was added to the eluent postcolumn in a 1:1 ratio. The mixed flow was split before being introduced into the mass spectrometer, and 5% of the sample was analyzed. The rest was manually collected at the split for further off-line analyses. Reprinted with permission from Schindler *et al.* (1995).

molecular masses, charge assignment of the ions becomes an important issue. However, only instruments with orthogonal-acceleration time-of-flight (oaTOF) or Fourier transform ion cyclotrone resonance (FTICR) analyzers afford sufficiently high resolution to resolve ions of higher charge states. Under low resolution conditions, the site-specific carbohydrate heterogeneity, the very source of some of our problems, may aid data interpretation. When trifluoroacetic acid (TFA) is the ion-pairing agent in the RP–HPLC separations, glycopeptides of the same amino acid sequence elute very close to each other, the most hydrophilic ones displaying slightly shorter retention times. Thus, one can recognize the glycopeptide ions from the heterogeneity. The mass differences will reflect the variations in the number of antennae, sialic acids, mannoses, and lactosamine units or reflect the presence or absence of fucose, depending on the structural type of carbohydrates. For example, complex oligosaccharides frequently display different numbers of sialic acid residues. The residue weight of sialic acid is 291.26 Da (average). The doubly charged ion series of differently sialylated glycoforms would differ by ~146 Da, while the mass difference between the triply charged ions would be ~97 Da and ~73 Da for quadruply charged ions.

Data shown in Fig. 1 were acquired on a quadrupole mass spectrometer that afforded sufficient resolution for resolving doubly charged ions but not ions of higher charge states (Schindler et al., 1995). Fraction 7 in Fig. 1 featured unresolved ions at m/z 1096.7 (70%), 1126.0 (18%), 1297.4 (15%), 1370.4 (100%), and 1406.8 (25%)—the relative abundances are listed in parentheses. As discussed earlier, the 73 Da mass difference between ions at m/z 1297.4 and 1370.4 suggests a charge of 4+, indicating heterogeneity in sialylation. Similarly, the ion at m/z 1406.8 could be 4+ charged, bearing an additional fucose residue ($4 \times 36.4 = ~146$ Da). Based on these assumptions, the just-mentioned ions could be considered as 4+ charged species. Thus, ions at m/z 1096.7 and 1126.0 could be assigned as 5+ charged ions representing the same molecules as m/z 1370.4 and 1406.8, (A and B), respectively. Checking the less abundant ions reveals an ion at m/z 1206.2 (5%) or 91 Da less then 1297.4 (C and D). For 4+ charged ions, this mass difference indicates heterogeneity in the number of HexHexNAc units (combined average residue weight of 365.33 Da). The charge state assignments of these ions is further supported by the presence of their 5+ charged counterparts at m/z 965.3 (2%) and 1038.6 (5%), respectively. In the 5+ ion series, an ion can also be found at m/z 1169.8 (2%) [(E): 73 Da higher than the major species at m/z 1096.7]. This mass difference at 5+ charge suggests the presence of additional antennae. Considering the potential glycosylation sites of the proteins and the addition of a sialylated complex carbohydrate, the molecules in this fraction were identified as the LCAT tryptic peptide, [16]AELSNHTRPVILVPGCLGNQLEAK[39], modified with disialo biantennary (C), diasialo triantennary (D), triasialo triantennary (A), trisialo fucosylated triantennary (B), and trisialo tetraantennary (E) complex oligosaccharides (Schindler et al., 1995).

Unfortunately, TFA has an adverse effect on the mass spectrometric sensitivity in electrospray ionization. Thus, in most current high-sensitivity LC–MS experiments, it has been replaced by formic acid. Such solvent systems [H_2O/acetonitrile/0.1% formic acid or H_2O/EtOH:PrOH (5:2)/ 0.1% formic acid] provide comparable chromatographic resolution for most peptides. However, the different ion-pairing agent has a profound effect on the retention of some acidic species: these elute later, in reversed order, in comparison to TFA-containing mobile phases. For example, in formic acid-containing solvent systems, a phosphorylated peptide will elute later than its unmodified counterpart (Wang *et al.*, 2001). Similarly, the different glycoforms are frequently separated, and those with the most sialic acids will elute latest. The coelution of a series of glycopeptides in a tryptic digest of the 50 kDa subunit of recombinant Factor VIII seems to contradict this statement (Fig. 2). Interestingly, Asn-239 of this protein is modified with oligomannose structures as well as sialylated and truncated complex carbohydrates, and the neutral and negatively charged glycoforms are all present in this fraction (Table I).

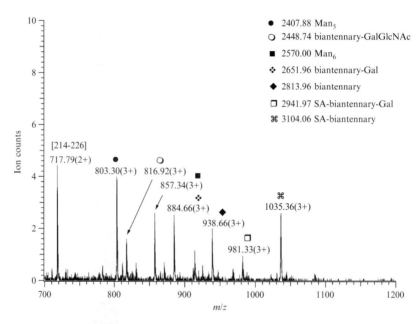

FIG. 2. Electrospray mass spectrum of a chromatographic peak (31.2 min–32 min) from the LC–MS analysis of a tryptic digest of the 50 kDa subunit of recombinant Factor VIII. Approximately 10 pmole of the digest was fractionated by RP–HPLC on a capillary (180 μm × 50 mm) C-18 column. Solvent A was 0.1% formic acid in H_2O. Solvent B was 0.1% formic acid in ethanol:propanol (5:2). The column was equilibrated at 1% B, and the amount of the organic solvent was linearly increased to 60% in 55 min. A Mariner ESI-oa-TOF mass spectrometer (Applied Biosystems, Framingham, MA) served as the detector. Different glycoforms of [231]MHTVNGYVN*R[240] eluted in this fraction, as listed in Table I. Glycoforms with neutral oligosaccharides feature hydrolysis of Asn-235 (+1 Da).

TABLE I
GLYCOFORMS OF [231]MHTVNGYVN*R[240] OF THE 50 KDA SUBUNIT OF RECOMBINANT
FACTOR VIII

Oligosaccharides	MH^+ calculated (monoisotopic mass)
Oligomannose structures	
$Man_5GlcNAc_2$	2406.99
$Man_6GlcNAc_2$	2569.05
Complex carbohydrates	
truncated (-GalGlcNAc) biantennary ($GalGlcNAcMan_3GlcNAc_2$)	2447.91
truncated (-Gal) biantennary ($GalGlcNAc_2Man_3GlcNAc_2$)	2651.10
biantennary ($Gal_2GlcNAc_2Man_3GlcNAc_2$)	2813.15
truncated (-Gal) monosialo biantennary ($Neu5AcGalGlcNAc_2Man_3GlcNAc_2$)	2942.20
monosialo biantennary ($Neu5AcGal_2GlcNAc_2Man_3GlcNAc_2$)	3104.25

All the neutral structures, however, displayed masses 1 Da higher than the calculated values (Table I). This experiment was carried out on an orthogonal acceleration time-of-flight mass spectrometer (Mariner, Applied Biosystems, Foster City, CA, www.appliedbiosystems.com) that afforded a mass accuracy of 100 ppm. Thus, this mass discrepancy must reflect structural changes, most likely the hydrolysis of Asn-235 to Asp. Figure 3 shows that the "normal" peptide bearing the neutral oligosaccharides eluted earlier, only the additional acidic residue brought the differently charged glycoforms together. Still, in this experiment, the glycopeptides eluted sufficiently close together to aid data interpretation. The retention time differences, however, may be significant as illustrated with the different glycoforms of the fetuin tryptic peptide [127]LC*PDC*PLLAP LNDSR[141] (C* represents S-β-4-ethylpyridyl Cys) (Fig. 4).

Confirmation of Glycopeptide Assignments

Endoglycosidases. Mass-based structure assignment usually requires independent confirmation. The identity of glycopeptides can be confirmed by endoglycosidase digestion (O'Neill, 1996). As mentioned earlier, PNGase F will free the peptides by removing most of the *N*-linked oligosaccharides. Since the modified Asn will be hydrolyzed, the molecular mass observed after PNGase F digestion should be a 1 Da higher per glycosylation site than the mass predicted for the original sequence. PNGase A, another

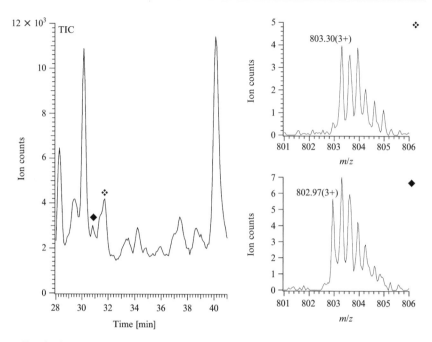

FIG. 3. A comparison showing the different retention of "normal" and Asn-235 hydrolyzed ^{231}MHTVNGYVN*R^{240} bearing neutral carbohydrates. These data are from the experiment described for Fig. 2. Part of the total ion current (TIC) chromatogram is shown with the appropriate peaks labeled and the isotope clusters of the Man$_5$-glycoform from 30.6–31.2 min (lower panel) and from 31.2–32 min (upper panel) showing the modified peptide.

member of this enzyme family, will release oligosaccharides containing α1,3-linked core fucose that are resistant to PNGase F digestion. Endo-β-N-acetylglucosaminidases D, F1, F2, F3, and H that cleave between the core GlcNAc residues also may be employed in these studies. Since the resulting peptides still bear a GlcNAc unit, the molecular mass observed after digestion should be higher than the predicted mass by 203 Da/glycosylation site for any given sequence stretch. Endo D, like PNGase F, will release all classes of N-linked carbohydrates. Endo H shows some specificity cleaving only oligomannose or hybrid structures. Endo F1 shows the same specificity as Endo H, while Endo F2 releases oligomannose and biantennary structures, and Endo F3 is partial to biantennary and triantennary oligosaccharides. Sequential application of Endo-β-N-acetylglucosaminidases and PNGase F may yield some information not only on site

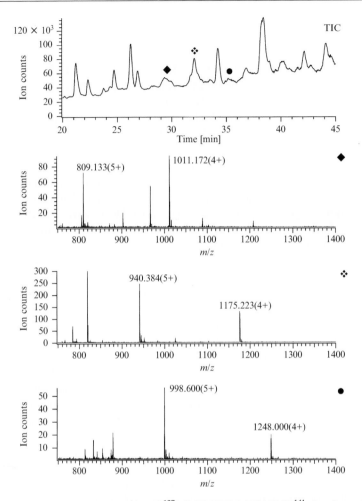

FIG. 4. Chromatographic separation of [127]LC*PDC*PLLAPLNDSR[141] glycoforms during an LC-MS analysis of a bovine fetuin tryptic digest. Approximately 400 fmole of the digest was fractionated by RP–HPLC on a 75 μm × 50 mm Pepmap column. Solvent A was 0.1% formic acid in H_2O. Solvent B was 0.1% formic acid in acetonitrile. The column was equilibrated at 5% B, and the amount of organic solvent was increased linearly to 40% in 40 min. A QSTAR-Pulsar QqTOF mass spectrometer (MDS Sciex, Concord, ON, Canada) served as the detector. The panels show the TIC trace with the fractions of interests labeled and show in elution order the corresponding mass spectra of the disialo biantennary, trisialo triantennary, and tetrasialo triantennary glycoforms. The instrument afforded sufficient resolution for charge state determination. Only the monoisotopic masses of the glycopeptides discussed are shown in the spectra. The calculated MH^+ values for the different glycoforms are 4041.68, 4697.91, and 4989.01. The mass measurement error was <10 ppm. (C* represents S-β-4-ethylpyridyl Cys.)

occupancy but some structural information on the oligosaccharides at any given site. For example, sequential Endo H/PNGase F digestion will "label" every oligomannose or hybrid site with a retained GlcNAc, while the complex sites are indicated by the 1 Da mass increase. CID experiments can readily identify these residues, even if the peptide contains multiple glycosylation sites (Tyagarajan *et al.*, 1997).

Exoglycosidases. While endoglycosidase digestions reveal the identity of the peptide modified and yield some information on the structure of the oligosaccharide, exoglycosidases may identify the sugar units as well as their linkage positions. Exoglycosidases may display specificity for cleaving only one kind of sugar and only in a single (or sometimes multiple) linkage position(s). A wide array of such enzymes are available commercially (Seikaguku Corporation, Tokyo, Japan, http://www.seikagaku.co.jp; QA-bio, San Mateo, CA, http://glycotools.9a-bio.com). The oligosaccharide structures of the LCAT glycopeptides discussed previously were assigned from data obtained by exoglycosidase digestions and consecutive mass measurements. For example, the presence of an additional HexHexNAc unit (component E) may reflect the elongation of antennae (polylactosamine structures) as well as the introduction of a new one. To differentiate between these structures, first the terminal sialic acids were removed using a neuraminidase (sialidase) enzyme, and then the glycopeptides were digested with *Streptococcus pneumoniae* β-galactosidase. This enzyme cleaves Gal$\beta1 \rightarrow 4$GlcNAc/GalNAc linkages. Obviously, the elongation of the existing antennae would not produce an additional terminal Gal. Thus, for a polylactosamine structure, the mass decrease observed should be 3×162 Da, while a tetra-antennary oligosaccharide would feature a 4×162 Da mass loss after this treatment, which was indeed what happened (Schindler *et al.*, 1995). For the "*in-situ*" analysis of *N*-linked carbohydrates, such exoglycosidase digestions followed by mass measurements represent the most promising approach (Settineri *et al.*, 1996; Stimson *et al.*, 1999; Sutton *et al.*, 1994; Tyagarajan *et al.*, 1997).

Collision-Induced Dissociation

For glycopeptide identifications, fragmentation information obtained by MS/MS experiments should also be sufficient. Collision-induced dissociation spectra of glycopeptides reveal the identity of the modified peptide as well as yield information on the carbohydrate structure (Stimson *et al.*, 1999). Data presented here were acquired from a tryptic digest of a recombinant bacterial enzyme, a carboxypeptidase G2 of *Pseudomonas aeruginosa*, expressed in *Pichia pastoris* (Medzihradszky *et al.*, 2004). This digest was first studied by LC–MS, and the 204-positive peaks were mostly indicative of the presence of tryptic peptides with a Gly-Lys C-terminus. The LC–MS results indicated extensive nonspecific cleavages, perhaps due to autolysis. Thus, it was suspected that the glycopeptides were also extensively cleaved, and that, in combination with the expected carbohydrate heterogeneity, lowered the relative amount of each species. To overcome this problem, the digest was fractionated by RP–HPLC using a microbore column, and the manually collected fractions were concentrated and subjected to electrospray analysis using nanospray sample introduction. The experiments were carried out on a QqTOF mass spectrometer (QSTAR-Pulsar, MDS Sciex, Concord, ON, Canada) that afforded ~10,000 resolution and, thus, permitted charge-state determination as well as accurate mass measurements. Multiple fractions revealed coeluting glycoforms, with molecular masses differing by 162 Da, the residue weight of a hexose residue. This was expected since yeast proteins feature oligomannose structures (Tanner and Lehle, 1987). However, no proper structure could be assigned to most of these masses because the size of the oligosaccharides and the peptide cleavage sites were not known. Thus, the suspected glycopeptide ions were subjected to low energy CID experiments (Figs. 5 and 6).

In general, the CID spectra of glycopeptides are rich in abundant ions formed *via* glycosidic bond cleavages. These fragments may be B-type oxonium ions (Domon and Costello, 1988) of the nonreducing end of the carbohydrate or Y-type ions with charge retention on the peptide. One can picture Y-series formation as removing the sugars one by one, so these fragments are characterized by the Hex-unit losses indicated. (From these CID data, the identity or the linkage of the carbohydrates cannot be determined.) In Fig. 5, one can follow the "trail" of carbohydrate losses, from m/z 1549.5(2+) to m/z 1144.97(2+) and determine that this glycopeptide contained at least eight mannoses. The next doubly charged ion in the series m/z 1043.5 is 203 Da lower, the result of a HexNAc cleavage. Its mass corresponds to ^{270}AG(GlcNAc)NVSNIIPASATLNADVR^{288}, a modified predicted tryptic peptide (the protonated complete peptide is labeled as y_{19}). All *N*-linked glycopeptides feature the full-length peptide bearing a single GlcNAc as an abundant fragment (Fig. 6) (Stimson *et al.*, 1999). Gas-phase deglycosylation of the peptide occurs to some extent (see ions marked with an asterisk in Fig. 5); however, this process is not as dominant as in *O*-glycosylated molecules (Alving *et al.*, 1999; Hanisch *et al.*, 1998; Medzihradszky *et al.*, 1996). Shorter glycopeptides usually do not feature any peptide fragments, but this tryptic sequence produced some, mostly C-terminal fragments sufficient to confirm the identity of the peptide. As was expected, the peptide bond cleavage at the N-terminus of the

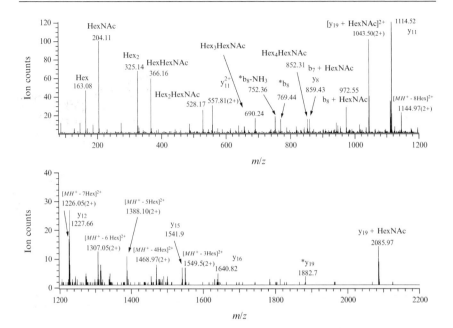

FIG. 5. Low-energy CID of ^{270}AG(Man$_8$GlcNAc$_2$)^{288}NVSNIIPASATLNADVR of recombinant carboxypeptidase G2 of *Pseudomonas aeruginosa* expressed in *Pichia pastori*. The spectrum was obtained from a manually collected and concentrated HPLC fraction of a tryptic digest on a QSTAR-Pulsar QqTOF tandem mass spectrometer (MDS Sciex) with nanospray sample introduction. Monoisotopic masses are listed. B-type carbohydrate ions are labeled showing their composition. Nomenclature for carbohydrate fragmentation: Domon and Costello (1988). Y-type ions are labeled indicating the number of hexoses cleaved. Peptide ions are labeled according to the nomenclature: Biemann (1990). Asterisks indicate gas-phase deglycosylation.

Pro residue yielded an abundant ion at m/z 1114.52 (y_{11}) (Medzihradszky *et al.*, 1994). Interestingly, its N-terminal counterpart, b_8, was also detected in both modified and gas-phase deglycosylated forms at m/z 972.55 and 769.44, respectively.

Since most of the glycopeptides observed in these experiments were formed via nonspecific cleavages, they lacked basic residues and readily formed Na or K adducts. These adduct ions were frequently more abundant than the protonated species and were selected for CID experiments. Figure 6 shows the CID spectrum of a triply charged ion at m/z 882.29. The isotope distribution of the precursor ion indicated two overlapping components. In such cases, an Asn $\rightarrow$ Asp or Gln $\rightarrow$ Glu conversion is

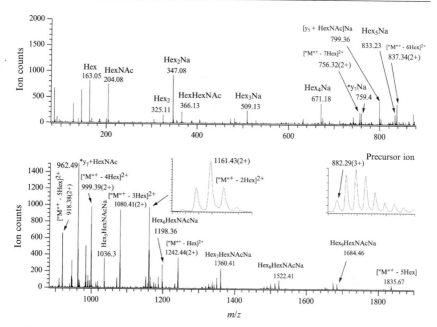

FIG. 6. Low-energy CID spectrum acquired from an overlapping ion cluster corresponding to [MHNa₂]³⁺ of ²²⁰QV(Man₁₀GlcNAc₂)NIT²²⁴ and [MH₂Na]³⁺ of ²²⁰QV(Man₉GlcNAc₂) NITGK²²⁶* of recombinant carboxypeptidase G2 of *Pseudomonas aeruginosa* expressed in *Pichia pastori*. Ions are labeled as in Fig. 5. The inserts show the precursor ion and a representative member of the Y-ion series and also illustrate the excellent resolution this instrument affords even in MS/MS mode. Monoisotopic masses are listed, but only the more abundant ion is labeled in fragments that represent both molecules. A detailed fragmentation pattern of the longer glycopeptide is presented in Scheme 1.

the most likely culprit. While following the trail of carbohydrate losses, this suspicion seemed to be confirmed, all of the Y-type fragments displayed an isotope profile in which the second isotope was more abundant than the monoisotopic peak (Fig. 6). The Y-type ions readily revealed the presence of at least seven mannoses, but then the subtraction of 203 Da (HexNAc) and different numbers of 162 Da (Hex) units did not yield any of the abundant fragment ions, as expected for the GlcNAc-modified peptide. When examining the other fragment masses, it became obvious that members of the B-ion series up the nine mannose residues were detected as sodiated ions (Fig. 6). If the carbohydrate size determined this way and the presence of the metal ion were taken into consideration, the modified peptide ion would be at *m/z* 961/962 (MNa⁺-9x Hex-HexNAc-Na+H).

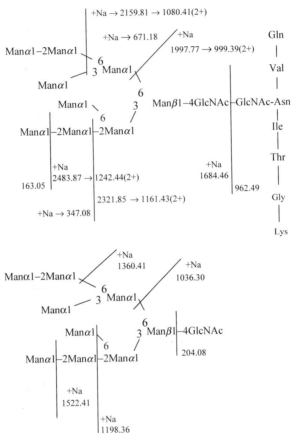

SCHEME 1. Fragments Observed in the Low Energy CID Spectrum of $[MH_2Na]^{3+}$ of Glycopeptide QV (Man$_8$GlcNAc$_2$)NITGK.[a]

[a]See Fig. 6.

Note: Carbohydrate structure based on Montesino, R., Garcia, R., Quintero, O., Cremata, J., A. (1998). Variation in N-linked oligosaccharide structures on heterologous proteins secreted by the methylotrophic yeast Pichia pastoris. Protein Expr. Purif. 14, 197–207. Not all the fragments observed are listed. Fragments of the same carbohydrate compositions may form via different cleavages.

Indeed, there was an abundant ion at m/z 962.49 that was also detected as gas-phase deglycosylated and as Na-adduct species at m/z 759.40 and 984.48, respectively, but in all three forms, the lower mass isotope was missing! The observed m/z 759.40 corresponds to QVNITGK of the protein sequence, for which the hydrolysis of the Gln residue would yield a 1 Da higher value. Thus, the overlapping component must be a different peptide represented by the only abundant ion not accounted for (i.e., m/z 799.36). This mass did not fit to any GlcNAc-modified peptide sequence in the protein studied. Careful examination of the original MS data revealed species with molecular weights 22 and 44 Da lower than this glycopeptide, suggesting a precursor ion corresponding to $[MHNa_2]^{3+}$. Thus, the ion at m/z 799.36 was eventually identified as the Na adduct of ^{220}QV(GlcNAc)NIT^{224}. In summary, the 882.29 monoisotopic mass corresponded to $[MHNa_2]^{3+}$ of ^{220}QV(Man$_{10}$GlcNAc$_2$)NIT^{224}, while the second peak of the cluster represented also the monoisotopic ion for $[MH_2Na]^{3+}$ of ^{220}QV(Man$_9$GlcNAc$_2$)NITGK^{226}. A detailed fragmentation pattern of the longer glycopeptide is presented in Scheme 1.

Based on the fragmentation pattern described above, multistage CID analysis can be and has been utilized to determine both the carbohydrate structure and the amino acid sequence of the peptide (Demelbauer *et al.*, 2004; Wada *et al.*, 2004). It also has been reported that infrared multiphoton dissociation (IRMPD), an alternative fragmentation technique offered by FTICR–MS, was utilized for the structural analysis of glycopeptides isolated from two-dimensional-gel fractionated proteins of human cerebrospinal fluid (Hakansson *et al.*, 2003). IRMPD also induces glycosidic bond cleavages preferentially.

Quantitative Assignment of Site Occupancy

PNGase F cleaves the β-aspartylglycosylamine linkage and leaves a telltale 1 Da mass increase behind: the glycosylated Asn is transformed into Asp. Thus, data can be obtained on site occupancy analyzing a tryptic digest before and after PNGase F digestion, even if not all the glycopeptides are "found" in the first experiment. This approach may help to identify previously overlooked species when a site is not 100% occupied. For example, ^{513}Ser-Leu-Asn-Asp-Ser-Thr-Asn-Arg^{520} of the B domain of recombinant Factor VIII, which contains two potential glycosylation sites, was first assigned as doubly glycosylated. However, LC–MS analysis of the PNGase F-digested peptides revealed that in about 15% of the protein, only one of the Asn-s was modified (Fig. 7) (Medzihradszky *et al.*, 1997). Since the hydrolysis of Asn residues has been reported as posttranslational modification as well as a side reaction occurring during sample preparations (see discussion of glycopeptides of the 50 kDa subunit of recombinant Factor VIII), a comparative study is necessary. As an alternative, PNGase F digestion could be performed in O^{18}-labeled water. After such digestion,

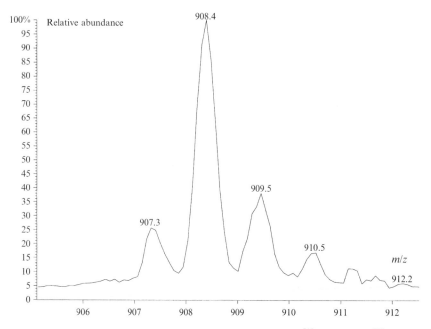

FIG. 7. The protonated molecular ions of tryptic glycopeptides [513]SLNDSTNR[520] of the B domain of recombinant Factor VIII after enzymatic deglycosylation with PNGase F. The MH^+ of the unmodified peptide is 906.4. The 1 Da molecular mass increase indicates that only one of the Asn residues was hydrolyzed by the enzyme (i.e., only one of the glycosylation sites was occupied, while the 2 Da mass increase shows that both potential sites were modified. Reprinted with permission from Medzihradszky et al. (1997).

all the originally glycosylated peptides would feature doublets separated by 2 Da for their MH^+ ions (Gonzalez et al., 1992).

Quantitative Assignments of Heterogeneity

If different glycoforms coelute during LC–MS analysis, the relative abundances of the ions representing the different species can be used for quantitation. It has been shown that results obtained this way are in good agreement with results obtained by other methods, such as high-pH anion exchange chromatography or NMR (Medzihradszky et al., 1993, 1994). However, if the glycoforms are separated as illustrated with the fetuin-peptide glycoforms, even estimation of the relative amounts of the different glycoforms becomes questionable because the ion abundances observed may be influenced differently by coelution with other peptides or differences in solvent composition and other factors.

MALDI

As mentioned earlier, MALDI–MS of unseparated enzymatic digests, commonly used for other purposes, cannot be utilized reliably for glycosylation studies. Thus, fractionation of the digests is required prior to the analysis. In addition, glycosidic bonds are more susceptible than the peptide bonds to fragmentation upon MALDI ionization. It has been reported that the most commonly used matrix for peptide analysis, α-cyano-4-hydroxycinnamic acid, induces strong metastable fragmentation in glycoproteins as well as in glycopeptides (Huberty et al., 1993; Karas et al., 1995). This phenomenon is especially apparent for sialylated carbohydrates. "Cooler" matrices, such as 2,5-dihydroxybenzoic acid (Huberty et al., 1993), 3-hydroxypicolinic acid (Karas et al., 1995), and 2,6-dihydroxy-acetophenone (Pitt and Gorman, 1996), are recommended as alternatives. Obviously for fractionation for MALDI, TFA may be applied in the solvents as an ion-pairing reagent. Thus, most glycoforms will coelute, and glycopeptide identification may be aided by the observed carbohydrate heterogeneity. The structural assignments should be confirmed using the same methods as with electrospray ionization discussed previously. MALDI analysis of the collected HPLC fractions has been used successfully for glycoprotein studies (Huberty et al., 1993; Ploug et al., 1998; Schuette et al., 2001; Sutton et al., 1994; Tyagarajan et al., 1997).

Summary

Off-line or on-line fractionation of the properly selected enzymatic digests of glycoproteins coupled with MS will yield a wealth of information. From the masses measured, the size of the carbohydrate(s) at any given site can be determined. The identity of the glycopeptides may be confirmed by mass measurements before and after endoglycosidase digestion. Similarly, MS/MS experiments may confirm the site assignments and reveal composition information on the oligosaccharides that in most cases can also be deduced from the masses themselves. Exoglycosidase digestions combined with consecutive mass measurements will provide linkage information. Under well-defined conditions, MS can also be used in a semiquantitative manner for both assigning the degree of site occupation as well as determining the relative amount of different glycoforms. The newest instrumentation provides superior information using only a few picomoles of sample or less for analysis (Figs. 4, 5, and 6).

Acknowledgments

I thank my colleagues and friends: Dr. Michael A. Baldwin for his helpful suggestions concerning this manuscript, David A. Maltby for providing the fetuin data and keeping our mass spectrometers in excellent condition, Dr. Daniel Spencer for the G2 sample, and Dr. Marc Besman for the Factor VIII protein. My work was supported by NIH grants NCRR RR01614, RR01296, RR014606, and RR015804 to the UCSF Mass Spectrometry Facility, Director A.L. Burlingame.

References

Alving, K., Paulsen, H., and Peter-Katalinić, J. (1999). Characterization of O-glycosylation sites in MUC2 glycopeptides by nanoelectrospray QTOF mass spectrometry. *J. Mass Spectrom.* **34**, 395–407.

Biemann, K. (1990). Nomenclature for peptide fragment ions (positive-ions). *Meth. Enzymol.* **193**, 886–887.

Bloom, J. W., Madanat, M. S., and Ray, M. K. (1996). Cell line and site-specific comparative analysis of the N-linked oligosaccharides on human ICAM-1des454–532 by electrospray ionization mass spectrometry. *Biochemistry* **35**, 1856–1864.

Bunkenborg, J., Pilch, B. J., Podtelejnikov, A. V., and Wisniewski, J. R. (2004). Screening for N-glycosylated proteins by liquid chromatography mass spectrometry. *Proteomics* **4**, 454–465.

Carr, S. A., Huddleston, M. J., and Bean, M. F. (1993). Selective identification and differentiation of N- and O-linked oligosaccharides in glycoproteins by liquid chromatography–mass spectrometry. *Protein Science* **2**, 183–196.

Charlwood, J., Skehel, J. M., and Camilleri, P. (2000). Analysis of N-linked oligosaccharides released from glycoproteins separated by two-dimensional gel electrophoresis. *Anal. Biochem.* **284**, 49–59.

Dage, J. L., Ackermann, B. L., and Halsall, H. B. (1998). Site localization of sialyl Lewis(x) antigen on alpha1-acid glycoprotein by high-performance liquid chromatography–electrospray mass spectrometry. *Glycobiology* **8**, 755–760.

Demelbauer, U. M., Zehl, M., Plematl, A., Allmaier, G., and Rizzi, A. (2004). Determination of glycopeptide structures by multistage mass spectrometry with low-energy collision-induced dissociation: Comparison of electrospray ionization quadrupole ion trap and matrix-assisted laser desorption/ionization quadrupole ion trap reflectron time-of-flight approaches. *Rapid Commun. Mass Spectrom.* **18**, 1575–1582.

Domon, B., and Costello, C. E. (1988). A systematic nomenclature for carbohydrate fragmentations in fab-MS/MS spectra of glycoconjugates. *Glycoconj. J.* **5**, 397–409.

Doucey, M. A., Hess, D., Blommers, M. J., and Hofsteenge, J. (1999). Recombinant human interleukin-12 is the second example of a C-mannosylated protein. *Glycobiology* **9**, 435–441.

Duffin, K. L., Welply, J. K., Huang, E., and Henion, J. D. (1992). Characterization of N-linked oligosaccharides by electrospray and tandem mass spectrometry. *Anal. Chem.* **64**, 1440–1448.

Fu, D., Chen, L., and O' Neill, R. A. (1994). A detailed structural characterization of ribonuclease B oligosaccharides by 1H NMR spectroscopy and mass spectrometry. *Carbohydr. Res.* **261**, 173–186.

Gallego, R. G., Blanco, J. L. J., Thijssen-van Zuylen, C. W. E. M., Gotfredsen, C. H., Vosho, H., Duus, J. O., Schachner, M., and Vliegenhart, J. F. G. (2001). Epitope diversity of

N-glycans from bovine peripheral myelin glycoprotein P0 revealed by mass spectrometry and nano probe magic angle spinning 1H NMR spectroscopy. *J. Biol. Chem.* **276,** 30834–30844.

Gonzalez, J., Takao, T., Hori, H., Besada, V., Rodriguez, R., Padron, G., and Shimonishi, Y. (1992). A method for determination of *N*-glycosylation sites in glycoproteins by collision-induced dissociation analysis in fast atom bombardment mass spectrometry: Identification of the positions of carbohydrate-linked asparagine in recombinant alpha-amylase by treatment with peptide-*N*-glycosidase F in 18O-labeled water. *Anal. Biochem.* **205,** 151–158.

Guttman, A. (1997). Multistructure sequencing of *N*-linked fetuin glycans by capillary gel electrophoresis and enzyme matrix digestion. *Electrophoresis* **18,** 1136–1141.

Guttman, A., and Ulfelder, K. W. (1997). Exoglycosidase matrix-mediated sequencing of a complex glycan pool by capillary electrophoresis. *J. Chromatogr. A.* **781,** 547–554.

Hagglund, P., Bunkenborg, J., Elortza, F., Jensen, O. N., and Roepstorff, P. (2004). A new strategy for identification of *N*-glycosylated proteins and unambiguous assignment of their glycosylation sites using HILIC enrichment and partial deglycosylation. *J. Proteome Res.* **3,** 556–566.

Hakansson, K., Emmett, M. R., Marshall, A. G., Davidsson, P., and Nilsson, C. L. (2003). Structural analysis of 2-D-gel-separated glycoproteins from human cerebrospinal fluid by tandem high-resolution mass spectrometry. *J. Proteome Res.* **2,** 581–588.

Hakomori, S. (2002). Glycosylation defining cancer malignancy: New wine in an old bottle. *Proc. Natl. Acad. Sci. USA* **99,** 10231–10233.

Hanisch, F. G., Green, B. N., Bateman, R., and Peter-Katalinić, J. (1998). Localization of *O*-glycosylation sites of MUC1 tandem repeats by QTOF–ESI mass spectrometry. *J. Mass Spectrom.* **33,** 358–362.

Harada, H., Kamei, M., Tokumoto, Y., Yui, S., Koyama, F., Kochibe, N., Endo, T., and Kobata, A. (1987). Systematic fractionation of oligosaccharides of human immunoglobulin G by serial affinity chromatography on immobilized lectin columns. *Anal. Biochem.* **164,** 374–381.

Hartmann, S., and Hofsteenge, J. (2000). Properdin, the positive regulator of complement, is highly C-mannosylated. *J. Biol. Chem.* **275,** 28569–28574.

Harvey, D. J. (2001). Identification of protein-bound carbohydrates by mass spectrometry. *Proteomics* **1,** 311–328.

Hironaka, T., Furukawa, K., Esmon, P. C., Yokota, T., Brown, J. E., Sawada, S., Fournel, M. A., Kato, M., Minaga, T., and Kobata, A. (1993). Structural study of the sugar chains of porcine factor VIII—Tissue- and species-specific glycosylation of factor VIII. *Arch. Biochem. Biophys.* **307,** 316–330.

Hofsteenge, J., Muller, D. R., de Beer, T., Loffler, A., Richter, W. J., and Vliegenhart, J. F. G. (1994). New type of linkage between a carbohydrate and a protein: C-glycosylation of a specific tryptophan residue in human RNase Us. *Biochemistry* **33,** 13524–13530.

Holmes, E. H., Greene, T. G., Tino, W. T., Boynton, A. L., Aldape, H. C., Misrock, S. L., and Murphy, G. P. (1996). Analysis of glycosylation of prostate-specific membrane antigen derived from LNCaP cells, prostatic carcinoma tumors and serum from prostate cancer patients. *The Prostate Suppl.* **7,** 25–29.

Huberty, M. C., Vath, J. E., Yu, W., and Martin, S. A. (1993). Site-specific carbohydrate identification in recombinant proteins using MALD–TOF MS. *Anal. Chem.* **65,** 2791–2800.

Huddleston, M. J., Bean, M. F., and Carr, S. A. (1993). Collisional fragmentation of glycopeptides by electrospray ionization LC–MS and LC–MS/MS: Methods for selective detection of glycopeptides in protein digests. *Anal. Chem.* **65,** 877–884.

Kapron, J. T., Hilliard, G. M., Lakins, J. N., Tenniswood, M. P. R., West, K. A., Carr, S. A., and Crabb, J. W. (1997). Identification and characterization of glycosylation sites in human serum clusterin. *Protein Sci.* **6,** 2120–2133.

Karas, M., Bahr, U., Strupat, K., Hillenkamp, F., Tsarbopoulos, A., and Pramanik, B. N. (1995). Matrix dependence of metastable fragmentation of glycoproteins in MALDI–TOF mass spectrometry. *Anal. Chem.* **67,** 675–679.

Kawasaki, N., Itoh, S., Ohta, M., and Hayakawa, T. (2003). Microanalysis of N-linked oligosaccharides in a glycoprotein by capillary liquid chromatography/mass spectrometry and liquid chromatography/tandem mass spectrometry. *Anal. Biochem.* **316,** 15–22.

Kumar, H. P. M., Hague, C., Haley, T., Starr, C. M., Besman, M. J., Lundblad, R., and Baker, D. (1996). Elucidation of *N*-linked oligosaccharide structures of recombinant human factor VIII using fluorophore-assisted carbohydrate electrophoresis. *Biotechnol. Appl. Biochem.* **24,** 207–216.

Kuster, B., Krogh, T. N., Mortz, E., and Harvey, D. J. (2001). Glycosylation analysis of gel-separated proteins. *Proteomics* **1,** 350–361.

Landberg, E., Huang, Y., Stromqvist, M., Mechref, Y., Hansson, L., Lundblad, A., Novotny, M. V., and Pahlsson, P. (2000). Changes in glycosylation of human bile-salt-stimulated lipase during lactation. *Arch. Biochem. Biophys.* **377,** 246–254.

Ling, V. A., Guzzetta, W., Canova-Davis, E., Stults, J. T., Hancock, W. S., Covey, T. R., and Shushan, B. I. (1991). Characterization of the tryptic map of recombinant DNA-derived tissue plasminogen activator by high-performance liquid chromatography–electrospray ionization mass spectrometry. *Anal. Chem.* **63,** 2909–2915.

Medzihradszky, K. F., Spencer, D. I. R., Begent, R. H. J., and Chester, K. A. (2004). Glycoforms obtained by expression in *Pichia pastoris* improve cancer targeting potential of a recombinant antibody-enzyme fusion protein. *Glycobiology* **14,** 27–37.

Medzihradszky, K. F. (2002). Characterization of site-specific glycosylation. *Methods Mol. Biol.* **194,** 101–125.

Medzihradszky, K. F., and Burlingame, A. L. (1994). The advantages and versatility of a high-energy collision-induced dissociation-based strategy for the sequence and structural determination of proteins. *Meth.: Comp. Meth. Enzym.* **6,** 284–303.

Medzihradszky, K. F., Gillece-Castro, B. L., Hardy, M. R., Townsend, R. R., and Burlingame, A. L. (1996). Structural elucidation of *O*-linked glycopeptides by high-energy, collision-induced dissociation. *J. Am. Soc. Mass Spectrom.* **7,** 319–328.

Medzihradszky, K. F., Settineri, C. A., Maltby, D. A., and Burlingame, A. L. (1993). Post-translational modifications of recombinant proteins determined by LC–electrospray mass spectrometry and high performance tandem mass spectrometry. In " Techniques in protein chemistry IV" (R. H. Angeletti, ed.), pp. 117–125. Academic Press, San Diego, CA.

Medzihradszky, K. F., Maltby, D. A., Hall, S. C., Settineri, C. A., and Burlingame, A. L. (1994). Characterization of protein *N*-glycosylation by reversed-phase microbore liquid chromatography/electrospray mass spectrometry, complementary mobile phases and sequential exoglycosidase digestion. *J. Am. Soc. Mass Spectrom.* **5,** 350–358.

Medzihradszky, K. F., Besman, M. J., and Burlingame, A. L. (1997). Structural characterization of site-specific *N*-glycosylation of recombinant human factor VIII by reversed-phase high-performance liquid chromatography–electrospray ionization mass spectrometry. *Anal. Chem.* **69,** 3986–3994.

Nemansky, M., Thotakura, N. R., Lyons, C. D., Ye, S., Reinhold, B. B., Reinhold, V. N., and Blithe, D. L. (1998). Developmental changes in the glycosylation of glycoprotein hormone free alpha subunit during pregnancy. *J. Biol. Chem.* **273,** 12068–12076.

O'Neill, R. A. (1996). Enzymatic release of oligosaccharides from glycoproteins for chromatographic and electrophoretic analysis. *J. Chromatogr. A.* **720,** 201–215.

Papac, D. I., Jones, A. J. S., and Basa, L. J. (1997). Matrix-assisted laser desorption/ionization time-of-flight mass spectrometry of oligosaccharides separated by high pH anion-exchange chromatography. *In* "Techniques in glycobiology" (R. R. Townsend and A. T. Hotchkiss, eds.), pp. 33–52. Marcel Dekker, New York.

Patel, T. P., and Parekh, R. B. (1994). Release of oligosaccharides from glycoproteins by hydrazinolysis. *Methods Enzymol.* **230**, 57–66.

Pitt, J. J., and Gorman, J. J. (1996). Matrix-assisted laser desorption/ionization time-of-flight mass spectrometry of sialylated glycopeptides and proteins using 2,6-dihydroxyacetophenone as a matrix. *Rapid Commun. Mass Spectrom.* **10**, 1786–1788.

Pless, D. D., and Lennarz, W. J. (1977). Enzymatic conversion of proteins to glycoproteins. *Proc. Natl. Acad. Sci. USA* **74**, 134–138.

Ploug, M., Rahbek-Nielsen, H., Nielsen, P. F., Roepstorff, P., and Dano, K. (1998). Glycosylation profile of a recombinant urokinase-type plasminogen activator receptor expressed in Chinese hamster ovary cells. *J. Biol. Chem.* **273**, 13933–13943.

Rudd, P. M., and Dwek, R. A. (1997). Rapid, sensitive sequencing of oligosaccharides from glycoproteins. *Cur. Op. Biotech.* **8**, 488–497.

Rush, R. S., Derby, P. L., Smith, D. M., Merry, C., Rogers, G., Rohde, M. F., and Katta, V. (1995). Microheterogeneity of erythropoietin carbohydrate structure. *Anal. Chem.* **67**, 1442–1452.

Schindler, P. A., Settineri, C. A., Collet, X., Fielding, C. J., and Burlingame, A. L. (1995). Site-specific detection and structural characterization of the glycosylation of human plasma proteins lecithin: Cholesterol acyltransferase and apolipoprotein D using HPLC–electrospray mass spectrometry and sequential glycosidase digestion. *Protein Sci.* **4**, 791–803.

Schuette, C. G., Weisberger, J., and Sandhoff, K. (2001). Complete analysis of the glycosylation and disulfide bond pattern of human beta-hexosaminidase B by MALDI–MS. *Glycobiology* **11**, 549–556.

Settineri, C. A., and Burlingame, A. L. (1996). Structural characterization of protein glycosylation using HPLC–electrospray ionization mass spectrometry and glycosidase digestion. *Meth. Mol. Biol.* **61**, 255–278.

Stack, R. J., and Sullivan, M. T. (1992). Electrophoretic resolution and fluorescence detection of *N*-linked glycoprotein oligosaccharides after reductive amination with 8-aminonaphthalene-1,3,6-trisulphonic acid. *Glycobiology* **2**, 85–92.

Staudacher, E., Altmann, F., Glossl, J., Marz, L., Schachter, H., Kammerling, J. P., Hard, K., and Vliegenhart, J. F. G. (1991). GDP-fucose: Beta-N-acetylglucosamine [Fuc to (Fuc alpha 1—-6GlcNAc)-Asn-peptide)alpha 1–3]-fucosyltransferase activity in honeybee (*Apis mellifica*) venom glands. The difucosylation of asparagine-bound *N*-acetylglucosamine. *Eur. J. Biochem.* **199**, 745–751.

Stephens, E., Sugars, J., Maslen, S. L., Williams, D. H., Packman, L. C., and Ellar, D. J. (2004). The *N*-linked oligosaccharides of aminopeptidase N from *Manduca sexta*: Site localization and identification of novel *N*-glycan structures. *Eur. J. Biochem.* **271**, 4241–4258.

Stimson, E., Hope, J., Chong, A., and Burlingame, A. L. (1999). Site-specific characterization of the *N*-linked glycans of murine prion protein by high-performance liquid chromatography–electrospray mass spectrometry and exoglycosidase digestions. *Biochemistry* **38**, 4885–4895.

Stroop, C. J. M., Weber, W., Gerwig, G. J., Nimtz, M., Kammerling, J. P., and Vliegenhart, J. F. G. (2000). Characterization of the carbohydrate chains of the secreted form of the human epidermal growth factor receptor. *Glycobiology* **10**, 901–917.

Sutton, C., O' Neill, J., and Cottrell, J. (1994). Site-specific characterization of glycoprotein carbohydrates by exoglycosidase digestion and laser desorption mass spectrometry. *Anal. Biochem.* **218**, 34–46.

Suzuki, N., Khoo, K. H., Chen, H. C., and Lee, Y. C. (2001). Isolation and characterization of major glycoproteins of pigeon egg white: Ubiquitous presence of unique *N*-glycans containing Gal-alpha1–4Gal. *J. Biol. Chem.* **276**, 23221–23229.

Tanner, W., and Lehle, L. (1987). Protein glycosylation in yeast. *Biochim. Biophys. Acta* **906**, 81–99.

Townsend, R. R., and Hardy, M. R. (1991). Analysis of glycoprotein oligosaccharides using high-pH anion exchange chromatography. *Glycobiology* **1**, 139–147.

Tyagarajan, K., Forte, J. G., and Townsend, R. R. (1996). Exoglycosidase purity and linkage specificity: Assessment using oligosaccharide substrates and high-pH anion-exchange chromatography with pulsed amperometric detection. *Glycobiology* **6**, 83–93.

Tyagarajan, K., Lipniunas, P. H., Townsend, R. R., and Forte, J. G. (1997). The *N*-linked oligosaccharides of the beta-subunit of rabbit gastric H,K-ATPase: Site localization and identification of novel structures. *Biochemistry* **36**, 10200–10212.

Varki, A. (ed.) (1999). Essentials of glycobiology Spring Harbor Laboratory Press, Cold Spring Harbor, N.Y.

Wada, Y., Nishikawa, A., Okamoto, N., Inui, K., Tsukamoto, H., Okada, S., and Taniguchi, N. (1992). Structure of serum transferrin in carbohydrate-deficient glycoprotein syndrome. *Biochem. Biophys. Res. Com.* **189**, 832–836.

Wada, Y., Tajiri, M., and Yoshida, S. (2004). Hydrophilic affinity isolation and MALDI multiple-stage tandem mass spectrometry of glycopeptides for glycoproteomics. *Anal. Chem.* **76**, 6560–6565.

Wang, X., Medzihradszky, K. F., Maltby, D. A., and Correia, M. A. (2001). Phosphorylation of native and heme-modified CYP3A4 by protein kinase C: A mass spectrometric characterization of the phosphorylated peptides. *Biochemistry* **40**, 11318–11326.

Watzlawick, H., Walsh, M. T., Yoshioka, Y., Schmid, K., and Brossmer, R. (1992). Structure of the *N*- and *O*-glycans of the A-chain of human plasma alpha-2HS-glycoprotein as deduced from the chemical compositions of the derivatives prepared by stepwise degradation with exoglycosidases. *Biochemistry* **31**, 12198–12203.

Yamashita, K., Koide, N., Endo, T., Iwaki, Y., and Kobata, A. (1989). Altered glycosylation of serum transferrin of patients with hepatocellular carcinoma. *J. Biol. Chem.* **264**, 2415–2423.

Zamze, S., Harvey, D. J., Chen, Y. J., Guile, G. R., Dwek, R. A., and Wing, D. R. (1998). Sialylated *N*-glycans in adult rat brain tissue—A widespread distribution of disialylated antennae in complex and hybrid structures. *Eur. J. Biochem.* **258**, 243–270.

Zhang, H., Li, X. J., Martin, D. B., and Aebersold, R. (2003). Identification and quantification of *N*-linked glycoproteins using hydrazide chemistry, stable isotope labeling, and mass spectrometry. *Nat. Biotechnol.* **21**, 660–666.

[7]   Methods in Enzymology: *O*-Glycosylation of Proteins

By JASNA PETER-KATALINIĆ

Abstract

Cell surface and extracellular proteins are *O*-glycosylated, where the most abundant type of *O*-glycosylation in proteins is the GalNAc attachment to serine (Ser) or threonine (Thr) in the protein chain by an a-glycosidic linkage. Most eukaryotic nuclear and cytoplasmic proteins modified by a-linked *O*-GlcNAc to Ser or Thr exhibit reciprocal *O*-GlcNAc glycosylation and phosphorylation during the cell cycle, cell stimulation, and/or cell growth. Less-investigated types of *O*-glycosylation *are* *O*-fucosylation, *O*-mannosylation, and *O*-glucosylation, but they are functionally of high relevance for early stages of development and for vital physiological functions of proteins. Glycosaminoglycans are a-linked to proteoglycans via a xylose-containing tetrasaccharide, represented by linear chains of repetitive disaccharides modified by carboxylates and *O*- or/and *N*-linked sulfates.

Analysis of *O*-glycosylation by mass spectrometry (MS) is a complex task due to the high structural diversity of glycan and protein factors. The parameters in structural analysis of *O*-glycans include determination of (i) *O*-glycosylation attachment sites in the protein sequence, (ii) the type of attached monosaccharide moiety, (iii) a core type in the case of GalNAc *O*-glycosylation, (iv) the type and size of the oligosaccharide portion, (v) carbohydrate branching patterns, (vi) the site of monosaccharide glycosidic linkages, (vii) the anomericity of glycosidic linkages, and (viii) covalent modifications of the sugar backbone chains by carbohydrate- and noncarbohydrate-type of substitutents.

Classical and novel analytical strategies for identification and sequencing of *O*-glycans by MS are described. These include methods to analyze *O*-glycans after total or partial release from the parent protein by chemical or enzymatic approach or to analyze *O*-glycosylated peptides by mapping and sequencing from proteolytic mixtures. A recombination process of multiply charged glycopeptides with electrons by electron capture dissociation Fourier transform ion cyclotrone resonance (FTICR)–MS has been introduced and is instrumental for nonergodic polypeptide backbone cleavages without losses of labile glycan substituents. A method for *O*-glycoscreening under increased sensitivity and efficient sequencing as a combination of an on-line coupling of capillary electrophoresis separation,

METHODS IN ENZYMOLOGY, VOL. 405   0076-6879/05 $35.00
    DOI: 10.1016/S0076-6879(05)05007-X

as well as an automated MS–tandem MS (MS/MS) switching under variable energy conditions collision-induced dissociation (CID) protocol, is beneficial for determination of *O*-acetylation and oversulfation (Bindila *et al.*, 2004a; Zamfir *et al.*, 2004a). *O*-glycomics by robotized chip-electrospray/ionization (ESI)–MS and MS/MS on the quadrupole time-of-flight (QTOF) and FTICR analyzers, accurate mass determination, and software for assignment of fragmentation spectra represent essentials for high-throughput (HTP) in serial screenings (Bindila *et al.*, 2004b; Froesch *et al.*, 2004; Vakhrushev *et al.*, 2005). Dimerization of intact *O*-glycosylated proteins can be investigated by matrix-assisted laser desorption/ionization–time-of-flight (MALDI–TOF)–MS after blotting.

Introduction: *O*-Glycosylation as a Common Posttranslational Modification of Proteins

Since 1805, glycoproteins have been studied by chemists and physiologists to reveal their structure and function (Montreuil, 1995). Already in 1865, Eichwald Ernst provided the first evidence on *O*-glycans showing that various mucins are composed of a genuine protein and a covalently linked sugar. During the past 20 years, in the late twentieth and early twenty-first centuries, a cannon of experimental methods in structural biology exploded, mostly due to technical advances in the field of recombinant DNA technology on one side and of development of sensitive instrumentation for analysis on the other. Although it is estimated that at least 50% of all mammalian proteins contain glycosylation at some point during their existence for regulation of cell traffic, cell differentiation, and cell/cell interactions, the function of glycosylation is in most cases not well understood (Apweiler *et al.*, 1999). The sugar chains contribute to structural and modulatory properties of proteins but also to recognition by other molecules, endogenous receptors, and exogenous agents. The quality control in the secretory pathways to regulate the protein expression during the cell life and differentiation in endoplasmatic reticulum (ER) is dependent on folding and assembly, processes in which *N*-glycosylation plays an essential role (Ellgaard *et al.*, 1999), whereas a general role for *O*-glycosylation has not yet been defined. Detailed investigation on biosynthesis and regulation of the pathways were carried out by a number of research groups (Brockhausen, 1995). It has been known for many years that most cell surface and extracellular proteins are *O*-glycosylated. During the last decade (1995–2005), it has been possible to demonstrate that most proteins in the nucleus and cytoplasm are also participating in *O*-glycosylation reactions by an *O*-Glc moiety and that this process could be a regulatory one (Hart, 1997). Another type of *O*-glycosylation called *O*-fucosylation, detected in the blood coagulation Factor VII (FVII) in its first epidermal growth factor-like domain (EGF-1), is found to be crucial for its binding to the tissue factor (TF) (Kao *et al.*, 1999).

Mass spectrometry (MS) has played a key role in the structural elucidation of protein glycosylation (Laine, 1990). To analyze *O*-glycosylation of proteins by MS is a complex task due to the high structural diversity of glycan chains and of the protein factors. The formation of carbohydrate chains does not occur according to a template-type of mechanism like these of the nucleic acids and proteins; accordingly, their structure cannot be directly correlated to the genetic code. To determine the full structure of glycoform(s) on single glycosylation sites is considered to be crucial for functional proteomic and glycomic types of studies.

The parameters in structural analysis of *O*-glycans (Egge, 1987) include determination of (i) utilization of potential *O*-glycosylation sites in the protein sequence for sugar attachment, (ii) the type of the monosaccharide moiety attached to a protein via *O*-glycosylation, (iii) one of eight possible core types in the case of GalNAc *O*-glycosylation, (iv) the type and size of the oligosaccharide chains attached to the core monosaccharide, (v) branching patterns of carbohydrate chains, (vi) the site of monosaccharide glycosidic linkages, (vii) the anomericity of the building blocks, and (viii) covalent modifications of the sugar backbone chains by carbohydrate- and noncarbohydrate-type of substituents.

Common Types of O-Glycosylation

O-GalNAc-Type Glycosylation. The most abundant type of *O*-glycosylation in proteins is the GalNAc-type, where the GalNAc is linked to serine (Ser) or threonine (Thr) in the protein chain by an α-glycosidic linkage. In the group of O-GalNAc glycosylated proteins, mucins are the most ubiquitous. Interest in mucin-associated carbohydrate structures arose primarily in the cancer research, in relation to the cell adhesion and metastasis, with the distribution of *O*-glycans in human tissues and sera being relevant for the diagnostic and prognostic aspects (Baldus and Hanisch, 2000). According to DNA analysis, mucins of the type MUC1 to MUC8 were shown to exhibit large domains of repeated peptides in tandem form that can be classified to distinct chromosomal sites. Mucin-linked *O*-glycans exhibit a high degree of structural variation already on the level of the core types and of chain elongations arising from the core extension (Fig. 1). In preparations from the same biological material, long and linear neutral oligosaccharide chains containing up to seven *N*-acetyllactosamine repeat units can be detected (Hanisch *et al.*, 1989) beside short glycoforms terminated by sialylation (Hanisch *et al.*, 1990). The carbohydrate chains of the

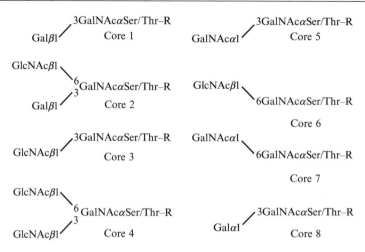

Fig. 1. Eight types of *O*-glycan core structures.

GalNAc-type *O*-glycans can be modified by noncarbohydrate substituents as detected in ovarian glycoproteins modified by sulfation and identified as ligands for E-selectin (Yuen *et al.*, 1992).

O-GlcNAc-Type Glycosylation. It was found that in most eukaryotic systems, nuclear and cytoplasmic proteins are modified with a β-linked *O*-GlcNAc to Ser or Thr (Hart, 1997). These proteins, involved in regulation processes of the cell, can exhibit reciprocal *O*-GlcNAc glycosylation and phosphorylation during the cell cycle, cell stimulation, and/or cell growth (Hart *et al.*, 1996). For MS studies, the major challenge in *O*-GlcNAc glycosylation studies is a determination of occupancy of multiple potential glycosylation sites, but the stoichiometry of single proteins is highly variable and that of the single glycosylation site is generally much lower than one (Person *et al.*, 2001).

O-Fucosylation. *O*-fucosylation is an unusual posttranslational modification detected in some proteins with interesting roles in physiological processes, such as coagulation, cell signaling, and metastasis. Although the exact function of the modification is still unclear, the number of proteins found to be modified is increasing. *O*-fucosylation is apparently phylogenetically preserved because it seems to occur on selected proteins of the primitive eukaryote *Dictyostelium discoideum* (Mreyen *et al.*, 2000). In human proteins, *O*-fucosylation was detected in the EGF-like modules in platelet and recombinant thrombospondin-1 and was discovered to be extended by a β1-3-linked glucose unit (Hofsteenge *et al.*, 2001). *O*-fucosylated motifs are common to clotting proteins Factor VII (Kao *et al.*, 1999) and Factor XII (Harris *et al.*, 1992) as well as the tissue plasminogen activator (Harris *et al.*, 1991). The major challenge for MS studies is the determination of the *O*-fucosylation site and the type of the sugar chain extension with up to four sugar units. The first step of glycosylation is performed by *O*-fucose-specific glycotransferase(s); one of them, the Fringe protein, was recently detected in *Drosophila* (Moloney *et al.*, 2000a) and in mammalian systems (Bruckner *et al.*, 2000).

O-*Mannosylation.* Mannose α-linked to serine or threonine was first found in yeast, where it can be elongated by a number of additional Man units, like in high-mannose *N*-glycosylation (Ballou, 1990). In all fungi investigated so far, *O*-glycans are attached to the amino acid chain via the Man unit. *O*-mannosylation in fungi is followed by additional glycosylation steps concerning the extension of the glycan chains by a number of Gal units (Leitao *et al.*, 2003). *O*-mannosylation has also been discovered in rat brain glycopeptides, where the mannose-linked glycan chain can be extended up to eight monosaccharide units and can be modified by sialylation, glucuronation/sulfation, and/or fucosylation (Chai *et al.*, 1999). The major challenge in *O*-mannosylation studies by MS is the determination of the attachment site in the amino acid chain and the determination of the type of hexose diastereomers as building blocks in the sugar chain.

O-*Glucosylation.* *O*-glucosylation was detected on proteins involved in blood clotting, such as Factor VII (Bjoern *et al.*, 1991) and Factor IX (Reimer, 1993), and on human Notch-1 protein (Moloney *et al.*, 2000b). The focus in glycosylation studies is to obtain clear evidence on the *O*-glucosylation next to *O*-fucosylation sites, which can be located in close proximity to each other. Besides, the β-linked *O*-Glc can serve as a core for short oligosaccharide chains by extension with further monosaccharide units.

Phosphoglycosylation. GlcNAc, Fuc, and Man can be indirectly linked to the protein at serine via a phosphate diester bridge, a glycosylation postulated to play a role in intracellular sorting and trafficking. In the protozoan, parasite *Leishmania* Manα1-P-Ser, also elongated by other sugars, has been found (Ilg *et al.*, 1996). Multiple *O*-glycoforms were found on the spore coat protein SP96 in *Dictyostelium discoideum*, where the Fucα1-3GlcNAcα1-P-Ser was the major species (Mreyen *et al.*, 2000).

O-*Glycosaminoglycan-Type Glycosylation.* Glycosaminoglycan (GAG) (synonym: mucopolysaccharide) chains are linked to proteoglycans via a β-linked xylose to Ser or Thr. GAG chains are linear and contain repetitive disaccharide units containing negatively charged functional groups, like carboxylate, *O*-, and *N*-linked sulfates. Distinct species, characterized according to the type of the basic repeating unit, are chondroitin sulfate, dermatan sulfate, heparan sulfate, and keratan sulfates I and II. In the protein-binding region, the carbohydrate structure is conserved for chondroitin sulfate, dermatan sulfate, and heparan sulfate and includes two galactose units and one glucuronic acid.

The major challenge for the structural analysis of GAG samples is the determination of the chain size, which is difficult due to the high number of carboxylate and sulfate functional groups. Such samples contain usually diverse cations in tight ion pairs, making careful desalting necessary (Vongchan et al., 2005). In intact proteoglycans, different types of GAG chains of length dispersity could be present, other than other types of glycosylation, such as N-glycosylation (Enghild et al., 1999) and GPI anchor-type glycosylation.

Collagen-Type Glycosylation. In the collagen-like modules, the hydroxylated lysine can be glycosylated by a β-linked galactose via an O-glycosidic linkage. In most collagen-like modules, usually only a short sugar chain of the disaccharide Glcα1-2Gal is present.

MS Analysis of O-Glycans Released from the Parent Protein

MS can be applied to the analysis of purified components or complex mixtures of O-glycans released from the parent protein. In this case, the analysis can be performed in the same way like that of any other saccharide. Analytical strategies for detection and sequencing of O-glycans from biological sources, which provide a general platform for understanding of ionization and fragmentation of native and derivatized samples, were developed in the 1980s using fast atom bombardment (FAB)–MS and liquid secondary ion (LSI)–MS. These basic principles are mostly applicable to electrospray/ionization (ESI)–MS and matrix-assisted laser desorption/ionization (MALDI)–MS, desorption methods that contributed to major developments in biological MS a few years later. Improved sensitivity and versatility of ionization and fragmentation devices in new mass spectrometers provide an excellent platform for implementation of novel, rapid strategies of analysis.

Release of O-*Glycans from the Parent Protein*

O-glycans are traditionally detached from the parent protein by β-elimination using strong bases, sodium hydroxide, or hydrazine. Depending on reaction conditions used, both N- and O-glycans can be released during hydrazinolysis (Patel et al., 1993), where selective isolation of O-glycans from proteins, which carry both N- and O-glycosylation, is not always achieved. Besides, the reducing terminus can become chemically modified under the reaction conditions. Classical β-elimination by sodium hydroxide under reductive conditions (Carlson, 1968) offers additional options for chemical derivatizations by tagging the reduced end. The

nonreductive release of *O*-linked oligosaccharides from mucin glycoproteins by β-elimination is also possible, frequently carried out as the first step
in synthesis of neoglycolipids, which are substrates carrying the defined
glycan structure that are covalently linked to a lipid portion, as those used
recently for identification of novel E-selectin ligands (Chai *et al.*, 1997). The
β-elimination of *O*-linked glycans takes place during the permethylation
reaction of intact glycoproteins, which requires high pH values. This procedure is of practical value for determination of the overall *O*-glycosylation
status of both glycoproteins and glycopeptides, as applied for human glycopeptides from urine to confirm the type of the glycan (Linden *et al.*,
1989). During this reaction, structural changes in the parent protein
are, however, inevitable, bringing this procedure out of consideration for
identification and sequencing of unknown proteins.

An enzymatic release of *O*-glycans from the parent protein is limited
by the high substrate specificity of *O*-glycosidases available, as shown for
the one from *Diplococcus pneumoniae*, which cleaves only the disaccharide
Galβ1-3GalNAc (Ravanat *et al.*, 1994). To obtain short oligosaccharide
blocks from the starting glycoproteins with long oligosaccharide chains,
other type of hydrolases, such as endoglycosidases, can be used; this is an
approach used for eliminative cleavage of GAGs by chondroitinase ABC in
decorin and biglycan (Zaia and Costello, 2001) and bikunin (Patel *et al.*,
1993), respectively.

MALDI–TOF–MS and ESI–MS of Released O-Glycans

MALDI–TOF–MS became the method of choice for mapping complex
mixtures of released neutral native and derivatized glycans due to its high
sensitivity and its relative tolerance to salts. The same preferences
concerning the choice of MALDI matrix for native and derivatized samples
can be applied as for the oligosaccharide analysis. In standard preparations
of *O*-glycan mixtures obtained by β-elimination from glycoproteins, a
MALDI–TOF molecular ion map could provide crucial data for determination of glycan expression. This general concept was applied to
detect components of mucin-type *O*-glycan mixture from the frog *Rana
temporaria* oviduct glycoprotein, in which 13 molecular species were detected in preparation obtained by alkaline borohydride release, ion
exchange chromatography, high-performance liquid chromatography
(HPLC) separation, and permethylation (Coppin *et al.*, 1999).

An instrumental advantage to carry out high-sensitivity analysis at the
accurate mass determination was demonstrated on a FTICR analyzer using
positive ion MALDI desorption of the highly heterogeneous neutral
O-glycan mixture from the frog *Xenopus laevis*, where 12 previously

unknown molecular species were detected in the mixture (Tseng et al., 1997). Negative-ion mode MALDI–TOF was the method of choice for mapping sialylated O-glycan mixtures containing mucin-type sialyl Lewis[x] oligosaccharides, products from a complete enzymatic synthesis that were derivatized at the reduced end with p-nitrophenyl group. More complex setup was designed to detect single components in the complex mixture of heparin-like GAGs by adding defined basic peptides to the GAG mixtures for the formation of noncovalent complexes prior to the MALDI–TOF analysis (Juhasz and Biemann, 1994; Venkataraman et al., 1999).

Fragmentation analysis of O-glycans is frequently performed by ESI desorption. Other than the possibility of performing in-source fragmentation by raising the orifice potential, low-energy CID can be carried out in ion trap (IT), triple quadrupole (Q3), quadrupole time-of-flight (QTOF), and FTICR analyzers. In some cases, it might be convenient to analyze permethylated samples, which can deliver bond cleavages, specific for the sequence and for the linkage type. On the other side, alkali-labile substituents may jeopardize analysis data. The prokaryotic O-glycan heptasaccharide, obtained under controlled base-catalyzed cleavage conditions, was submitted to ESI–IT MS/MS as a permethylated substrate (Reinhold et al., 1995). Sulfated complex alditol mixtures from pig colon mucin glycoproteins, separated by liquid chromatography (LC) on amino-bonded column as native substrates, had to be desalted on a porous graphitized carbon column prior to ESI–MS analysis. Assignment of the carbohydrate sequence and topology of the sulfate moieties was obtained by in-source fragmentation (Thomsson et al., 1999). For assignment of the GalNAc core, anomericity and modifications by Rha moieties in *Aneurinibacillus thermoaerophilus* DSM 10155 S-layer glycoprotein-derived substrates were analyzed directly by LC–ESI–MS and by nuclear magnetic resonance (NMR) (Wugeditsch et al., 1999).

Determination of O-Glycosylation Sites by ESI–QTOF–MS/MS

High potential of QTOF analyzers for direct determination of GalNAc O-glycosylation sites has been demonstrated in the analysis of a number of complex proteins from biological material, available in only low picomolar amounts. In a single MS/MS experiment, sequence of the carbohydrate and of the peptide chains as well as the glycosylation site(s) can be determined (Fig. 2) (Hanisch et al., 1998). Optimization of the conditions for collisional dissociation in the hexapole collisional cell is a crucial step in this procedure because of relative bond lability in the attachment region, resulting in the loss of sugar moiety without leaving any structural mark on the peptide backbone. A systematic fragmentation study of singly and doubly O-glycosylated MUC2 peptides by nanoelectrospray/ionization

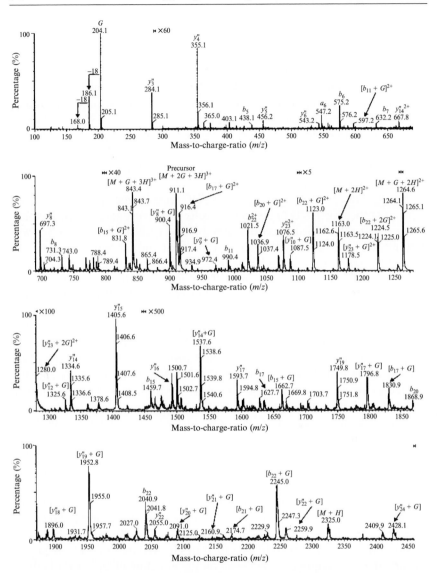

FIG. 2. (+)ESI-QTOF-MS/MS of the triply charged ion at m/z 911.1 corresponding to GalNAc-disubstituted TAP25-2 peptide. The y_n and b_m sequence ions derived from C-terminal and N-terminal, respectively, are assigned in accordance to the peptide sequence and their glycosylation status. G = GalNAc. Nomenclature of the fragment ions follows the rules established by Roepstorff and Fohlman (1984). "Localization of *O*-glycosylation sites of MUC1 tandem repeats by QTOF ESI Mass Spectrometry", F-G Hanisch *et al.*, *J. Mass Spectrom.*, **33,** 358–362. 1998 Copyright John Wiley & Sons Limited. Reproduced with permission [45].

(nanoESI)–QTOF–tandems/MS has been carried out, where distinct fragmentation patterns were obtained for isobaric glycopeptides with different glycosylation sites and different carbohydrate chains, respectively (Alving *et al.*, 1999). The M surface protein from human hepatitis B virus is in its pre-S2 domain and is *N*- and *O*-glycosylated. The *O*-glycosylated peptide of 30 amino acids, containing a single GalNAc-type *O*-glycan for one out of ten possible glycosylation sites, was fragmented in a nanoESI–QTOF–MS/MS analysis using the triply charged molecular ion as a precursor. Thr-37 was identified as a glycosylation site carrying a GalGalNAc disaccharide (Schmitt *et al.*, 1999), as one of five sites predicted by the NetOGlyc 2.0 Prediction Server (Hansen *et al.*, 1998).

Site-specific *O*-glycosylation of MUC1 tandem repeat peptides from secretory mucin of T47D breast cancer cells was investigated in the fraction obtained by affinity isolation on immobilized BC3 antibody. MUC1 was partially deglycosylated by enzymatic treatment and submitted to proteolysis by clostripain to obtain PAP20 peptides and to be mapped by ESI–QTOF to define the maximal number of GalNAc moieties. In a single nanoESI–QTOF–MS/MS experiment, the glycosylation sites of the PAP20–HexNAc2 were determined, but a polymorphism by amino acid replacements in its conserved tandem repeat region was also detected (Hofsteenge *et al.*, 2001). The attachment site of an unusual *O*-GlcNAc-type pentasaccharide modification in the cytoplasmic F-box-binding protein SKP1 in *Dictyostelium discoideum* was analyzed by ESI–QTOF–MS/MS using the triply charged molecular ion as a precursor and confirmed by Edman degradation to be a HyPro-143 (Teng-Umnuay *et al.*, 1998). Glycosylation analysis of a multidomain protein thrombospondin-1 was carried out in conjunction to its biological activity in cell adhesion, motility, and growth and was localized in the three 60 amino acid modules. The *O*-fucosylation of these modules has been investigated by nanoESI–QTOF–MS/MS, and for the first time, the structure of the *O*-linked disaccharide Glc-Fuc-*O*-Ser/Thr and its attachment sites were successfully identified by direct structural MS analysis (Fig. 3) (Hofsteenge *et al.*, 2001). Only after optimization of experimental conditions in the collisional cell, such as gas pressure, collisional energy, and acquisition time, a reasonable coverage of glycosylated peptide sequence ions, rendering unequivocal evidence on the attachment site, was obtained (Macek *et al.*, 2001).

Procedure

The glycopeptide sample was dissolved in methanol/0.4% trifluoroacetic acid (1:1, v:v) 4–8 pmol/μL and loaded into a capillary. A potential of 1.1 kV was applied to the capillary tip, and the cone voltage was set to 40 V.

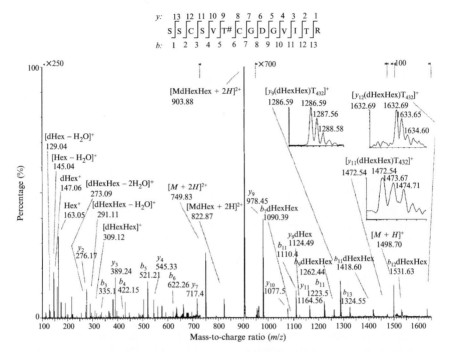

FIG. 3. (+)ESI-QTOF-MS/MS of the doubly charged molecular ion of the glycopeptide at *m/z* 903.88 corresponding to T₄₉-Ch-2. Inset fragmentation scheme of the T₄₉-Ch-2 glycopeptide. The assignment of fragment ions is in accordance with Roepstorff and Fohlman 1984. Reproduced by permission of The American Society for Biochemistry And Molecular Biology from "C-Mannosylation and *O*-Fucosylation of the Thrombospondin Type 1 Module", Hofsteenge *et al.*, 2001, 276, 9, 6485–6498 [52].

Under these conditions, the estimated flow of the sample into the analyzer was around 50 nL/min. For fragmentation analysis, a doubly charged precursor ion was selected in the quadrupole analyzer and partially fragmented in the hexapole collision cell, with the pressure of collision gas (Ar) 2.7×10^{-5} mbar and collision energy typically at 20 eV. Data acquisition was optimized to give the highest possible resolution and signal-to-noise ratio even in the case of low-abundance signals. Up to 2000 TOF pulses were accumulated for each spectrum shown. NaI and/or deglycosylated *b* and *y* ion series of the analyzed peptide were used as mass standards for instrument calibration. The mass accuracy of all measurements was within mass-to-charge ratio (*m/z*) 0.2 (Macek *et al.*, 2001).

A proteomic-type approach has been introduced for *O*-GlcNAc glycosylation analysis from intact proteins. LC–QTOF–MS/MS was successfully applied to identify *O*-GlcNAc glycosylation sites in α-crystallin proteolytic digests obtained either from solution or by in-gel digestion. Peptide mixture was loaded on a C18 microcolumn (300-μm inner diameter (I.D.) $\times$ 5 cm) and on-line introduced into the ion source by distal coated tips. Mass spectra were acquired by automatic switching between survey MS and MS/MS modes. Collisional energies were adapted to the size of the peptides from 28 to 32 V up to m/z 900 as well as 30 to 35 V from 900 on (Chalkley and Burlingame, 2001).

Determination of Glycosylation Sites by PSD–MALDI–TOF–MS

PSD–MALDI–TOF fragment ion analysis under standard conditions in α-cyano-4-hydroxycinnamic acid as a matrix renders rather complex pictures, providing enough ions for assignment of glycosylation sites if, in particular, prolines as preferable cleavage sites are present in the chain. However, the amount of deglycosylated fragment ions and internal peptide chain cleavages is significant. Formation of b and y peptide fragment ions in TAP25 MUC1 mono-, di-, and triglycosylated species was dominant although a significant degree of neutral losses (CO, H_2O, NH_3) was observed (Goletz et al., 1997). Fragmentation analysis by PSD–MALDI–TOF was studied in O-glycosylation of MUC4 peptides with blocked N and C termini. Fragmentation patterns of the glycosylated MUC4 peptides were compared to those of the nonglycosylated to study the influence of sugar constitutents to the fragmentation behavior and ion abundance. Fragmentation patterns obtained were compared to those in nanoESI–MS/MS on the triple quadrupole in low-energy CID and in the ion trap analyzer (Alving, 1998). The glycosylation site on the Thr10 in the N-terminal O-glycosylated peptide from a boar sperm surface protein was identified by PSD–MALDI–TOF (Bezouska et al., 1999).

Determination of the O-Glycosylation Status by MALDI–TOF and ESI–MS

In many laboratories, MALDI–TOF and ESI–MS are used for determination of the GalNAc O-glycosylation status in peptides, where complementary methods like Edman sequencing, modified for detection of GalNAc-linked Ser and Thr, provide evidence about the glycosylation sites. For studies of epigenetic regulatory mechanisms ruling the initiation of O-glycosylation in vitro, incorporation of the GalNAc moieties in the peptides containing up to 25 amino acids and five to six potential glycosylation sites by the recombinant GalNAc transferase was monitored by MALDI–TOF, and the sequence of utilization of glycosylation sites determined by Edman sequencing (Hanisch et al., 1999). Of practical importance is the

MALDI–TOF glycosylation analysis, carried out in parallel in the linear and reflectron modes, where the difference in mass in these two experiments can indicate the extent of glycosylation. As indicators for all types of *O*-glycosylation, appropriate oxonium ions, at *m/z* 204 for HexNAc, *m/z* 163 for Hex, and *m/z* 147 for dHex, can serve for single ion monitoring detection of glycosylated peptides in complex mixtures. Due to lability of the glycosidic bond between the core sugar and the Ser/Thr site, these ions can be detected in positive ion ESI–MS. For such type of studies, the cone voltage in the ESI ion source can be raised to achieve higher abundance of indicator ions. Glycopeptides carrying *O*-GlcNAc were identified by microbore HPLC on-line with triple quadrupole ESI–MS after in-gel digestion of their parent protein by monitoring oxonium ions at *m/z* = 204 for the *O*-GlcNAc and *m/z* = 366 for the *O*-GlcNAc-Gal, *in-vitro* galactosylated ones (Haynes and Aebersold, 2000).

Identification of *O*-Glycosylation in Proteins by In-gel Alkylaminylation

The first step in the analysis of proteoms is generally two-dimensional gel electrophoresis, followed by in-gel proteolysis of the protein bands and the mass spectrometric identification of the eluted fragments. Most proteins are posttranslationally *N*- or/and *O*-glycosylated, where those with particularly high carbohydrate content are difficult to be detected in gels because of either a low level of Coomassie blue color staining or the diffuse appearance of spots in the gel. The other aspect is the sterical hindrance at proteolysis of glycoproteins, in particular within domains with clustered carbohydrate chains where the number of mismatched proteolytic cleavage sites can be rather high. For *N*-glycosylated proteins, a standard useful protocol for in-gel deglycosylation by endo-*N*-glycosidase peptide *N*-glycosidase F (PGNase F) for glycomics-type investigations is already available. Due to the lack of endoglycosidases without substrate specificity restrictions for enzymatic *O*-deglycosylation, new protocols are required for glycoproteomics in *O*-linked carbohydrates. Under strong alkaline conditions for β-elimination, which is a classical chemical method for detachment of *O*-glycans from protein, a structural integrity of the protein is not maintained. Alternatively, *O*-glycans can be detached from the protein backbone under acidic conditions by trifluoromethanesulfonic acid, which does not severely affect the protein backbone; however, the reaction is rather difficult to be controlled concerning the degree of deglycosylation on one side and the exclusion of the moisture during the reaction on the other (Gerken *et al.*, 1992). Several refined modifications of the β-elimination reaction have been tested by different authors; these include the use of ammonia as a reagent leading to the introduction of the amino group at the glycosylation site (Rademaker *et al.*, 1998) or use of ethylamine for improved yields of the detached sugar portion (Chai *et al.*, 1997). The use of alkylamine RNH_2 as a reagent is superior to the use of ammonia due to the clear-cut labeling of the former glycosylation site, where $\Delta m = RNH - 1$ is the mass increment for the glycosylation marker pro site instead of $\Delta m = -1$ mass increment for the substitution of the hydroxy group on serine or theronine by amino group (Fig. 4) (Hanisch *et al.*, 2001). The aminoalkyl site-specific marker does not interfere with the enzymatic proteolysis for protein identification, does not compromise ionization of modified peptides in complex proteolytic mixtures during the MALDI or ESI desorption and can be easily located in the peptide chain using tandem MS sequencing.

Procedure for In-gel O-Deglycosylation

For the procedure of in-gel deglycosylation (Hanisch *et al.*, 2001), Coomassie blue-stained bands were cut out of the gel, washed with 50% aqueous methanol for 1 h, and dried in the Speedvac (Savant, Düsseldorf, Germany). The dried gel piece was taken up in 70% aqueous ethylamine or 40% aqueous methylamine and incubated at 50° for 18 or 6 h, respectively. The liquid phase containing the liberated glycans was removed.

The dried gel piece was submitted to tryptic digestion under standard conditions. The eluted peptides were analyzed by MALDI–TOF mapping. The *O*-glycosylation site-specific label contributed to the sensitivity of the peptide detection due to the alkylamino group introduced by a factor 3–5, in comparison to the starting *O*-glycosylated peptide.

The same conditions for β-elimination/alkylamination can be used for the *O*-deglycosylation procedure in solution. The degree of deglycosylation reaction and the insertion of the alkylamino marker can be directly monitored by MALDI–TOF due to the stability of the new modification.

Identification of *O*-Glycosylation Sites in Peptides by
    ECD–FTICR–MS

Direct MS determination of *O*-glycosylation sites have been performed by use of post-source decay (PSD) and collision-induced dissociation (CID). A problem to be overcome is the optimization of the fragmentation experiment to keep the carbohydrate chain glycosidic linkage to the peptide backbone intact because under standard conditions, the most labile bonds within glycopeptide ions (i.e., loss of sugar residues) occur faster than cleavage of peptide bonds. Electron capture dissociation (ECD) has been introduced as a new low-energy dissociation method (Cooper *et al.* 2005; Zubarev, 1998). The recombination process of a multiply charged species

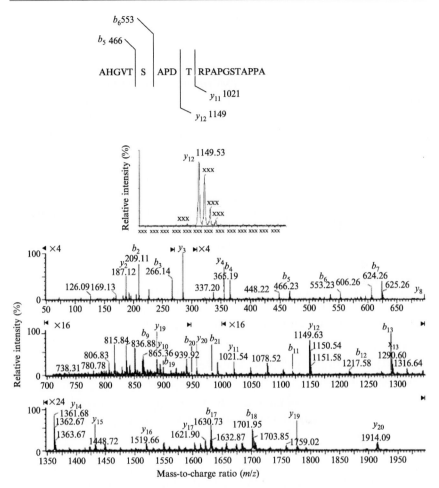

FIG. 4. (+)ESI–QTOF–MS/MS of ethylaminated MUC1 glycopeptide A13 (carrying GalNAc at Thr10, average molecular mass of 2162.3 Da). Sequence coverage: 100%. *Y* series ions indicate the glycosylation site at Thr10. Inset: Enlarged pattern of the y_{12} ion at *m/z* 1149.53 indicating the substitution site by a shift of +27 Th due to the presence of ethylamine group; amino acid sequence of the A13 peptide. Nomenclature of the fragment ions follows the rules established by Roepstorff and Fohlman (1984). Reprinted from *Anal. Biochem.* 290, Hanisch FG *et al.*, "Glycoprotein identification and localization of *O*-glycosylation sites by mass spectrometric analysis of deglycosylated/alkylaminated peptide fragments" 47–59, Copyright 2001, with permission from Elsevier [62].

with an electron of low kinetic energy was shown to generate a nonergodic polypeptide backbone cleavage without a loss of labile substituents (Mirgorodskaya *et al.*, 1999).

Fragmentation behavior of MUC2 glycopeptides, obtained either by solid-phase synthesis or by tryptic digest of glycoproteins, was studied by three general MS/MS fragmentation modes for FTICR analyzers, sustained off-resonance irradiation (SORI), infrared multiphoton dissociation (IRMPD), and ECD experiments (Mormann, 2001, 2004). Although mechanistically distinct, fragmentation processes of SORI and IRMPD gave rise to comparable fragmentation patterns. As expected, both IRMPD and SORI of the stored waveform inverse fourier transform (SWIFT)-isolated singly charged ions were shown to lose easily sugar groups. In one case, the loss of CO_2 was also observed as a side fragmentation pathway.

In course of the ECD experiment, however, by recombination of the SWIFT-isolated doubly charged glycopeptide ions with electrons of low kinetic energy, radical cations are generated to allow subsequent fragmentation. Due to the nonergodic character of this process, the cleavage of the polypeptide backbone producing intense c and $z^{\cdot}$ fragment ions without loss of the sugar substituents is observed. The series of ions obtained from the ECD experiments together with the sequence information obtained from the low-energy CID spectra allow the determination of the O-glycosylation site leading to a complete characterization of glycopeptides, like in the case of O-fucosylated properdin (Fig. 5) (Mormann, 2001, 2004).

Capillary Electrophoresis and MALDI–MS for Identification of O-Glycosylation

Complete profiling of complex glycan and glycopeptide mixtures from biological sources is hardly achievable by conventional methods. The difficulty demands for improved approaches and analytical instrumentation able to characterize glycosylation patterns or determine the structure of individual mixture components. The coupling of separation devices to either ESI–MS or MALDI–MS should contribute toward comprehensive characterization of glycosylation patterns (Duteil *et al.*, 1999; Medzihradszky *et al.*, 1998; Monsarrat *et al.*, 1999). The off-line CE–MALDI approach was very useful in studies of mucin carbohydrate chain elongation studies performed on the MUC5AC gene products, expressed mainly in the respiratory and gastric tracts, incorporating up to six GalNAc moieties *in vitro* (Soudan *et al.*, 1999).

Glycoscreening of O-linked Glycans by On-line Sheathless CE-Tandem ESI–QTOF–MS

By combined CE–MS technique, not only can molecular masses be measured with good accuracy, but also with employing MS/MS, specific fragment ions may be generated from individual components in

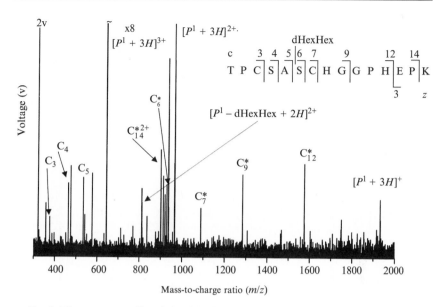

FIG. 5. Electron capture dissociation FTICR mass spectrum of the triply protonated properdin derived O-fucosylated glycopeptide ions $[P^1+3H]^{3+}$. Nomenclature of the fragment ions follows the rules established by Roepstorff and Fohlman (1984). Reprinted and adapted from *Int. J. Mass Spectrom.*, 234, M. Mormann *et al.*, "Structural studies on protein O-fucosylation by electron capture dissociation", 11–21, Copyright 2003, with permission from Elsevier [16].

O-glycopeptide mixtures, which can be off-line CE separated to deduce the molecular structure (Zamfir *et al.*, 2000). On-line coupling of CE and MS described more than 10 years ago (Olivares, 1987) was dramatically improved in terms of sensitivity by the recent development of microESI and nanoESI ion sources. A newly developed methodology for O-glycoscreening was applied to a mixture of O-glycosylated peptides obtained by partial purification from urine of patients suffering from a hereditary N-acetylhexosaminidase deficiency known as Schindler's disease (Peter-Katalinić, 1994). The steps include (i) the use of a home-built sheathless CE/ESI–MS interface, (ii) high pH of the buffer for CE separation, (iii) CE buffer compatible for MS of carbohydrates, (iv) use of negative ion MS/MS, and (v) the high-speed automated "on-the-fly" MS–MS/MS switching abilities of the QTOF–MS (Fig. 6) (Zamfir and Peter-Katalinić, 2001).

Procedure for the On-line CE–QTOF–MS/MS

For the on-line CE–QTOF–MS/MS procedure (Zamfir *et al.*, 2002), the sample was submitted to the experiment in 50 mM ammonium acetate/ammonia pH 12.0 in a concentration of 0.75 mg/ml buffer and 40% MeOH.

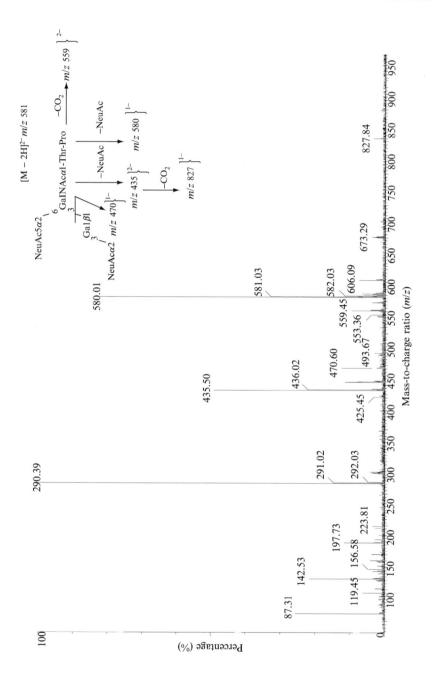

On-line CE–ESI analysis was performed in negative ion mode under a -1.1-kV potential applied to the CE microspray tip, which produced the appropriate ESI performance and gave an overall potential difference of $(30 + 1.1) = 31.1$ kV along the CE column. By applying the injection pressure for 8 s, approximately 25 nl out of 5 μl original volume was injected.

In the single high-speed automated switching "on-the-fly" experiment from MS to tandem MS mode, four precursor ions of interest can be fragmented, delivering diagnostic fragment ions even under the restrictive acquiring time conditions imposed by the MS/MS to MS switching back criteria.

In the MS/MS spectra of the singly charged Neu5AcGalGal NAc-Ser and the doubly charged tetrasaccharide Neu5Ac$_2$GalGalNAc-Ser Y- at m/z 469.6 and C-type at m/z 673.8 Da were dominant ions. In analogy, the same C-type ions complemented by the Y-type ions at m/z 435.5 and 827.8 Da resulted from CID decay of a doubly charged pseudomolecular ion at m/z 581.0, which is an indicative for the Neu5Ac$_2$GalGalNAc-Thr-Pro composition of this species (Fig. 6). The design and methodology of the sheathless CE–ESI/QTOF interface and the employed methodology turned out to be the best suited for glycomics, providing high coverage of detection and identification by data-dependent CE–MS analysis and MS to MS/MS switching, in view of the provided continuous signal stability of the microsprayer and the detection and acquisition speed of the QTOF instrument.

Characterization of GAG Oligosaccharides by CE–tandem ESI-QTOF–MS

For structural characterization of chondroitin/dermatan sulfate (CS/DS) glycosaminoglycan (GAG) oligosaccharides, several factors are in general limiting the accessibility for MS analysis; these factors include low ionization yield for longer chain GAGs, in-source desulfation, and generation of multiply charged species adequate for fragmentation analysis. Therefore, specific protocols were required, under which these factors can be avoided or reduced. In the case of a GAG complex mixture from bovine aorta obtained after β-elimination and enzymatic depolymerization by chondroitin B lyase, the approach of off-line CE and (-)ESI–MS and

---

Fig. 6. On-line sheathless CE/$(-)\mu$ESI-QTOF-automatic switching MS/MS of the doubly charged ion at m/z 581.0 corresponding to Neu5Ac$_2$HexHexNAc-Thr-Pro migrating at min 17.83. Fragmentation scheme of the doubly charged ion at m/z 581.0 corresponding to Neu5Ac$_2$HexHexNAc-Thr-Pro deduced from the spectrum. "Glycoscreening by on-line sheathless capillary electrophoresis electrospray ionization quadrupole time-of-flight tandem mass spectrometry, Zamfir *et al.*, 2001, 22, 11, 2448–2457. Copyright 2001. Reproduced and adapted by permission of John Wiley & Sons [73].

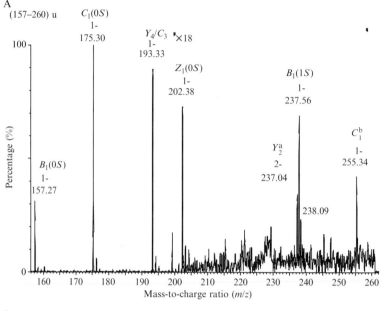

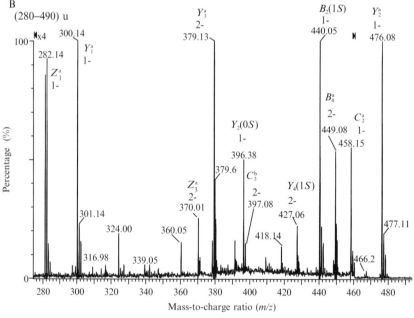

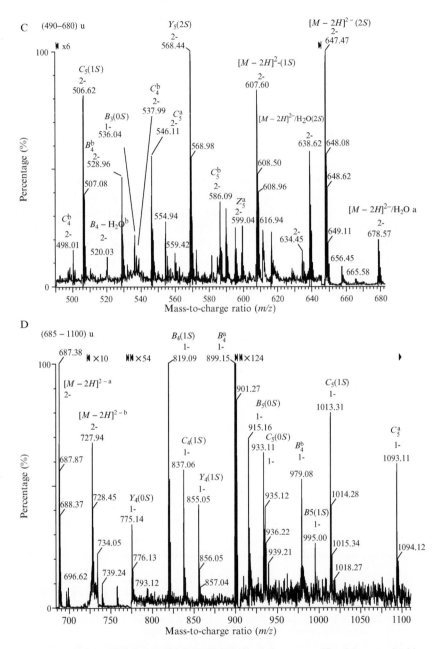

FIG. 7. Off-line CE/(−)nanoESI-QTOF-MS/MS of the pentasulfated hexasaccharide – glycosaminoglycan detected detected in the MS1 of the first CE fraction as triply charged ion

MS/MS (Zamfir *et al.*, 2002) has been optimized first. Hexa- to decasaccharide components of different sulfation grade containing a double bond at the nonreducing side of the molecule could be desorbed from the buffer in the (-)ESI–MS without loss of sulfate and then submitted to fragmentation by (-)-mode ESI–MS/MS to allow a complete sequence assignment of single components (Fig. 7). Oversulfation of CS/DS domains in human fibroblasts decorin has been explored by the slightly modified approach of the off-line CE–ESI–MS/MS. A novel tandem MS protocol of low-energy CID under variable energy condition (CID–VE) was developed to identify sulfation distribution in up to 12-mer GAG oligosaccharide components in the mixture; this novel protocol was demonstrated to be highly efficient for generation of sequence ions in the case of the pentasulfated hexasaccharide, representing a potential domain for interaction with growth factors, like FGF-2 (Zamfir *et al.*, 2003). Using the homemade sheathless interface, the human decorin CS/DS oligosaccharide mixture from HEK293 transfected cells, obtained by chondroitin B lyase partial digestion, was analyzed by on-line CE–MS (Zamfir *et al.*, 2004a). In the MS mode, it was possible to detect larger oligosaccharides up to 22-mer carrying different numbers of sulfate groups. Using the already optimized protocol by autoMS/MS switching in the on-line procedure, an eicosasaccharide with 11 sulfates has been sequenced and the position of the additional sulfate determined. General protocols for CE–MS of glycoconjugates have also been reviewed (Zamfir and Peter-Katalinić, 2004).

O-Glycomics: Chip-ESI–MS and MS/MS on the QTOF and FTICR, Accurate Mass Determination, and Software

According to the complexity of structural parameters relevant for biological activity (discussed in Chapter 1 of this volume), a performance higher than for proteomics is to be expected from MS in *O*-glycomics. The screening results can be useful only if the prerequisites of high-throughput glycan mapping and sequencing under accurate mass determination can be fulfilled. Microfluidics, automatization/robotization, and software for

---

at *m/z* 511.38. (A) *m/z* 157–260, (B) *m/z* 280–490, (C) *m/z* 490–680, (D) *m/z* 685–1100. Nomenclature of fragment ions is in accordance to the rules established by Domon and Costello (1988). [a]Regular sequence ions, [b]oversulfated sequence ions, (nS) indicates the sequence ions derived by desulfation, where n = number of sulfate groups; S = sulfate group. Reprinted and adapted from A. Zamfir *et al.*, "Structural investigation of chondroitin/dermatan sulfate oligosaccharides from human skin fibroblast decorin", Glycobiology, 2003, 13, 11, 733–742, by permission of Oxford University Press [74].

assignment of MS data are presently the most relevant basic technical requirements to be introduced and coupled to the best possible mass analyzers in terms of sensitivity and versatility of experiment design. Progress has been already achieved by introducing the Advion Nanomate robotized dispenser and chip to glycoscreening by coupling it to the QTOF (Zamfir *et al.*, 2004b) and FTICR (Froesch *et al.*, 2004) mass analyzers. The first fully automated chip-based mass spectrometry approach for complex carbohydrate system analysis was applied to the urine analysis of patients suffering from hereditary diseases. A number of already known and previously unknown *O*-linked glycoforms were detected in the urine of patients with Schindler's disease by this approach, whereas in the urine of patients with congential disorder of glycosylation (CDG), a mixture of *N*- and *O*-glycans could be partially *ad hoc* identified using the in-house-developed computer algorithm (Vakhrushev *et al.*, 2005).

The admission of samples obtained by CE preseparation to the (-)ESI–QTOF via the Nanomate system was shown to provide advantages in comparison to the capillary-based admission due to the higher ionization yield, decreased in-source decay, and stable spray over large time slots, allowing high-quality fragmentation experiments on single precursor ions in the mixture (Bindila *et al.*, 2004a). The overlapping of ions at mapping in MS1 due to the different charge states and the high number of glycoconjugate ions (originated from the diversity in sialylation, core type, chain length, and additional modifications by amino acids, peptides, sulfates, and phosphates) can be overcome by using a high-resolution FTICR instrument (Vakhrushev *et al.*, 2005). The coupling of the Nanomate to the FTICR for high-performance glycoscreening and sequencing has been carried out by constructing a new system to allow the alignment with the ESI source and has been applied to glycoscreening of urine fractions from a patient and a healthy person, allowing detection of intact sialylated glycoforms by mapping and detection of their sialylation status by SORI–CID. To clarify branching patterns of the disialylated Ser-linked hexasaccharide automated SORI–CID, MS^2 was performed selecting a precursor ion from the mixture of 20 components using the same setup (Fig. 8) (Froesch *et al.*, 2004).

A second type of microfluidic device probed for high-throughput *O*-glycomics on the QTOF instrument is a planar or thin microsprayer made from synthetic polymer material containing microchannels of 120-μm width with gold-coated microelectrodes and an integrated reservoir (Zamfir *et al.*, 2005). A number of relevant parameters for high throughput/high sensitivity were observed to be improved using this setup in comparison to the nanoelectrospray admission mode; the improvements included significant decrease of the in-source decay, increase of the ionization yield, a stable spray over long timeslots, a higher salt tolerance, and an ability to perform

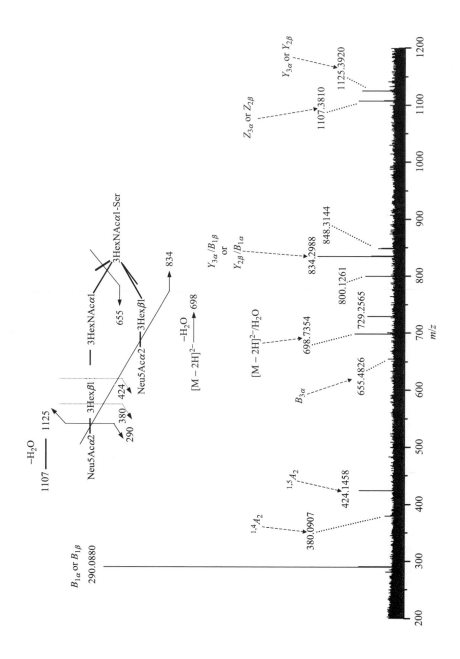

automatic MS to MS/MS switching under versatile CID experimental designs using one of the glycoconjugate mixtures from the healthy human urine. Under the high-throughput MS/MS conditions, the determination of the 2–3 versus 2–6 Neu5Ac linkage by ring cleavage ions, as described before for gangliosides (Meisen *et al.*, 2003), was possible.

Coupled to the high-resolution FTICR, this chip has been tested with different O-glycoconjugate mixtures from healthy individuals' as well as patients' urine fractions to reveal the presence of novel glycoforms not detected before due to partial overlapping of the nominally same or similar molecular ions of different charge states and constitution or due to the low abundance in the mixture (Bindila *et al.*, 2004b). In conclusion, high-throughput O-glycomics can be achieved by combined efforts toward development of microfluidic devices, in particular if combined with efficient MS analyzers like the QTOF and FTICR. This platform must be strongly supported by a biological O-glycan database for identification (Cooper *et al.*, 1999) and supplemented by software by which a *de novo* assignment of complex glycoforms, even isobaric structures, can be achieved. When introducing the O-glycomics system to molecular medicine, focus will be directed toward development of databases for human body fluid profiling and appropriate software for automatic assignment of fragmentation spectra, obtained under optimized MS conditions and accurate mass determination (Vakhrushev *et al.*, 2005; Zamfir *et al.*, 2004b).

Identification of the Noncovalent O-Glycoprotein Dimer by Blot IR-MALDI-MS

Glycophorin A (GpA), an integral transmembrane protein from the human erythrocyte membrane, contains in its N-terminal 72 amino acids stretch as well as 15 O- and one N-glycosylation sites, expressing glycans with different degrees of sialylation (Fukuda *et al.*, 1987; Cooper *et al.*, 1999). In the case of maximal glycosylation, the molecular mass of the monomeric glycoprotein is calculated to be 30 kDa. A newly designed experiment for analysis of a dimeric protein as a noncovalent complex under MALDI conditions was shown for the first time, using IR–MALDI–MS as a

---

FIG. 8. (−)ChipESI-FTICR SORI-CID MS^2 of the doubly charged ion at *m/z* 707.7419 corresponding to Neu5Ac$_2$Hex$_2$HexNAc$_2$-Ser detected in the glycopeptide mixture from urine of healthy individual. Inset: Fragmentation scheme of the Ser-linked hexasaccharide. Nomenclature of fragment ions follows the rules established by Domon and Costello (1988). "Coupling of fully automated chip electrospray to Fourier transform ion cyclotron resonance mass spectrometry for high-performance glycoscreening and sequencing", M. Froesch *et al.*, 2004, Rapid Commun. Mass Spectrom. 18, 3084–3092. Copyright 2004 John Wiley & Sons Ltd. Reproduced and adapted with permission [79].

"softer" desorption technique than the UV-MALDI (Meisen, 1997). The dimeric GpA, very stable under the conditions of the SDS–PAGE separation, with electroblotting and desorption/ionization directly from the Western blot analysis, was identified in parallel by MS and in Western blot using monoclonal antiGpA antibody. The molecular ion of the dimer was about 6 kDa lower than estimated by the SDS–PAGE using the standard marker proteins. Moreover, molecular ions with two, three, and four positive charges were detected with declining abundance.

Procedure for Detection of Noncovalent GpA Dimer by SDS–PAGE, Western Blot Analysis, and IR–MALDI–MS

For the detection of noncovalent GpA dimer procedure (Meisen, 1997), GpA with MM blood group activity was isolated from pooled human blood according to a modified procedure of Hanahan and Ekholm (1974). After SDS–PAGE according to Laemmli (1970) and Schägger and Jagow (1987), on 0.75-mm thick slab gels, the protein was electrotransferred onto nitrocellulose sheets (0.45 μm, Schleicher and Schüll, Dassel, Germany,) at 95 V for 105 min. GpA was detected by Western blot using monoclonal anti-GpA antibody. For immunodetection, the blots were incubated in 3% bovine serum albumin in PBS-T (50 mM KH$_2$PO$_4$, 50 mM Na$_2$HPO$_4$, 150 mM NaCl, pH 7.2, 0.05% Tween 20) for at least 2 h, followed by incubation with a 1:1000 dilution of anti-glycophorin A (Sigma, Deisenhofen, Germany,) in 1% BSA/PBS-T for 3 h at room temperature. After thorough washing, a 1:2000 dilution of alkaline phosphatase conjugated anti-mouse-IgM antibody (Sigma) in 1% BSA/PBS-T was added for 3 h. After washing with TBS (50 mM Tris-HCl, 150 mM NaCl, pH 7.5), the blots were developed with 37.5 μl BCIP (50 mg/ml DMF) and 50 μl NBT [76.92 mg/ml DMF, 70% (v/v)] in 10 ml TBS. Molecular mass of the stained GpA band was roughly estimated by comparison to commercial standards to be at 66 kDa. No equilibrium between monomeric and dimeric form was observed in the blot. The membrane was incubated in the solution of succinic acid as an IR-MALDI matrix. The GpA band was cut out of the blot and glued to the target. In the IR-MALDI–MS, the molecular $[M + H]^+$ ion was found at $m/z = 59860 \pm 50$ (Fig. 9) (Meisen, 1997).

MALDI–MS was carried out with a laboratory-built time-of-flight instrument at the University of Münster using 12-keV ion energy and a single-stage reflection followed by a secondary-electron multiplier (EMI 9643, Electron Tube Ltd., Ruislip, England) equipped with a conversion dynode using 15-keV postacceleration. As an IR laser, an Er-YAG laser (Schwartz Electrooptics SEO 123, Orlando, FL; 2.94 μm, 150-ns pulse duration) was used.

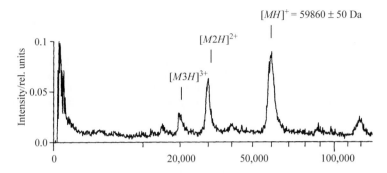

FIG. 9. IR-MALDI MS of the Glycophorin A desorbed directly from the Westen-Blot. A band at 66 kDa has been stained by the monoclonal antibody directed against the *N*-terminus of the glycoprotein in the Western blot analysis indicating the identity of Glycophorin A. Reprinted with permission from Doctoral Thesis of Dr. Iris Meisen (1997).

Determination of the Overall *O*-Glycosylation of a Glycoprotein by Permethylation

Making use of alkaline lability of the *O*-glycosidic linkage a rapid protocol has been developed for detaching all *O*-glycans from the protein or peptide (Ravanat *et al.*, 1994) by β-elimination and permethylation as a one-pot reaction according to known protocols (Ciucana, 1984). The mixture of permethylated *O*-glycans can be obtained by extraction and submitted to MS for rapid mapping and sequencing.

Preparative Procedure for Permethylation of the Intact Glycoprotein

The amount of 100 μg of each sample was dissolved in 200-μl DMSO containing 1 mg/ml glucose as the carrier. Then 30 mg of dried, powdered NaOH and 200 μl of MeI were added. After 2 h, another 200 μl of MeI were added. After a complete reaction time of 4 h at room temperature, the permethylated glycans were extracted by adding 7 ml of $CHCl_3$ and then they were placed in a centrifuge at 3000 rpm (1620 g) for 10 min. The supernatant was dried under N_2 atmosphere, dissolved in 500 μl of $CHCl_3$/MeOH 1:1 (v/v) and separated from remaining reagents on a small self-packed column of Sephadex LH-20. The resulting fractions were dried under N_2 atmosphere and analyzed by MS. The original reaction was developed for the FAB–MS analysis and can be scaled down to the 1–5-μg starting material if a carrier saccharide (mostly glucose) is added.

Acknowledgments

The ongoing engagement in projects cited in this chapter by present and past members of my group and external collaborations is greatly acknowledged. A generous financial support for the 9.4 T FTICR (Apex II, Bruker Daltonik, Bremen, Germany) was provided by Deutsche Forschungsgemeinschaft (Project PE 415/14). The QTOF mass spectrometer (Waters/Micromass) was purchased by the HbfG grant by the Land of Nordrhine-Westfalia, Germany. The technical help with the manuscript was provided by L. Bindila (University of Münster).

References

Alving, K., Korner, R., Paulsen, H., and Peter-Katalinić, J. (1998). Nanospray–ESI low-energy CID and MALDI post-source decay for determination of O-glycosylation sites in MUCU peptides. *J. Mass Spectrom.* **33**, 1124–1133.

Alving, K., Paulsen, H., and Peter-Katalinić, J. (1999). Characterization of O-glycosylation sites in MUC2 glycopeptides by nanoelectrospray QTOF mass spectrometry. *J. Mass Spectrom.* **34**, 395–407.

Apweiler, R., Hermjakob, H., and Sharon, N. (1999). On the frequency of protein glycosylation, as deduced from analysis of the SWISS-PROT database. *Biochim. Biophys. Acta* **1473**, 4–8.

Baldus, S. E., and Hanisch, F. G. (2000). Biochemistry and pathological importance of mucin-associated antigens in gastrointestinal neoplasia. *Adv. Cancer Res.* **79**, 201–248.

Ballou, C. E. (1990). Isolation, characterization, and properties of *Saccharomyces cerevisiae mnn* mutants with nonconditional protein glycosylation defects. *Methods Enzymol.* **185**, 440–470.

Bezouska, K., Sklenar, J., Novak, P., Halada, P., Havlicek, V., Kraus, M., Ticha, M., and Jonakova, V. (1999). Determination of the complete covalent structure of the major glycoform of DQH sperm surface protein, a novel trypsin-resistant boar seminal plasma O-glycoprotein related to pB1 protein. *Protein Sci.* **8**, 1551–1556.

Bindila, L., Almeida, R., Sterling, A., Allen, M., Peter-Katalinić, J., and Zamfir, A. (2004a). Off-line capillary electrophoresis/fully automated nanoelectrospray chip quadrupole time-of-flight mass spectrometry and tandem mass spectrometry for glycoconjugate analysis. *J. Mass Spectrom.* **39**, 1190–1201.

Bindila, L., Froesch, M., Lion, N., Vukelić, Z., Rossier, J. S., Girault, H. H., Peter-Katalinić, J., and Zamfir, A. D. (2004b). A thin chip microsprayer system coupled to Fourier transform ion cyclotron resonance mass spectrometry for glycopeptide screening. *Rapid Commun. Mass Spectrom.* **18**, 2913–2920.

Bjoern, S., Foster, D. C., Thim, L., Wiberg, F. C., Christensen, M., Komiyama, Y., Pedersen, A. H., and Kisiel, W. (1991). Human plasma and recombinant factor VII. Characterization of O-glycosylations at serine residues 52 and 60 and effects of site-directed mutagenesis of serine 52 to alanine. *J. Biol. Chem.* **266**, 11051–11057.

Brockhausen, I. (1995). Biosynthesis of O-glycans of the N-acetylgalactosamine α-Ser/Thr linkage type. *In* "Glycoproteins" (J. F. G. V. J. Montreuil and H. Schachter, eds.). Chapter 5, Biosynthesis 3. Elsevier, Amsterdam; New York.

Bruckner, K., Perez, L., Clausen, H., and Cohen, S. (2000). Glycosyltransferase activity of fringe modulates Notch–Delta interactions. *Nature* **406**, 411–415.

Carlson, D. M. (1968). Structures and immunochemical properties of oligosaccharides isolated from pig submaxillary mucins. *J. Biol. Chem.* **243**, 616–626.

Chai, W., Feizi, T., Yuen, C. T., and Lawson, A. M. (1997). Nonreductive release of O-linked oligosaccharides from mucin glycoproteins for structure/function assignments as neoglycolipids: Application in the detection of novel ligands for E-selectin. *Glycobiology* **7**, 861–872.

Chai, W., Yuen, C. T., Kogelberg, H., Carruthers, R. A., Margolis, R. U., Feizi, T., and Lawson, A. M. (1999). High prevalence of 2-mono- and 2,6-di-substituted manol-terminating sequences among O-glycans released from brain glycopeptides by reductive alkaline hydrolysis. *Eur. J. Biochem.* **263**, 879–888.

Chalkley, R. J., and Burlingame, A. L. (2001). Identification of GlcNAcylation sites of peptides and alpha-crystallin using Q-TOF mass spectrometry. *J. Am. Soc. Mass Spectrom.* **12**, 1106–1113.

Ciucana, I., and Kerek, F. (1984). A simple and rapid method for the permethylation of carbohydrates. *Carbohydr. Res.* **131**, 209–217.

Cooper, C. A., Wilkins, M. R., Williams, K. L., and Packer, N. H. (1999). BOLD–a biological O-linked glycan database. *Electrophoresis* **20**, 3589–3598.

Cooper, H. J., Hakansson, K., and Marshall, A. G. (2005). The role of electron capture dissociation in biomolecular analysis. *Mass Spectrom. Rev.* **24**, 201–222.

Coppin, A., Maes, E., Morelle, W., and Strecker, G. (1999). Structural analysis of 13 neutral oligosaccharide-alditols released by reductive beta-elimination from oviducal mucins of *Rana temporaria*. *Eur. J. Biochem.* **266**, 94–104.

Domon, B., and Costello, C. E. (1988). Structure elucidation of glycosphingolipids and gangliosides using high-performance tandem mass spectrometry. *Biochemistry* **27**, 1534–1543.

Duteil, S., Gareil, P., Girault, S., Mallet, A., Feve, C., and Siret, L. (1999). Identification of heparin oligosaccharides by direct coupling of capillary electrophoresis/ionspray–mass spectrometry. *Rapid Commun. Mass Spectrom.* **13**, 1889–1898.

Egge, H., and Peter-Katalinić, J. (1987). Fast-atom-bombardment mass spectrometry for structural elucidation of glycoconjugates. *Mass Spectrom. Rev.* **6**, 331–393.

Ellgaard, L., Molinari, M., and Helenius, A. (1999). Setting the standards: Quality control in the secretory pathway. *Science* **286**, 1882–1888.

Enghild, J. J., Thogersen, I. B., Cheng, F., Fransson, L. A., Roepstorff, P., and Rahbek-Nielsen, H. (1999). Organization of the inter-alpha-inhibitor heavy chains on the chondroitin sulfate originating from Ser(10) of bikunin: Post-translational modification of IalphaI-derived bikunin. *Biochemistry* **38**, 11804–11813.

Froesch, M., Bindila, L. M., Baykut, G., Allen, M., Peter-Katalinić, J., and Zamfir, A. D. (2004). Coupling of fully automated chip electrospray to Fourier transform ion cyclotron resonance mass spectrometry for high-performance glycoscreening and sequencing. *Rapid Commun. Mass Spectrom.* **18**, 3084–3092.

Fukuda, M., Lauffenburger, M., Sasaki, H., Rogers, M. E., and Dell, A. (1987). Structures of novel sialylated O-linked oligosaccharides isolated from human erythrocyte glycophorins. *J. Biol. Chem.* **262**, 11952–11957.

Gerken, T. A., Gupta, R., and Jentoft, N. (1992). A novel approach for chemically deglycosylating O-linked glycoproteins. The deglycosylation of submaxillary and respiratory mucins. *Biochemistry* **31**, 639–648.

Goletz, S., Thiede, B., Hanisch, F. G., Schultz, M., Peter-Katalinić, J., Muller, S., Seitz, O., and Karsten, U. (1997). A sequencing strategy for the localization of O-glycosylation sites of MUC1 tandem repeats by PSD–MALDI mass spectrometry. *Glycobiology* **7**, 881–896.

Hanahan, D. J., and Ekholm, J. E. (1974). The preparation of red cell ghosts (membranes). *Methods Enzymol.* **31**, 168–172.

Hanisch, F. G., Green, B. N., Bateman, R., and Peter-Katalinić, J. (1998). Localization of O-glycosylation sites of MUC1 tandem repeats by QTOF–ESI mass spectrometry. *J. Mass Spectrom.* **33,** 358–362.

Hanisch, F. G., Jovanovic, M., and Peter-Katalinić, J. (2001). Glycoprotein identification and localization of O-glycosylation sites by mass spectrometric analysis of deglycosylated/alkylaminylated peptide fragments. *Anal. Biochem.* **290,** 47–59.

Hanisch, F. G., Muller, S., Hassan, H., Clausen, H., Zachara, N., Gooley, A. A., Paulsen, H., Alving, K., and Peter-Katalinić, J. (1999). Dynamic epigenetic regulation of initial O-glycosylation by UDP-N-acetylgalactosamine: Peptide N-acetylgalactosaminyltransferases. Site-specific glycosylation of MUC1 repeat peptide influences the substrate qualities at adjacent or distant Ser/Thr positions. *J. Biol. Chem.* **274,** 9946–9954.

Hanisch, F. G., Peter-Katalinić, J., Egge, H., Dabrowski, U., and Uhlenbruck, G. (1990). Structures of acidic O-linked polylactosaminoglycans on human skim milk mucins. *Glycoconj. J.* **7,** 525–543.

Hanisch, F. G., Uhlenbruck, G., Peter-Katalinić, J., Egge, H., Dabrowski, J., and Dabrowski, U. (1989). Structures of neutral O-linked polylactosaminoglycans on human skim milk mucins. A novel type of linearly extended poly-N-acetyllactosamine backbones with Gal beta(1-4)GlcNAc beta(1-6) repeating units. *J. Biol. Chem.* **264,** 872–883.

Hansen, J. E., Lund, O., Tolstrup, N., Gooley, A. A., Williams, K. L., and Brunak, S. (1998). NetOglyc: Prediction of mucin type O-glycosylation sites based on sequence context and surface accessibility. *Glycoconj. J.* **15,** 115–130.

Harris, R. J., Leonard, C. K., Guzzetta, A. W., and Spellman, M. W. (1991). Tissue plasminogen activator has an O-linked fucose attached to threonine-61 in the epidermal growth factor domain. *Biochemistry* **30,** 2311–2314.

Harris, R. J., Ling, V. T., and Spellman, M. W. (1992). O-linked fucose is present in the first epidermal growth factor domain of factor XII but not protein C. *J. Biol. Chem.* **267,** 5102–5107.

Hart, G. W. (1997). Dynamic O-linked glycosylation of nuclear and cytoskeletal proteins. *Annu. Rev. Biochem.* **66,** 315–335.

Hart, G. W., Kreppel, L. K., Comer, F. I., Arnold, C. S., Snow, D. M., Ye, Z., Cheng, X., Della Manna, D., Caine, D. S., Earles, B. J., Akimoto, Y., Cole, R. N., and Hayes, B. K. (1996). O-GlcNAcylation of key nuclear and cytoskeletal proteins: Reciprocity with O-phosphorylation and putative roles in protein multimerization. *Glycobiology* **6,** 711–716.

Haynes, P. A., and Aebersold, R. (2000). Simultaneous detection and identification of O-GlcNAc-modified glycoproteins using liquid chromatography–tandem mass spectrometry. *Anal. Chem.* **72,** 5402–5410.

Hofsteenge, J., Huwiler, K. G., Macek, B., Hess, D., Lawler, J., Mosher, D. F., and Peter-Katalinić, J. (2001). C-mannosylation and O-fucosylation of the thrombospondin type 1 module. *J. Biol. Chem.* **276,** 6485–6498.

Ilg, T., Stierhof, Y. D., Craik, D., Simpson, R., Handman, E., and Bacic, A. (1996). Purification and structural characterization of a filamentous, mucin-like proteophosphoglycan secreted by *Leishmania* parasites. *J. Biol. Chem.* **271,** 21583–21596.

Juhasz, P., and Biemann, K. (1994). Mass spectrometric molecular-weight determination of highly acidic compounds of biological significance via their complexes with basic polypeptides. *Proc. Natl. Acad. Sci. USA* **91,** 4333–4337.

Kao, Y. H., Lee, G. F., Wang, Y., Starovasnik, M. A., Kelley, R. F., Spellman, M. W., and Lerner, L. (1999). The effect of O-fucosylation on the first EGF-like domain from human blood coagulation factor VII. *Biochemistry* **38,** 7097–7110.

Laemmli, U. K. (1970). Cleavage of structural proteins during the assembly of the head of bacteriophage T4. *Nature* **227,** 680–685.

Laine, R. A. (1990). Glycoconjugates: Overview and strategy. *Methods Enzymol.* **193,** 539–553.

Leitao, E. A., Bittencourt, V. C., Haido, R. M., Valente, A. P., Peter-Katalinić, J., Letzel, M., de Souza, L. M., and Barreto-Bergter, E. (2003). Beta-galactofuranose-containing *O*-linked oligosaccharides present in the cell wall peptidogalactomannan of *Aspergillus fumigatus* contain immunodominant epitopes. *Glycobiology* **13,** 681–692.

Linden, H. U., Klein, R. A., Egge, H., Peter-Katalinić, J., Dabrowski, J., and Schindler, D. (1989). Isolation and structural characterization of sialic-acid-containing glycopeptides of the *O*-glycosidic type from the urine of two patients with an hereditary deficiency in alpha-N-acetylgalactosaminidase activity. *Biol. Chem. Hoppe Seyler* **370,** 661–672.

Macek, B., Hofsteenge, J., and Peter-Katalinić, J. (2001). Direct determination of glycosylation sites in *O*-fucosylated glycopeptides using nanoelectrospray quadrupole time-of-flight mass spectrometry. *Rapid Commun. Mass Spectrom.* **15,** 771–777.

Medzihradszky, K. F., Besman, M. J., and Burlingame, A. L. (1998). Reverse-phase capillary high-performance liquid chromatography–high-performance electrospray ionization mass spectrometry: An essential tool for the characterization of complex glycoprotein digests. *Rapid Commun. Mass Spectrom.* **12,** 472–478.

Meisen, I. (1997). Strukturuntersuchungen an aus Erthrozytenmembranen isoliertem Glycophorin Amit immunochemischen und massen Spektrometrischen Methoden. Ph.D. dissertation, University of Bonn, Germany.

Meisen, I., Peter-Katalinić, J., and Muthing, J. (2003). Discrimination of neolacto-series gangliosides with alpha2-3- and alpha2-6-linked N-acetylneuraminic acid by nanoelectrospray ionization low-energy collision-induced dissociation tandem quadrupole TOF MS. *Anal. Chem.* **75,** 5719–5725.

Mirgorodskaya, E., Roepstorff, P., and Zubarev, R. A. (1999). Localization of *O*-glycosylation sites in peptides by electron capture dissociation in a Fourier transform mass spectrometer. *Anal. Chem.* **71,** 4431–4436.

Moloney, D. J., Panin, V. M., Johnston, S. H., Chen, J., Shao, L., Wilson, R., Wang, Y., Stanley, P., Irvine, K. D., Haltiwanger, R. S., and Vogt, T. F. (2000a). Fringe is a glycosyltransferase that modifies Notch. *Nature* **406,** 369–375.

Moloney, D. J., Shair, L. H., Lu, F. M., Xia, J., Locke, R., Matta, K. L., and Haltiwanger, R. S. (2000b). Mammalian Notch1 is modified with two unusual forms of *O*-linked glycosylation found on epidermal growth factor-like modules. *J. Biol. Chem.* **275,** 9604–9611.

Monsarrat, B., Brando, T., Condouret, P., Nigou, J., and Puzo, G. (1999). Characterization of manno-oligosaccharide caps in mycobacterial lipoarabinomannan by capillary electrophoresis–electrospray mass spectrometry. *Glycobiology* **9,** 335–342.

Montreuil, J. (1995). Glycosyltransferases involved in the synthesis of *N*-glycan antennae. *In* "Glycoproteins" (J. F. G. V. J. Montreuil and H. Schachter, eds.). Chapter 5, Biosynthesis 2C. Elsevier, Amsterdam; New York.

Mormann, M., Macek, B., Gonzalez de Peredo, A., Hofsteenge, J., and Peter-Katalinić, J. (2004). Structural studies on protein *O*-fucosylation by electron capture dissociation. *Int. J. Mass Spectrom.* **234,** 11–21.

Mormann, M., Meisen, I., Hakansson, K., Quenzer, T., Emmett, M., Marshall, A. G., and Peter-Katalinić, J. (2001). Characterization of *Muc* glycopeptides by nanoelectrospray low-energy CID and electron capture dissociation. *Proc. 49th ASMS Conf. MS Allied Top,* 27–31 May, Chicago, IL.

Mreyen, M., Champion, A., Srinivasan, S., Karuso, P., Williams, K. L., and Packer, N. H. (2000). Multiple O-glycoforms on the spore coat protein SP96 in Dictyostelium discoideum. Fuc(alpha1–3)GlcNAc-alpha-1-P-Ser is the major modification. J. Biol. Chem. 275, 12164–12174.

Olivares, J. A., Nguyen, N., Yonker, C., and Smith, R. D. (1987). On-line mass spectrometric detection capillary zone electrophoresis. Anal. Chem. 59, 1230–1232.

Patel, T., Bruce, J., Merry, A., Bigge, C., Wormald, M., Jaques, A., and Parekh, R. (1993). Use of hydrazine to release in intact and unreduced form both N- and O-linked oligosaccharides from glycoproteins. Biochemistry 32, 679–693.

Person, M. D., Brown, K. C., Mahrus, S., Craik, C. S., and Burlingame, A. L. (2001). Novel interprotein cross-link identified in the GGH-ecotin D137Y dimer. Protein Sci. 10, 1549–1562.

Peter-Katalinć, J., Williger, K., Egge, H., Green, B. N., Hanisch, F.-G., and Schindler, D. (1994). The application of electrospray mass spectrometry for structural studies on a tetrasaccharide monopeptide from the urine of a patient with alpha-N-acetyl-hexosamini-dase deficiency. J. Carbohydr. Chem. 13, 447–456.

Rademaker, G. J., Pergantis, S. A., Blok-Tip, L., Langridge, J. I., Kleen, A., and Thomas-Oates, J. E. (1998). Mass spectrometric determination of the sites of O-glycan attachment with low picomolar sensitivity. Anal. Biochem. 257, 149–160.

Ravanat, C., Gachet, C., Herbert, J. M., Schuhler, S., Guillemot, J. C., Uzabiaga, F., Picard, C., Ferrara, P., Freund, M., and Cazenave, J. P. (1994). Rat platelets contain glycosylated and nonglycosylated forms of platelet factor 4. Identification and characterization by mass spectrometry. Eur. J. Biochem. 223, 203–210.

Reimer, K. B., Meldal, M., Kusumoto, S., Fukase, K., and Bock, K. (1993). Small-scale solid-phase O-glycopeptide synthesis of linear and cyclized hexapeptides from blood clotting factor IX containing o-(a-D-xyl-1-3-a-D-xyl-1-3-B-D-Glc)-Lser. J. Chem. Soc., Perkin Trans. I, 925.

Reinhold, B. B., Hauer, C. R., Plummer, T. H., and Reinhold, V. N. (1995). Detailed structural analysis of a novel, specific O-linked glycan from the prokaryote Flavobacterium meningosepticum. J. Biol. Chem. 270, 13197–13203.

Roepstorff, P., and Fohlman, J. (1984). Proposal for a common nomenclature for sequence ions in mass spectra of peptides. Biomed. Mass Spectrom. 11, 601.

Schägger, H., and von Jagow, G. (1987). Tricine-sodium dodecyl sulfate-polyacrylamide gel electrophoresis for the separation of proteins in the range from 1 to 100 kDa. Anal. Biochem. 166, 368–379.

Schmitt, S., Glebe, D., Alving, K., Tolle, T. K., Linder, M., Geyer, H., Linder, D., Peter-Katalinić, J., Gerlich, W. H., and Geyer, R. (1999). Analysis of the pre-S2 N- and O-linked glycans of the M surface protein from human hepatitis B virus. J. Biol. Chem. 274, 11945–11957.

Soudan, B., Hennebicq, S., Tetaert, D., Boersma, A., Richet, C., Demeyer, D., Briand, G., and Degand, P. (1999). Capillary zone electrophoresis and MALDI-mass spectrometry for the monitoring of in vitro O-glycosylation of a threonine/serine-rich MUC5AC hexadecapep-tide. J. Chromatogr. B Biomed. Sci. Appl. 729, 65–74.

Teng-Umnuay, P., Morris, H. R., Dell, A., Panico, M., Paxton, T., and West, C. M. (1998). The cytoplasmic F-box binding protein SKP1 contains a novel pentasaccharide linked to hydroxyproline in Dictyostelium. J. Biol. Chem. 273, 18242–18249.

Thomsson, K. A., Karlsson, N. G., and Hansson, G. C. (1999). Liquid chromatography–electrospray mass spectrometry as a tool for the analysis of sulfated oligosaccharides from mucin glycoproteins. J. Chromatogr. A 854, 131–139.

Tseng, K., Lindsay, L. L., Penn, S., Hedrick, J. L., and Lebrilla, C. B. (1997). Characterization of neutral oligosaccharide-alditols from *Xenopus laevis* egg jelly coats by matrix-assisted laser desorption Fourier transform mass spectrometry. *Anal. Biochem.* **250,** 18–28.

Vakhrushev, S., Mormann, M., and Peter-Katalinić, J. (2005). *Proteomics* 5. In press.

Venkataraman, G., Shriver, Z., Raman, R., and Sasisekharan, R. (1999). Sequencing complex polysaccharides. *Science* **286,** 537–542.

Vongchan, P., Warda, M., Toyoda, H., Toida, T., Marks, R. M., and Linhardt, R. J. (2005). Structural characterization of human liver heparan sulfate. *Biochim. Biophys. Acta.* **1721,** 1–8.

Wugeditsch, T., Zachara, N. E., Puchberger, M., Kosma, P., Gooley, A. A., and Messner, P. (1999). Structural heterogeneity in the core oligosaccharide of the S-layer glycoprotein from *Aneurinibacillus thermoaerophilus* DSM 10155. *Glycobiology* **9,** 787–795.

Yuen, C. T., Lawson, A. M., Chai, W., Larkin, M., Stoll, M. S., Stuart, A. C., Sullivan, F. X., Ahern, T. J., and Feizi, T. (1992). Novel sulfated ligands for the cell adhesion molecule E-selectin revealed by the neoglycolipid technology among O-linked oligosaccharides on an ovarian cystadenoma glycoprotein. *Biochemistry* **31,** 9126–9131.

Zaia, J., and Costello, C. E. (2001). Compositional analysis of glycosaminoglycans by electrospray mass spectrometry. *Anal. Chem.* **73,** 233–239.

Zamfir, A., Konig, S., Althoff, J., and Peter-Katalinć, J. (2000). Capillary electrophoresis and off-line capillary electrophoresis–electrospray ionization quadrupole time-of-flight tandem mass spectrometry of carbohydrates. *J. Chromatogr. A* **895,** 291–299.

Zamfir, A., and Peter-Katalinić, P. (2001). Glycoscreening by on-line sheathless capillary electrophoresis–electrospray ionization-quadrupole time-of-flight tandem mass spectrometry. *Electrophoresis* **22,** 2448–2457.

Zamfir, A., and Peter-Katalinić, P. (2004). Capillary electrophoresis-mass spectrometry for glycoscreening in biomedical research. *Electrophoresis* **25,** 1949–1963.

Zamfir, A., Seidler, D. G., Kresse, H., and Peter-Katalinić, J. (2002). Structural characterization of chondroitin/dermatan sulfate oligosaccharides from bovine aorta by capillary electrophoresis and electrospray ionization quadrupole time-of-flight tandem mass spectrometry. *Rapid Commun. Mass Spectrom.* **16,** 2015–2024.

Zamfir, A., Seidler, D. G., Kresse, H., and Peter-Katalinić, J. (2003). Structural investigation of chondroitin/dermatan sulfate oligosaccharides from human skin fibroblast decorin. *Glycobiology* **13,** 733–742.

Zamfir, A., Seidler, D. G., Schonherr, E., Kresse, H., and Peter-Katalinić, J. (2004a). On-line sheathless capillary electrophoresis/nanoelectrospray ionization-tandem mass spectrometry for the analysis of glycosaminoglycan oligosaccharides. *Electrophoresis* **25,** 2010–2016.

Zamfir, A., Vakhrushev, S., Sterling, A., Niebel, H. J., Allen, M., and Peter-Katalinić, J. (2004b). Fully automated chip-based mass spectrometry for complex carbohydrate system analysis. *Anal. Chem.* **76,** 2046–2054.

Zamfir, A. D., Lion, N., Vukelić, Z., Bindila, L., Rossier, J., Girault, H. H., and Peter-Katalinić, J. (2005). Thin chip microsprayer system coupled to quadrupole time-of-flight mass spectrometer for glycoconjugate analysis. *Lab Chip* **5,** 298–307.

Zubarev, R. A., Kelleher, N. L., and McLafferty, F. W. (1998). Electron capture dissociation of multiple charged protein cations. A nonergodic process. *J. Am. Chem. Soc.* **120,** 3265–3266.

[8] Analysis of Glycosylphosphatidylinositol Protein Anchors: The Prion Protein

By Michael A. Baldwin

Abstract

Membrane proteins constitute a substantial fraction of the human proteome. A small subgroup associates with membranes through the presence of a C-terminal lipid anchor that is joined to the protein via a phosphoglycan. The prion protein (PrP), an abnormally folded form that causes fatal neurodegeneration, is one example of a glycosylphosphatidylinositol (GPI)-anchored protein. Although GPI-anchored proteins were first recognized some 20 years ago (in the mid-1980s), relatively few GPI anchors have been analyzed in detail. Therefore, a description of the analysis of the PrP–GPI anchor using a variety of mass spectrometric methods is of interest even though some of the approaches adopted could be facilitated through the use of newer, more sensitive techniques.

Introduction

Many proteins, such as cell surface receptors and ion channels, carry out their functions in intimate association with lipid membranes. The majority of these have extensive hydrophobic regions within their amino acid chains, frequently allowing the protein to wind back and forth through the membrane several times. In many cases, the active sites of such molecules are in the aqueous environment of either the cytosol or the extracellular medium, with the hydrophobic domain merely acting as an anchor. Other proteins lacking hydrophobic peptide motifs have lipids attached that serve the same purpose of anchoring the active moiety to the membrane. The lipids are often attached directly to the amino acid chain, but a class of lipidated proteins identified during the 1980s has lipids linked to an oligosaccharide spacer, which is attached via phosphoethanolamine to the protein C terminus (Ferguson *et al.*, 1985). Because the lipids are in the form of alkylacylglycerol linked to the oligosaccharide through phosphoinositol, the entire appendage is generally referred to as a glycosylphosphatidylinositol (GPI) anchor.

An example of a GPI-anchored protein is the prion protein (PrP), identified by Prusiner and colleagues as the agent responsible for the fatal neurodegenerative diseases known as prion diseases (Bolton *et al.*, 1982). A major outbreak of mad cow disease and its transmission to humans in the

METHODS IN ENZYMOLOGY, VOL. 405   0076-6879/05 $35.00
DOI: 10.1016/S0076-6879(05)05008-1

form of new variant Creutzfeldt-Jakob disease greatly raised public awareness of these unique diseases (Prusiner, 1997). Because prion diseases can be transmitted by intracerebral inoculation of PrP in a specific conformation (PrP^{Sc}), intensive study has centered on this protein. PrP was first reported as a GPI-anchored protein on the basis of amino acid analysis carried out by acid hydrolysis and high-performance liquid chromatography (HPLC), which gave a peak consistent with the presence of ethanolamine. This assignment was subsequently confirmed by gas chromatography mass spectrometry (GC–MS) (Stahl et al., 1987). This result, together with the presence of two sites for N-linked glycosylation, explained why the protein of nominal mass ~23 kDa migrated at 33 to 35 kDa by sodium dodecyl sulfate–polyacrylamide gel electrophoresis (SDS–PAGE). The presence of the GPI anchor was also confirmed by the observation that after denaturation, PrP^{Sc} could be released from membranes by the action of an enzyme that cleaves the lipids from the anchor, namely phosphatidylinositol-specific phospholipase C (PI-PLC) (Stahl et al., 1990a).

All GPI anchors that have been fully characterized conform to a common pattern with the sugar glycan core containing the sequence Man$_3$-HexN-Inos phosphate, the nonreducing terminus of which is attached through one phosphoethanolamine to the C terminus of the protein. The inositol phosphate has a lipophilic tail containing the alkylacylglycerol. There is variability in the attachment of further sugar units, and a feature only identified in mammalian anchors is an additional phosphoethanolamine that may carry a third lipid chain (Ferguson et al., 1988). Prior to the analysis of PrP^{Sc}, the analytical methodology employed in characterizing GPI anchors was based on radiolabeling by hydrazinolysis, exoglycosidase digestion, and gel permeation chromatography. However, these approaches were not sufficiently sensitive for the small-scale analysis required for PrP^{Sc} isolated from infected hamster brain; thus, several new approaches were implemented for the structural characterization of this GPI anchor. At the time, the site of attachment of the GPI was not known although it had been predicted based on a comparative analysis of other GPI-anchored proteins (Ferguson and Williams, 1988). It also remained to be determined whether molecules that adopted the pathogenic conformation were covalently different from the normal cellular prion protein (PrP^{C}), such as in the chemical structure of the GPI.

Analysis of the C-Terminal Peptide GPI

The PrP 27–30 is the protease resistant core derived from limited proteolysis of infectious PrP^{Sc} isolated from the brains of hamsters artificially infected with scrapie, a disease of sheep (Bolton et al., 1982).

Compared with the mature protein, PrP 27–30 was known to be N-terminally truncated but C-terminally intact, with retention of the GPI anchor. It was shown to be largely insoluble in its native state, but after denaturation in 6 M guanidine hydrochloride, it could be solubilized by boiling in 0.1% SDS. As well as identifying the GPI anchor, it was necessary to confirm that the amino acid sequence conformed to that predicted by earlier N-terminal sequencing by Edman chemistry (Prusiner et al., 1984) and cloning of a cDNA (Oesch et al., 1985). To achieve this confirmation, the lipids would be removed by digestion with PI-PLC, and the protein would be digested with a protease; then, the peptides would be separated by HPLC and identified by MS. The strategy adopted was to select an enzyme that would produce a relatively small number of peptides, the smaller of which would be analyzed directly and the larger of which would be digested with a second enzyme for further analysis. Endoproteinase Lys-C cleaving at the C-terminal side of lysine was selected as the first enzyme. PrP 27–30 would then give nine fragments and full-length PrP would give twelve fragments ranging from single lysines to a 75-amino acid glycopeptide, His111-Lys185. In practice, not all sites would be cleaved with equal efficiency (e.g., Lys-Pro bonds, of which there are three in PrP, are relatively stable and digestion would likely be incomplete).

The digestion was carried out as planned, after which the SDS was precipitated with 1 M guanidine hydrochloride. After removing the precipitate by centrifugation, the peptides in the supernatant were separated over a C18 analytical column, and fractions were collected based on UV absorption at the dual wavelengths of 214 and 280 nm. A small aliquot from each fraction was surveyed by capillary electrophoresis and amino acid analysis. The peptide sequences could be predicted from the cDNA; consequently, the amino acid analysis gave valuable clues regarding the likely nature of the peptides in each fraction. In this way, a likely candidate for the fraction containing the C-terminal peptide was identified. MS analysis of the fractions was carried out by fast atom bombardment (FAB) (Barber et al., 1981) or liquid secondary ion mass spectrometry (LSI–MS) (Aberth et al., 1982), which were readily available at the time this study was initiated (1989). These methods involve the ablation of cationized species, such as protonated molecular ions from solution in a weakly acidic viscous liquid matrix of low volatility by bombardment with high-energy argon or xenon atoms or cesium ions. A common matrix for both techniques is 1:1 glycerol/thioglycerol acidified with 0.1% trifluoroacetic acid (TFA). In this study, the ions were mass analyzed in a double-focusing sector mass spectrometer. The upper mass limit was about 2500 Da and the resolving power 3000 Da; thus, peaks at adjacent mass numbers were clearly resolved, allowing mass measurement to ±0.2 Da. LSI–MS was ideal for studying the mid-sized peptides of about 5 to 20 residues, with seven of the nine predicted fragments from PrP 27–30 falling in this category. The 75-residue peptide was anticipated to be too large for this methodology, and a tetrapeptide Thr-Asn-Met-Lys was believed to be too small and hydrophilic for efficient detection of this species.

Despite significant success with most of the peptide analyses (Stahl *et al.*, 1993), the putative C-terminal peptide gave no signal by LSI–MS while still attached to the GPI. A number of possible approaches exist for removing parts of a GPI by chemical or enzymatic methods, as summarized in Fig. 1. It was decided to treat the sample with 50% aqueous hydrofluoric (HF) acid, a procedure that should hydrolyze the phosphodiester and phosphate linkages within the GPI and remove all but the ethanolamine from the peptide. The fraction was dried on a vacuum centrifuge, and the residue was incubated overnight with 50% aqueous HF at 4°. This process was followed by further drying with KOH pellets in the vacuum trap of the centrifuge to absorb the remaining HF. The residue was dissolved in LSI–MS matrix and probed by MS, revealing a peptide of mass-to-charge ratio (m/z) 1354.4. Assuming attachment of the GPI at Ser231 as previously predicted, this corresponded to the calculated mass for the expected

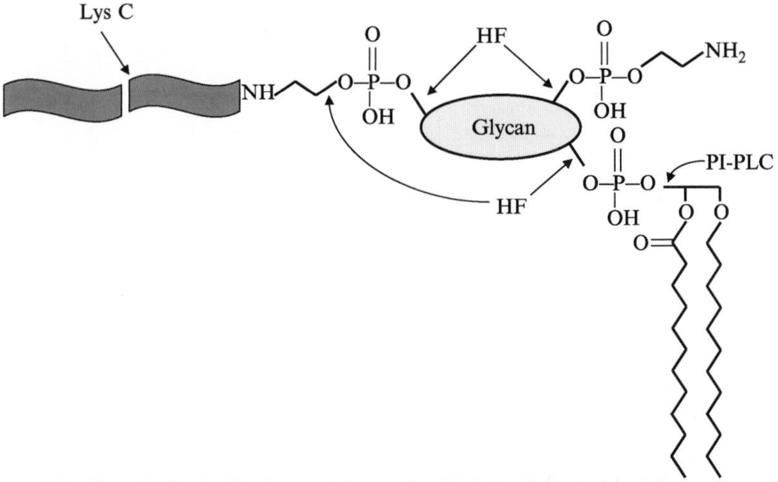

FIG. 1. Chemical and enzymatic treatment to reduce the size and complexity of the GPI anchor for MS analysis. PI-PLC removed the acylalkylglycerol lipids, and then endoproteinase Lys C cut the amino acid chain after Lys220, giving the C-terminal peptide attached to the phosphorylated glycan. Incubation with 50% aqueous HF was used to hydrolyze the phosphodiester bonds and to release the glycan and the peptide, which was separated by RP-HPLC and analyzed independently.

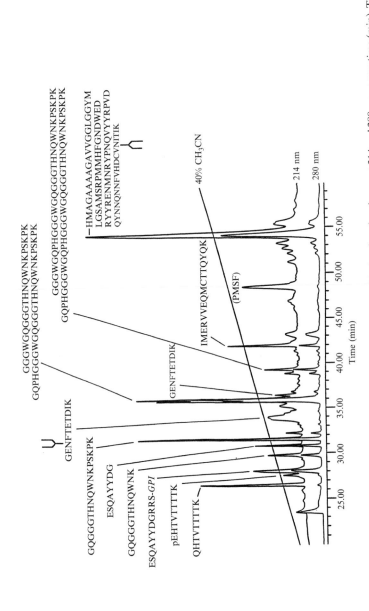

FIG. 2. RP-HPLC chromatogram of an endoproteinase Lys C digest of PrP 27–30, showing absorbance at 214 and 280 nm versus time (min). The gradient line shows acetonitrile content. The peptide compositions shown for each peak were derived from various MS experiments as described. Reprinted from Stahl et al. in Prusiner, S. B., Collinge, J., Powell, J., and Anderton, B. (Eds.) (1992). Prion diseases of humans and animals, pp. 361–379. Copyright Elsevier Science.

C-terminal peptide plus ethanolamine (i.e., ESQAYYDGRRS-Ea). In due course, all the peptides anticipated for the PrP sequence were identified by a combination of MS techniques, as shown in the chromatogram in Fig. 2 (Stahl *et al.*, 1993). The group of peptides included a minor component ($\sim$15%) at m/z 932.3 and was identified as a C-terminal peptide Glu221-Gly228 with no GPI anchor. This finding was not an anticipated Lys-C cleavage site and was attributed to a small fraction of PrP that was devoid of the GPI (Stahl *et al.*, 1990b).

At about the time this analysis was in progress, electrospray ionization (ESI) became available. This atmospheric ionization technique is optimized for the transformation of involatile and highly polar ions in solution to isolated, desolved gas-phase ions within the mass spectrometer (Fenn *et al.*, 1990). ESI is also better suited than LSI–MS for the analysis of larger species because they typically acquire multiple charges, thereby appearing in the mass spectrum at lower m/z than singly charged species. Using a quadrupole mass spectrometer, electrospray ionization mass spectrometry (ESI–MS) of an aliquot of the unhydrolyzed fraction, infused in 1:1 water/ acetonitrile with 1% acetic acid, revealed several molecular species. The deconvoluted spectrum is shown in Fig. 3. Certain components of the GPI could be predicted by analogy with other known structures; that is, the glycan core would be linked to the protein through phosphoethanolamine, a second phosphoethanolamine group would probably be present, and the core would contain at least three mannose residues linked to glucosamine

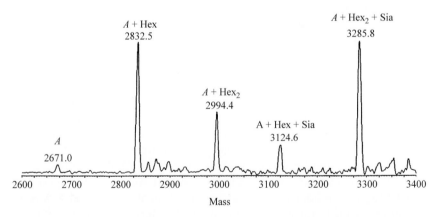

FIG. 3. Transformed ESI–MS spectrum of the heterogeneous GPI species attached to the C-terminal peptide Glu221-Ser231. Key: A = (Peptide 221–231); Ea-P-Hex$_3$-(HexNAc) (P-Ea)-HexN-Ino-P: Ea = ethanolamine, Hex = hexose, HexN = hexosamine, HexNAc = N-acetylhexosamine, Ino = inositol, P = phosphate, Sia = sialic acid. Reprinted from Baldwin, M., in Caughey, B. (2001). *Advances in Protein Chemistry* **57,** 29–54. Copyright Academic Press.

and phosphoinositol. Alkylacylglycerol had been removed by the action of PI-PLC. The MS analysis of oligosaccharides differs from that of peptides and proteins for which the basic building blocks are the 20 naturally occurring amino acids, each having a unique mass. The basic units of oligosaccharides are mostly isomers of only four or five unique compositions; thus, it is relatively easy to assign the composition of an oligosaccharide on the basis of the number of hexose units (mass 162), hexosamines (mass 161), N-acetylhexosamines (mass 203), and sialic acids (mass 291), among others. The various molecular species revealed by ESI–MS, differing in mass by combinations of 162 and 291 Da, were assigned carbohydrate compositions. The smallest was calculated to contain the peptide, two phosphoethanolamines, three hexoses, one hexosamine, one N-acetyl hexosamine, and phosphoinositol (the basic mammalian GPI with an additional N-acetylhexosamine). Larger forms contained a further one or two hexose units and, unexpectedly, sialic acid (N-acetylneuraminic acid), a sugar not previously reported as a component of a GPI (Stahl et al., 1992).

From these analyses, it was learned that the GPI was heterogeneous with at least five separate species, and its composition was consistent with known GPI structures but with sialic acid as a novel component. The heterogeneity was confirmed by various separative techniques, including capillary electrophoresis, Dionex high-performance anion exchange chromatography (HPAEC), and reversed-phase (RP)-HPLC at pH 7. Cleavage of sialic acid by neuraminidase was also demonstrated. ESI–MS showed that only species with at least four hexose units were sensitive to Jack bean α-mannosidase, and then only one residue was removed, consistent with the third mannose away from the glucosamine being the site of attachment to the protein. MS was not able to identify the particular isomeric form of each sugar unit (e.g., hexoses can be glucose, mannose, or galactose, differing in the configuration of the ring hydroxyl groups). However, the sugars were identified by acid hydrolysis of the GPI and separation of the various monosaccharides by HPAEC. Comparison with the monosaccharide standards confirmed three or four mannose units, the glucosamine, and sialic acid. N-acetylgalactosamine and galactose were also identified.

Using matrix-assisted laser desorption/ionization (MALDI) analysis of a C-terminal peptide to identify a GPI-anchor at Asn-77 without inositol-associated phospholipids, sialic acid was subsequently identified as a component of the GPI of a soluble form of CD59 (Meri et al., 1996). By using residue-specific exoglycosidases, chemical modification, and MALDI–MS, structures of seven different GPI-anchor variants were determined. As with PrP, variant forms of the anchor had deletions and/or extensions of one or more monosaccharide units. Also analogous to PrP, sialic acid linked to an *N*-acetylhexosamine-galactose arm was found in two GPI-anchor variants.

Branching Patterns in the GPI Glycan by Tandem MS

Oligosaccharides differ from peptides and proteins in that they are frequently branched. Branching patterns are usually determined by digestion with glycosidases with specific activities followed by chromatographic analysis of the products, comparing retention times with those of standards. Standards may not be available for studies on novel structures, however, and this approach requires the use of a relatively large amount of material. Nuclear magnetic resonance analysis can identify linkages within oligosaccharides, but this technique is substantially less sensitive. By contrast, MS is extremely sensitive, and it was anticipated that the fragmentation induced by collision-induced dissociation (CID) might reveal the branching patterns. Tandem MS (MS/MS) also has the ability to study individual selected species in unseparated mixtures, such as the heterogeneous GPI glycans. High-resolution MS/MS with high-energy CID was carried out on the glycans released by 50% HF and permethylated to enhance surface activity and volatility; this technique was also carried out to give a positively charged quaternary ammonium cation as the glucosamine became triply methylated, greatly enhancing the sensitivity for MS detection. After methylation, the masses of the protonated molecular ions for the most abundant species were measured as 1312.5, 1557.6, and 1761.6 Da, corresponding to the permethylated glycans lacking sialic acid. The MS/MS spectra shown in Fig. 4 represent fragmentation of these three species separated in the first mass analyzer of the MS/MS mass spectrometer and caused to undergo collisions with neutral gas atoms, the fragment ions being separated in the second mass analyzer (Baldwin *et al.*, 1990a).

The key to the interpretation of the tandem spectra was a series of ring cleavages across the individual hexose residues (so-called X ions) that were seen to terminate in an ion of *m/z* 510 for the glucosamine/inositol moiety. Ions representing the loss of successive sugars revealed which sugar was substituted (which was the branch site). Complementary information came from the oxonium ions (called B ions by analogy with the fragmentation of peptide ions) from charge-remote fragmentations giving single or multiple sugar units from the nonreducing end of each chain. These findings are shown in Fig. 5; the largest of the GPI species have sialic acid or *N*-acetyl neuraminic acid (NANA). The oxonium ions at *m/z* 376.1, 580.2, and 825.3 confirmed the presence of a linear sequence of ions corresponding to NANA-Hex-HexNAc.

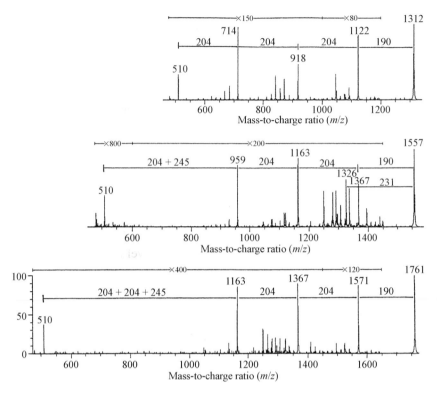

FIG. 4. Partial tandem mass spectra for the region above m/z 500 of permethylated GPI anchor glycans devoid of sialic acid, showing the fragmentation of m/z 1312, 1557, and 1761. High-energy CID measurements were carried with LSI–MS using a four-sector tandem mass spectrometer. Reprinted from Baldwin, M. (1990). *Analytical Biochemistry* **191**, 174–182. Copyright Academic Press.

Thus, tandem MS together with chromatographic and glycosidase data allowed the identification of the six species shown in Fig. 6 (Baldwin *et al.*, 1990b). The branching was clearly revealed by cleavages along each branch of a biantennary structure. One branch represented the normal GPI glycan core with a chain of either three or four hexose units attached to the GlcN-ino. The branch point was at the hexose immediately adjacent to the glucosamine with a chain of one, two, or three sugars in the sequence Sia-Gal-GalNAc. The calculated molecular masses listed in Fig. 6, which include the mass of the C-terminal peptide, correlated with the experimentally determined values in the ESI spectrum in Fig. 4.

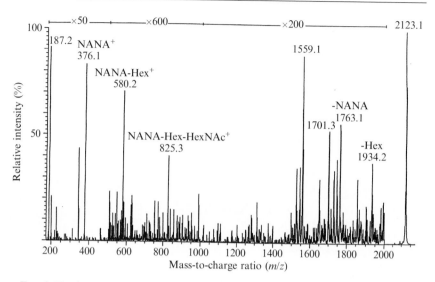

FIG. 5. Tandem mass spectrum of a permethylated GPI glycan of m/z 2123 containing sialic acid (NANA). Reprinted from Stahl, N., *et al.* (1992). *Biochemistry* **31**, 5043–5053. Copyright American Chemical Society.

Analysis of Intact PrP by MALDI–MS and ESI–MS

Enzyme digestion and peptide mapping are powerful methods for identifying the sites and the nature of any posttranslational modifications. However, measuring the intact molecular mass of the protein is the most effective way for ensuring that nothing was overlooked during the mapping experiments and for confirming the mass increment due to a posttranslational modification such as a GPI. ESI–MS and MALDI–MS are both capable of ionizing and accurately determining the mass of intact proteins. ESI–MS generally offers higher resolution, but for heterogeneous samples, such as multiply glycosylated proteins, the presence of an overlapping series of highly charged ions can complicate the spectral analysis. MALDI–MS may not resolve all the individual molecular species for a protein such as PrP, but it can provide the centroid or peak top mass of the unresolved peak profile.

Samples for MALDI–MS are dissolved in a solution containing an approximately 10^4 molar excess of a UV-absorbing matrix, normally an aromatic acid with additional conjugated double bonds, such as α4-hydroxy-α-cyanocinnamic acid (4-HCCA). Without detergents or denaturants that suppress the ionization process, membrane proteins such as PrP are generally insoluble in solvents that are suitable for the matrix compound.

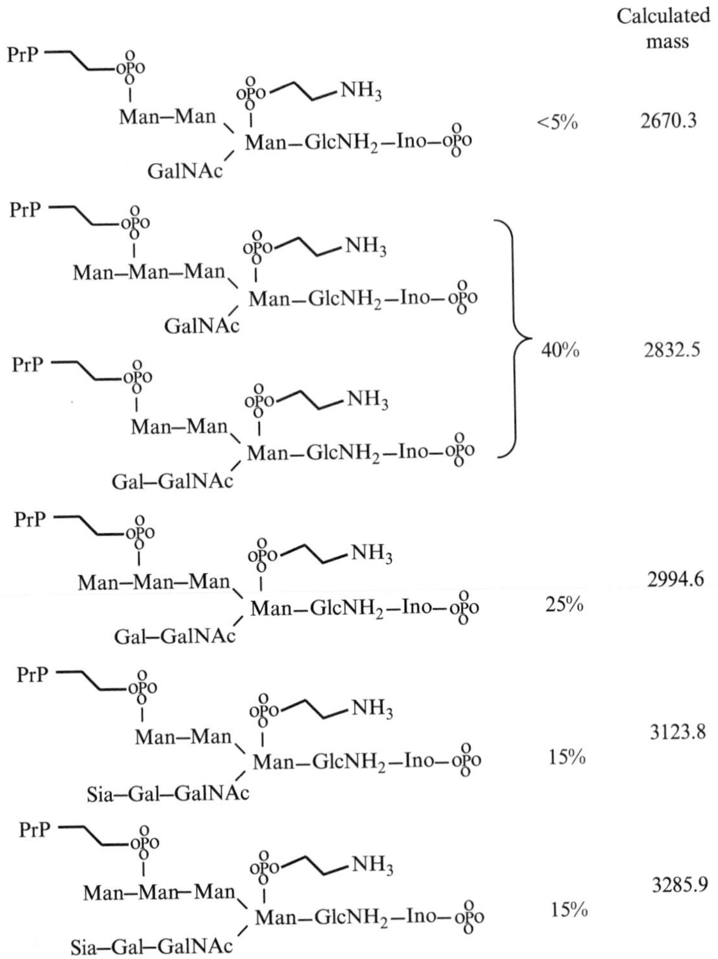

Calculated
mass

PrP —— $-O\overset{O}{\underset{O}{P}}O$
          $O\overset{O}{\underset{O}{P}}O$——NH$_3$
Man—Man $\overset{O}{\underset{O}{P}}$O
          Man—GlcNH$_2$—Ino—$O\overset{O}{\underset{O}{P}}O$ <5%   2670.3
GalNAc

PrP —— $-O\overset{O}{\underset{O}{P}}O$
          $O\overset{O}{\underset{O}{P}}O$——NH$_3$
Man—Man—Man
          Man—GlcNH$_2$—Ino—$O\overset{O}{\underset{O}{P}}O$
GalNAc

                                                              40%   2832.5

PrP —— $-O\overset{O}{\underset{O}{P}}O$
          $O\overset{O}{\underset{O}{P}}O$——NH$_3$
Man—Man
          Man—GlcNH$_2$—Ino—$O\overset{O}{\underset{O}{P}}O$
Gal—GalNAc

PrP —— $-O\overset{O}{\underset{O}{P}}O$
          $O\overset{O}{\underset{O}{P}}O$——NH$_3$
Man—Man—Man
          Man—GlcNH$_2$—Ino—$O\overset{O}{\underset{O}{P}}O$ 25%   2994.6
Gal—GalNAc

PrP —— $-O\overset{O}{\underset{O}{P}}O$
          $O\overset{O}{\underset{O}{P}}O$——NH$_3$
Man—Man   3123.8
          Man—GlcNH$_2$—Ino—$O\overset{O}{\underset{O}{P}}O$ 15%
Sia—Gal—GalNAc

PrP —— $-O\overset{O}{\underset{O}{P}}O$
          $O\overset{O}{\underset{O}{P}}O$——NH$_3$
Man—Man—Man   3285.9
          Man—GlcNH$_2$—Ino—$O\overset{O}{\underset{O}{P}}O$ 15%
Sia—Gal—GalNAc

FIG. 6. Structures of the various GPI anchors. The calculated masses include the C-terminal peptide and can be directly correlated with the masses shown in Fig. 3 Key: Man = mannose, Gal = galactose, GalNAc = N-acetyl galactosamine, GlcNH$_2$ = glucosamine, Ino = inositol, Sia = sialic acid. Reprinted from Baldwin, M., in Caughey, B. (2001). *Advances in Protein Chemistry* **57**, 29–54. Copyright Academic Press.

In the early 1990s, PrP was analyzed by MALDI in the laboratory of Dr. Brian Chait at Rockefeller University. PrP^{Sc} was treated with PNGase F to remove the N-linked sugars and reduce the heterogeneity, and the protein was then dissolved in hexafluoroisopropanol and mixed 1:1 with a solution of 5 mg/ml 4-HCCA in 2:5:2 chloroform/methanol/0.1% TFA. An amount of 1 mL of this mixture containing approximately 1 pmol of protein was deposited on the MALDI target, and the solvent was evaporated. This was irradiated and ionized with 354 nm radiation from a Nd-YAG laser, the ablated ions being mass analyzed by mass spectrometry, i.e., time-of-flight (MS–TOF). The mass spectrum showed broad peaks corresponding to singly, doubly, and triply charged ions, giving an average mass of ~25,350 Da. Taking account of the relative abundance of the different GPIs, the molecular masses for amino acids Lys23-Ser231 combined with the various GPI species (Fig. 6) and the likely GPI lipids, a weighted mean of 25,329 Da was calculated. Thus, there was relatively good agreement between the calculated and measured numbers, but it was not good enough to confirm or refute that no modification had been destroyed or overlooked (Baldwin *et al.*, 1993).

In these experiments, the resolving power of linear MALDI–MS was not sufficient to separate the various forms present due to the heterogeneity of the GPI anchor, hence the difficulty of measuring the molecular weight with high precision. ESI–MS gives higher resolution, but it relies upon the protein being soluble; PrP^C is more soluble than PrP^{Sc}, and Fig. 7 compares spectra of PrP^C obtained by MALDI–MS and ESI–MS. The protein was treated with PI-PLC to remove the lipids, thereby enhancing solubility in the ESI–MS buffer of 1:1 acetonitrile/water with 1% acetic acid, but the heterogeneous GPI glycan was left intact. *N*-linked oligosaccharides were removed with PNGase F because they were known to be highly heterogeneous. The upper spectrum (Fig. 7A) shows MALDI–MS data for hamster brain PrP^C mixed with myoglobin as an internal mass calibrant. The PrP peaks are broader than those of myoglobin and show unresolved structure. Using peak tops to define the mass of each ion, PrP^C appeared as a singly charged ion at 24,466.1, doubly charged at 12,236.9, and triply charged at 8161.6. The mean molecular mass calculated from these three species was 24,472.9 Da. Note that although it is generally more accurate to use centroids rather than peak tops, this is not true in the case of unresolved peaks arising from overlapping ions of different compositions. Due to much higher charge states, a spectrum of the same sample from ESI–MS in the lower panel (Fig. 7B) appears to be completely different. Peaks are observed for the attachment of between 20 and 32 protons, each charge state showing at least two major species. The raw data in Fig. 7B were deconvoluted to the molecular weight pattern shown in Fig. 7C. Here, the different glycoforms of the GPI are well-resolved and give a molecular weight for the major species of 24,474.0. The calculated value for the PrP sequence plus the major GPI glycoform is 24,474.5 Da, in excellent agreement with the ESI–MS result (0.002%) and in surprisingly good agreement with MALDI–MS, considering the poorly resolved nature of the MALDI spectrum.

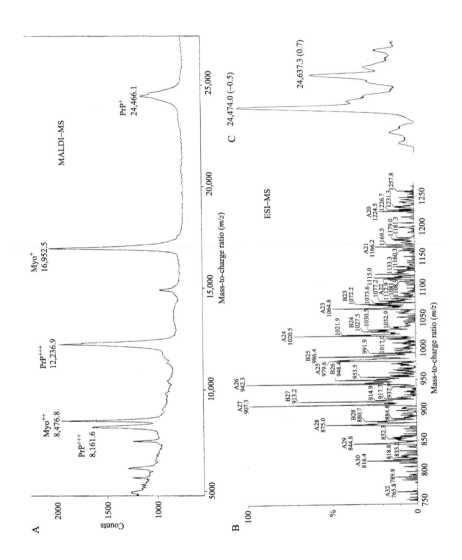

Unanswered Questions Concerning the GPI of PrP

A further difference between oligosaccharides and proteins arises from the presence of alternative sites for linking individual sugars together. The glycosidic bond from carbon-1 of the nonreducing residue in a disaccharide can be linked to positions 2, 3, 4, or 6 of the reducing residue. Furthermore, the glycosidic bond at carbon-1 has two possible configurations, giving either α- or β-anomers. Each of these differences may be crucial in terms of biochemical action, but MS cannot be applied easily to their analysis. Permethylation and hydrolysis followed by peracetylation and reduction with sodium borodeuteride give partially methylated alditol acetates specifically labeled at one end, which can be analyzed by GC–MS to reveal the linkage positions. Unfortunately, this strategy is difficult to prosecute with less than 5 to 10 nmole of homogeneous starting material. There are glycosidases that are specific for both anomericity and linkage position, and these glycosidases are valuable in dealing with a structure likely to conform to a previously identified pattern. In this way, it was shown that the third mannose in the chain, the site of attachment of the GPI to the protein, becomes susceptible to hydrolysis by α-mannosidase only after HF treatment. Interestingly, the middle mannose in the chain of three at the core of the GPI does not show the same sensitivity even though previous studies on GPIs from other proteins have shown α-1–6-linked mannose at this point. Another approach that could be explored to investigate linkage positions and anomericity is based on stereospecific reactions with metal ions (Gaucher and Leary, 1999; Konig and Leary, 1998).

Despite these limitations, the analysis of the GPI anchor of PrP^{Sc} by MS was a success, considering the limited amounts of material available and the relatively modest performance of the mass spectrometers available at the time of the experiment. Furthermore, some methods were developed that could have more general applicability. Future studies on any other GPI anchor would probably exploit ESI–MS to a greater degree, and the use of an instrument such as a quadrupole orthogonal acceleration time-of-flight (Qqoa-TOF) mass analyzer would allow more extensive CID experiments to be carried out with relative ease and efficiency. In

---

Fig. 7. Comparison of PrP^C spectra obtained by MALDI–MS and ESI–MS. (A) Mass spectra of PrP^C from MALDI–MS. (B) Mass spectra of PrP^C from ESI–MS. (C) The deconvoluted spectrum of raw data from Fig. 7B. For MALDI–MS, the mass scale was calibrated with myoglobin ions as an internal standard. For ESI–MS, the mass scale was calibrated externally using the doubly and singly charged ions of gramicidin S. Fig. 7C gives measured molecular weights for the major species that differ by one hexose. Deviations from the calculated values are shown in parentheses.

addition to identifying novel features of this GPI, the important aim of demonstrating whether there are differences between the GPIs of PrP^C and PrP^{Sc} was fulfilled, at least in part. Within the limitations of the information that could be derived from the very small amounts of PrP^C that could be isolated, which was much less than the available PrP^{Sc}, no differences were found (Stahl *et al.*, 1992). This is consistent with more recent evidence that prion diseases are attributable solely to different protein conformations (Cohen and Prusiner, 1998).

References

Aberth, W., Straub, K., and Burlingame, A. L. (1982). Secondary ion mass spectrometry with cesium ion primary beam and liquid target matrix for analysis of bioorganic compounds. *Anal. Chem.* **54,** 2029–2034.

Baldwin, M. A., Stahl, N., Reinders, L. G., Gibson, B. W., Prusiner, S. B., and Burlingame, A. L. (1990a). Permethylation and MS/MS of oligosaccharides having free hexosamine: A general strategy for analysis of GPI anchors. *Anal. Biochem.* **191,** 174–182.

Baldwin, M. A., Stahl, N., Burlingame, A. L., and Prusiner, S. B. (1990b). Structure determination of glycoinositol phospholipid anchors by permethylation and tandem mass spectrometry. *Meth.: Comp. Meth. Enzymol.* **1,** 306–314.

Baldwin, M. A., Wang, R., Pan, K.-M., Hecker, R., Stahl, N., Chait, B. T., and Prusiner, S. B. (1993). Matrix-assisted laser desorption/ionization mass spectrometry of membrane proteins: The scrapie prion protein. *In* "Techniques in Protein Chemistry" (R. Hogue Angeletti, ed.) Vol. IV. Academic Press, San Diego, CA.

Barber, M., Bordoli, R. S., Sedgewick, R. D., and Tyler, A. N. J. (1981). Fast atom bombardment of solids (F.A.B.): A new ion source for mass spectrometry. *J. Chem. Soc. Chem. Commun.* 325–326.

Bolton, D. C., McKinley, M. P., and Prusiner, S. B. (1982). Identification of a protein that purifies with the scrapie prion. *Science* **218,** 1309–1311.

Cohen, F. E., and Prusiner, S. B. (1998). Pathologic conformations of prion proteins. *Annu. Rev. Biochem.* **67,** 793–819.

Fenn, J. B., Mann, M., Meng, C. K., Wong, S. K., and Whitehouse, C. M. (1990). Electrospray ionization for mass spectrometry of large biomolecules. *Science* **246,** 64–71.

Ferguson, M. A., Haldar, K., and Cross, G. A. (1985). *Trypanosoma brucei* variant surface glycoprotein has a sn-1,2-dimyristyl glycerol membrane anchor at its COOH terminus. *J. Biol. Chem.* **260,** 4963–4968.

Ferguson, M. A., and Williams, A. F. (1988). Cell-surface anchoring of proteins via glycosyl-phosphatidylinositol structures. *Annu. Rev. Biochem.* **57,** 285–320.

Ferguson, M. A., Homans, S. W., Dwek, R. A., and Rademacher, T. W. (1988). Glycosyl-phosphatidylinositol moiety that anchors *Trypanosoma brucei* variant surface glycoprotein to the membrane. *Science* **239,** 753–759.

Gaucher, S. P., and Leary, J. A. (1999). Determining anomericity of the glycosidic bond in Zn (II)-diethylenetriamine-disaccharide complexes using MSn in a quadrupole ion trap. *J. Amer. Soc. Mass Spectrom.* **10,** 269–272.

Konig, S., and Leary, J. A. (1998). Evidence for linkage position determination in cobalt coordinated pentasaccharides using ion trap mass spectrometry. *J. Amer. Soc. Mass Spectrom.* **9,** 1125–1134.

Meri, S., Lehto, T., Sutton, C. W., Tyynela, J., and Baumann, M. (1996). Structural composition and functional characterization of soluble CD59: Heterogeneity of the oligosaccharide and glycophosphoinositol (GPI) anchor revealed by laser-desorption mass spectrometric analysis. *Biochem. J.* **316,** 923–935.

Oesch, B., Westaway, D. M., Wälchli, M., McKinley, M. P., Kent, S. B. H., Aebersold, R., Barry, R. A., Tempst, P., Teplow, D. B., Hood, L. E., Prusiner, S. B., and Weissmann, C. (1985). A cellular gene encodes scrapie PrP 27–30 protein. *Cell* **40,** 735–746.

Prusiner, S. B., Groth, D. F., Bolton, D. C., Kent, S. B., and Hood, L. E. (1984). Purification and structural studies of a major scrapie prion protein. *Cell* **38,** 127–134.

Prusiner, S. B. (1997). Prion diseases and the BSE crisis. *Science* **278,** 245–251.

Stahl, N., Borchelt, D. R., Hsiao, K., and Prusiner, S. B. (1987). Scrapie prion protein contains a phosphatidylinositol glycolipid. *Cell* **51,** 229–240.

Stahl, N., Borchelt, D. R., and Prusiner, S. B. (1990a). Differential release of cellular and scrapie prion proteins from cellular membranes by phosphatidylinositol-specific phospholipase C. *Biochemistry* **29,** 5405–5412.

Stahl, N., Baldwin, M. A., Burlingame, A. L., and Prusiner, S. B. (1990b). Identification of glycoinositol phospholipid-linked and truncated forms of the scrapie prion protein. *Biochemistry* **29,** 8879–8884.

Stahl, N., Baldwin, M. A., Hecker, R., Pan, K.-M., Burlingame, A. L., and Prusiner, S. B. (1992). Glycosylinositol phospholipid anchors of the scrapie and cellular prion proteins contain sialic acid. *Biochemistry* **31,** 5043–5053.

Stahl, N., Baldwin, M. A., Teplow, D. B., Hood, L. E., Gibson, B. W., Burlingame, A. L., and Prusiner, S. B. (1993). Structural studies of the scrapie prion protein using mass spectrometry and amino acid sequencing. *Biochemistry* **32,** 1991–2002.

[9] Comprehensive Mass Spectrometric Analysis of the 20S Proteasome Complex

By LAN HUANG and A. L. BURLINGAME

Abstract

The 20S proteasome is a multicatalytic protein complex that plays an important role in intracellular protein degradation from archaebacteria to eukaryotes. This complex is made up of two copies each of seven different alpha (α) and seven different beta (β) subunits arranged into four stacked rings ($\alpha_7\beta_7\beta_7\alpha_7$). Although the proteasome's cylindrical structure is conserved, the subunit composition of the 20S protein complex varies during the evolution, and the number of subunits increases from archaebacteria to mammals. To fully characterize the 20S proteasome subunit composition and understand the subunit functions, we, the authors of this chapter, have developed and employed various mass spectrometry (MS)-based approaches to generate a comprehensive profile of the 20S proteasomes from

METHODS IN ENZYMOLOGY, VOL. 405
Copyright 2005, Elsevier Inc. All rights reserved.

0076-6879/05 $35.00
DOI: 10.1016/S0076-6879(05)05009-3

rat liver and *Tropanosoma brucei*. We have identified 7 α and 10 β subunits, including 7 essential and 3 nonessential β subunits from rat 20S proteasome complex using two-dimensional (2-D) gel electrophoresis and tandem MS (MS/MS). In addition, multiple isoforms of most of the subunits were determined; indicating the composition of rat 20S proteasome complex was much more complicated than expected. Further analysis of the intact protein molecular weight of each subunit using LC–MS confirmed the heterogeneous population of the 20S proteasome and revealed that many of the experimental measured molecular weights do not correspond well with the theoretical values deduced from the sequences in protein databases. This finding is mostly due to the sequence errors in the protein databases and possible posttranslational modifications. Although the protein sequences of rat 20S proteasome are present in the databases, the sequences of the 20S proteasome from *T. brucei* were not available at the time when the analysis was carried out. To determine the subunit composition of the 20S proteasome from *T. brucei*, we developed a homology-based database searching tool to identify unknown proteins based on the novel sequences determined by *de novo* sequencing using MS/MS. As a result, 14 subunits (7 α and 7 β) were identified on the 2-D gel, which was later confirmed by the full-length sequences. Using the same approach, we also identified and characterized an activator protein, PA26, from *T. brucei*. The purified recombinant PA26 self-assembles into a heptamer ring, which can bind and activate the 20S proteasome from *T. brucei* as well as rat. Compared to the human PA28 complex, PA26 may be the prototype activator protein involved in proteasomal protein degradation. Therefore, the MS-based strategy developed here for identification of the known and unknown protein complexes can be generalized for the study of other protein complexes.

Introduction

The proteasome is an intracellular multifunctional proteinase that is responsible for most of the nonlysosomal protein degradation in both the nucleus and cytosol. It is involved in many important biological processes, such as cell cycle progression, apoptosis, DNA repair, chromosomal maintenance, transcriptional activation, metabolism, immune response, signal transduction, stress response, and antigen presentation (Coux *et al.*, 1996; Lupas and Baumeister, 1998; Voges *et al.*, 1999). The 20S proteasome is the catalytic core of the 26S proteasome complex responsible for ATP-dependent protein degradation of ubiquinated substrates. With a molecular weight of approximately 750 kDa, the eukaryotic 20S proteasome complex is composed of at least 14 subunits with molecular masses of 21 to 34 kDa and different charges (pI 3–10). These subunits fall into two categories, α and β, on the basis of their sequence similarities. The hollow cylindrical structure of the 20S proteasome has been conserved through evolution, as shown for the crystal structures of proteasomes from archaebacterium *Thermoplasma acidophilum* (Lowe *et al.*, 1995) and yeast *Saccharomyces cerevisiae* (Groll *et al.*, 1997), whereas the complexity of the subunit composition has changed during evolution. The 20S proteasomes from archaebacteria contain only one type each of α and β subunits. In contrast, eukaryotic proteasomes are more divergent, composed of seven α and seven β subunits. For the eukaryotic 20S proteasome, the cylindrical structure consists of four stacked heptameric rings, each of which is organized from seven homologous but nonidentical subunits. The two outer rings contain seven α subunits ($\alpha1$–$\alpha7$) that direct the assembly of the complex and form the "gate" through which substrates enter and products are released. The two inner rings are made up of seven β subunits ($\beta1$–$\beta7$), including three, $\beta1(Y)$, $\beta2(Z)$, and $\beta5(X)$, that are catalytically active in the mature enzyme complex to generate a catalytic center for various protease activities (Coux *et al.*, 1996; Groll *et al.*, 1997; Lupas and Baumeister, 1998; Schmidtke *et al.*, 1996; Voges *et al.*, 1999).

In mammalian cells, these three β subunits can be replaced by three nonessential subunits, $\beta1i(RING12/LMP2)$, $\beta2i(MECL-1)$, and $\beta5i$ (C1/LMP7), upon induction by the T-cell-derived antiviral cytokine interferon-γ to form the immunoproteasome and alter the proteolytic specificity as well as generate antigenic peptides for presentation by major histocompatibility complex (MHC) class I molecules (Coux *et al.*, 1996; Groll *et al.*, 1997; Kloetzel, 2001; Lowe *et al.*, 1995; Lupas and Baumeister, 1998; Stoltze *et al.*, 2000; Voges *et al.*, 1999). It is known that most β-type subunits are synthesized as proproteins that undergo posttranslational processing during proteasome maturation (Schmidtke *et al.*, 1996). This process may generate further subunit composition complexity. The total number of β-type subunits may be increased in multicellular organisms during evolution for adaptation to environmental stress (Tanaka, 1995). Furthermore, duplicated proteasomal genes, encoding testes-specific isoforms, have been cloned from *Drosophila melanogaster* (Yuan *et al.*, 1996). A family of 23 distinct proteasomal genes, 13 α- and 10 β-type genes, which can be classified into the 7 α- and 7 β-type subfamilies, was identified in *Arabidopsis thaliana* (Fu *et al.*, 1998). Even though nothing is known about the functional role of these duplicated genes in *D. melanogaster* and *A. thaliana*, the presence of multiple genes suggests that 20S proteasomes may exist in cells as heterogeneous populations with functional diversity. To clarify the individual roles of these subunits, elucidate their relationships, and characterize their putative posttranslational modifications and physiologically functional forms, it is essential to investigate the subunit compositional heterogeneity of 20S proteasome complex and the primary structures of each individual subunit. Therefore, we, the authors of this chapter, have undertaken various approaches to achieve a comprehensive characterization of the 20S proteasome complex using MS. The MS analyses of the 20S proteasomes from both rat liver and *T. brucei* have been carried out in our laboratory and will be discussed here.

Materials and Methods

Protein Preparation and Separation by Two-Dimensional Gel Electrophoresis

The 20S proteasomes from rat liver and 20S proteasomes from procyclic forms of *T. brucei* strain 427 (MiTat 1.4) were prepared as described in Hua *et al.* (1996). A 2-D gel electrophoresis was performed as described in the manufacturer's instructions with slight modifications (Amersham Pharmacia Biotech, Piscataway, NJ, http://internalmed.wustl.edu/divisions/enzymes/ComAP.HTM) (Huang *et al.*, 1999). To explain briefly, a mixture of 30–80 μl of the 20S proteasome in Tris-Borate (TBE) buffer (90 mM TBE, pH 8.0, 2 mM ethylenediaminetetracacetic acid or EDTA) containing 40% glycerol was mixed with 350–320 μl of the rehydration buffer [with a pH of 3–10 (0.36 meq ml^{-1}, 1:50 by volume)] containing 7 M urea, 2 M thiourea, 4% CHAPS, 65 mM dithiothreitol, Pharmalyte, and a trace (~0.01%) of bromophenol blue. The solution was used to swell the Immobiline DryStrips overnight (18 cm, pH 3–10, nonlinear) for isoelectrofocusing on the Multiphor II (Amersham Pharmacia Biotech). Using an EPS 3500 XL electrophoresis power supply (Amersham Pharmacia Biotech), proteins were focused at 20° according to the following voltage gradient program: 0–300 V, 600 Vh; 300–1000 V, 1000 Vh; 1000–3500 V, 3500 Vh; 3500 V, >50 kVh. After a standard equilibration step, proteins were further separated by sodium dodecyl sulfate–polyacrylamide gel electrophoresis (SDS–PAGE) using the IsoDalt system (Amersham Pharmacia Biotech). The separating gel was an 11% nongradient gel made with 30.8% Duracryl solution (ESA, Chelmsford, MA). Each gel was run at 50 mA for 9 to 10 h. The 20S proteasome subunits from rat liver were stained for 15 h in 0.1% Coomassie Blue R-380, 40% methanol, 10% acetic acid, and 50% water. Gels were destained with 40% methanol, 10% acetic acid, and 50% water. Other gels were stained by silver staining as with slight modification (Shevchenko *et al.*, 1996). To describe briefly, gels were fixed in 50%

methanol and 5% acetic acid overnight, washed with 50% methanol for 20 min, and washed twice with water for 15 min. The gels were then sensitized by 0.02% sodium thiosulfate for 2 min, washed again with water for 1 min twice, and treated with 1.5% silver nitrate for 20 min in a cold room (4°). The gels were further washed twice with water for 1 min before they were developed in a 2% sodium carbonate (Na_2CO_3) solution containing 0.04% Formalin. Coomassie-stained gels were stored in water, while silver-stained gels were stored in a solution of 1% acetic acid at 4° until analyzed.

Protein In-Gel Digestion. Protein spots were excised from either Coomassie-stained or silver-stained gels and digested as described (Shevchenko *et al.*, 1996). Gels were first cut to small pieces and washed three times with 25 mM ammonium biocarbonate (NH_4HCO_3) in 50% acetonitrile (CAN), and then the gels were dried by Speedvac and rehydrated in a 25 mM NH_4HCO_3 solution containing trypsin. The amount of trypsin used depended on the amount of protein loaded on the gel. The protein gel pieces were digested overnight at 37°. Peptides were then extracted by washing with high-performance liquid chromatography (HPLC) grade water followed by three washes with 50% ACN/5% trifluoroacetic acid (TFA) at room temperature. The combined supernatants were dried down by Speedvac and redissolved in 50% ACN/5% TFA prior to unseparated digests analysis. Some of the peptide extracts were further separated by reversed-phase (RP)-HPLC on a Vydac C18 column. The HPLC fractions were collected, concentrated, and analyzed.

Mass Spectrometric Analysis: Molecular Weight Measurements on Unseparated and Separated Tryptic Digests. Molecular weights of all tryptic peptides were determined by analyzing 1/20th of unseparated digests and/or 1/10th of each HPLC fraction using a matrix-assisted laser desorption/ionization (MALDI) delayed extraction reflectron time-of-flight (TOF) instrument (PerSeptive Biosystems, DE-STR mass spectrometer, Framingham, MA) equipped with a nitrogen laser (337 nm wavelength). Peptides were cocrystallized with equal volume of matrix consisting of saturated solution of α-cyano-4-hydroxycinnamic acid prepared in 50% ACN/1% TFA. All MALDI spectra were either externally calibrated using a standard peptide mixture or internally calibrated using trypsin autolysis products. The monoisotopic masses from all spectra recorded for a particular peptide are reported here.

Peptide Sequencing by Post-source Decay. After screening of the MALDI mass spectra of tryptic digests for each spot, selected peptides were subjected to post-source decay (PSD) analysis for partial amino acid sequence determination. This method detects the fragmentation product ions (metastable ions) of a selected mass value window containing the chosen precursor ion that occurs in the field-free region between the ion source and the reflectron.

Peptide Sequencing by a Quadrupole Orthogonal Acceleration Time-of-Flight Mass Spectrometer Equipped with a Nanoelectrospray Source. Peptide sequencing using MS/MS was performed on a prototype quadrupole orthogonal acceleration time-of-flight (Qqoa-TOF) mass spectrometer (Sciex, Toronto, Canada) equipped with a nanoelectrospray ionization (nanoESI) source (Protana A/S, Odense, Denmark), as described (Huang *et al.*, 1999). An approximately 2-μL sample was introduced into a metalized, drawnout silica capillary, the end of which was opened by touching against a metal surface in the source. A conventional mass spectrum was first obtained to identify peptide masses by multiple charging, and then tandem mass spectra of the peptide ions were collected by collision-induced dissociation (CID) using argon collision gas. The collision energy was selected independently for each precursor ion by visual inspection of the spectrum.

Database Searching

Protein Identification by Peptide Mass Mapping. All the measured tryptic peptide masses by MALDI–TOF were submitted to the MS-Fit program developed by Karl Clauser and Peter Baker in our laboratory (University of California, San Francisco, CA, http://prospector.ucsf.edu) (Clauser *et al.*, 1999). Basically, this program is designed to compare the experimental mass values with theoretical values from protein databases, calculated by applying the enzyme cleavage rules. By using an appropriate scoring algorithm, the closest match or matches can be identified.

Protein Identification by Peptide Sequencing. Although peptide mass mapping is fast and simple, its success can be compromised by the purity of the protein spot in one gel spot, errors in the sequence database, and the observation of too few peptides in the MS map from a given protein. Therefore, peptide sequencing using either PSD or nanoESI–MS/MS was carried out to confirm or establish unambiguous protein identifications. PSD or MS/MS spectra were analyzed by the MS-Tag program (University of California, San Francisco, CA), which is designed to match the fragment-ion tag data contained in a user's tandem mass spectrum to a peptide sequence in an existing database. Each fragment ion is characteristic of the sequence of the peptide under analysis and adds a constraint to database searching. Therefore, all ions present provide very high discriminating power for database searching. MS-Tag integrates all of these constraints by considering each fragment ion and parent ion mass independently to match a single peptide sequence in a genomic database. Proteins in each spot on a 2-D gel were identified with certainty when combining both MS-Fit and MS-Tag searching results.

When the protein sequences were not present in the databases, *de novo* sequencing was performed. The PSD or tandem mass spectra were interpreted manually as described in the results section of this chapter, and the novel peptide sequences established by tandem mass spectrometry were used to design oligonucleotide probes for cDNA cloning.

Identification of Unknown Proteins Using Homology Searches of Novel Peptide Sequences. The strategy for identification of unknown proteins using novel sequences is the same as described (Huang *et al.*, 2001). A static nonredundant database (as of 3/11/99 NcBInr) was downloaded and modified by removing sequences matching three of our query sets that had been submitted to the database following their identification early in this project (Huang *et al.*, 1999). MS-Pattern (University of California, San Francisco, CA) and the Gapped BLAST search engine (version 2.0a19 of BLASTP, Washington University, St. Louis, http://blast.wustl.edu/) were used as homology searching tools to find "best" homologues for unknown protein identification. Subsequent congruence analysis of MS-Pattern and Gapped BLAST search results were carried out for all sequence fragments from a given protein using MS-Shotgun (University of California, San Francisco, CA), a modification of the MS-Shotgun program (Pegg and Babbitt, 1999). Protein hits were sorted either by MS-Shotgun score or by *p* values. Furthermore, functional assignments inferred from the database search outputs were further evaluated using multiple alignments [using either Pileup (Feng and Doolittle, 1987) or ClustalW (Thompson *et al.*, 1994)] of top-scoring protein hits from each MS-Shotgun with a catenation search. Percentages of identities were calculated for these hits to identify a divergent set of sequences from available species when possible.

Results and Discussion

Part A: MS Analysis of 20S Proteasome Complex from Rat Liver

Initially, we investigated the 20S proteasome from rat liver simply because most of the rat proteasome subunits were present in the databases, and this complex could be used as a model to develop MS approaches for comprehensive characterization of macromolecular machines. Two strategies have been undertaken to study rat 20S proteasome complex subunit composition and heterogeneity, as shown in Fig. 1.

Identification and Characterization of Rat 20S Proteasomes Separated by 2-D Gel Electrophoresis. In Path 1 (Fig. 1), 20S proteasome complexes were purified from rat liver and separated by 2-D gel electrophoresis.

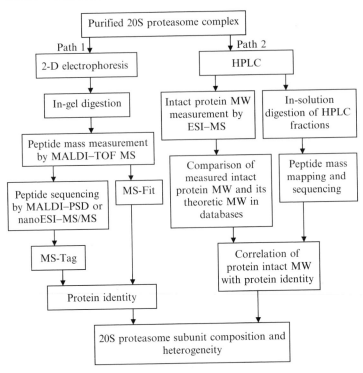

FIG. 1. Strategy for protein identification and characterization of 20S proteasome from rat liver.

Coomassie Blue staining revealed 38 spots, as shown in Fig. 2 (Huang *et al.*, 2000). All the visible protein spots were digested and analyzed using MALDI–TOF–MS, and the measured peptide masses were submitted into MS-Fit program for database searching. Selected peptides were analyzed by MALDI–PSD to obtain partial peptide sequence information to confirm protein identities. As a result, 14 essential subunits, including 7 alpha subunits [i.e., α1(IOTA), α2(C3), α3(C9), α4(C6-I), α5(ZETA), α6(C2), and α7(C8)] and 7 beta subunits [i.e., β1(Y/DELTA), β2(Z), β3(C10-II/THETA), β4(C7-I), β5(X), β6(C5), and β7(N3)] were identified by peptide mass mapping and peptide sequencing (Huang *et al.*, 1999, 2000). In addition, three nonessential β subunits of β1i(RING12/LMP2), β2i (MECL-1), and β5i(C1/LMP7) were also identified that could replace the other three catalytic β subunits β1(Y/DELTA), β2(Z), and β5(X), respectively, upon interferon γ induction to form immunoproteasome for antigen

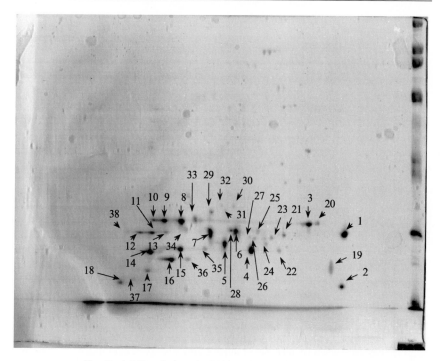

FIG. 2. A 2-D gel picture of 20S proteasome from rat liver.

presentation of MHC class I peptides. Clearly, the number of protein spots observed on 2-D gel of rat 20S proteasome complex is much higher than the number of proteasome subunits (14 for eukaryotes, 17 for mammalian cells) identified. MS analysis of the rest of the spots revealed that the 12 subunits of $\alpha1$(IOTA), $\alpha2$(C3), $\alpha3$(C9), $\alpha4$(C6-I), $\alpha6$(C2), $\alpha6$(C8), $\beta2$(Z), $\beta3$(C10-II/THETA), $\beta4$(C7-I), $\beta7$(N3), $\beta2i$(MECL-1), and $\beta5i$(C1/LMP7) were observed in more than one spot, suggesting that multiple forms of those subunits were present (Huang et al., 2000). The results are summarized in Table I.

The reason for the presence of multiple isoforms is not clear; perhaps the presence is a result of different localization of proteasome subunits with diverse functions, cDNA splicing, premature forms, and/or possible posttranslational modifications (Coux et al., 1996; Elenich et al., 1999; Haass and Kloetzel, 1989; Huang et al., 2000; Kloetzel, 2001; Kristensen et al., 1994; Lupas and Baumeister, 1998; Voges et al., 1999). Furthermore, several new peptides different from the sequences of existing rat

TABLE I

SUMMARY OF IDENTIFICATION AND CHARACTERIZATION OF RAT 20S PROTEASOME SEPARATED BY
2-D GEL ELECTROPHORESIS USING MS

Spot #	Subunits identified	Subfamily
1	ZETA	α
2	Ring12(Lmp2)	β
3, 20	C8	α
4, 22	RN3	β
5, 24	C10-II	β
6^a, 25, 26	Z^g	β
7^b, 30	IOTA	α
8^c, 29, 31, 33^d	C2	α
9, 10	C9	α
11, 12, 13, 28^e, 34^f	C6-1	α
14	C5	β
15, 35	C3	α
16, 36	C7-I	β
17	X	β
18, 37	C1(Lmp 7)	β
19	Y(DELTA)	β
21, 23	MECL-1^g	β
27	IOTA and C10-II	

[a] Peptides with sequences identical to subunit C6-1 were present in spot 6.
[b] Peptides with sequences identical to subunit Z were present in spot 7.
[c] Peptides with sequences identical to subunit C9 were present in spot 8.
[d] Spot 33 is a doublet.
[e] Peptides with sequences identical to subunit Z were present in spot 28.
[f] Peptides with sequences identical to subunit C2 were present in spot 34.
[g] Newly identified subunits.

proteasome subunits in the databases have also been found by peptide mass mapping and sequencing (Huang and Burlingame, unpublished data, 2000). The results indicate that sequence errors are present in the databases, which would complicate protein identification and characterization. In comparison to the human and mouse 20S proteasome complex separated by 2-D gel electrophoresis (Elenich *et al.*, 1999; Haass and Kloetzel, 1989; Kristensen *et al.*, 1994; Ni *et al.*, 1995; Thomson and Rivett, 1996), various numbers of isoforms have also been observed. Therefore, the heterogeneous composition established in this work appears to be a general phenomenon for eukaryotic 20S proteasomes.

Identification and Characterization of Rat 20S Proteasome Complex Separated by HPLC Coupled with On-Line MS Analysis. The results described previously that were obtained using 2-D gel electrophoresis/MS

(Strategy 1, Path 1) provide strong evidence that proteasome subunit composition is more complicated than expected. To further investigate the subunit composition heterogeneity, a second strategy (Path 2, Fig. 1) was employed. In this approach, we were seeking to correlate the subunit composition with the intact protein molecular weight to investigate the subunit heterogeneity and gain evidence for possible posttranslational modifications. The rat 20S proteasome was first separated by RP-HPLC, part of the HPLC eluent going straight to an on-line ESI–QTOF mass spectrometer for intact subunit molecular weight measurement, whereas the rest of the HPLC fractions were collected for subsequent protein identification using mass mapping and peptide sequencing. The proteins identified in each fraction would be correlated with the molecular weight measurement. If the expected molecular weight based on protein identity corresponded well with the measured molecular weight, it meant that the protein sequence deduced from cDNA was correct. Otherwise, either the error was present in the cDNA sequence, or the subunit was modified. The total ion chromatography of the LC–MS run is illustrated in Fig. 3, which is identical to the HPLC–UV trace.

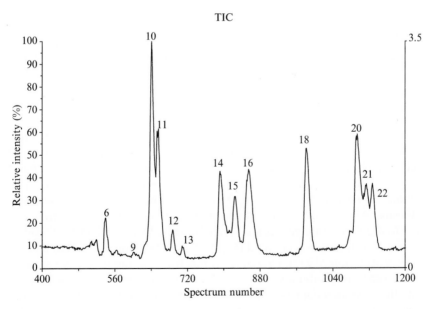

FIG. 3. Total ion current chromatography (TIC) of 20S proteasome complex from rat liver using LC–ESI QTOF–MS.

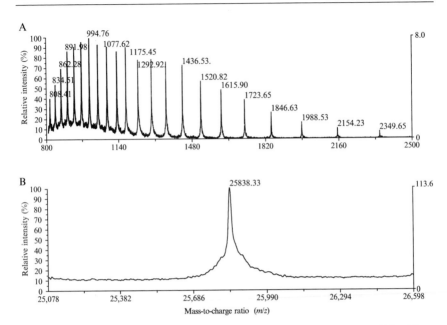

FIG. 4. (A) ESI spectrum of peak 10 in Fig. 3. (B) Transformed spectrum of Fig. 4A.

The rat 20S proteasome complex was separated into 13 major chromatographic peaks, corresponding to 26 measured molecular weights, which again revealed directly the composition heterogeneity present in rat 20S proteasomes (Huang and Burlingame, unpublished data, 2000). The HPLC fractions of the 13 peaks were collected and digested with trypsin and analyzed by MS for protein identification. An ESI spectrum of fraction 10 is shown in Fig. 4A, and the transformed ESI spectrum is shown in Fig. 4B, giving one dominant peak with measured molecular weight of 25838.33. The subunits in fraction 10 were identified as subunit C3, C2, and MECL-1 by peptide mass mapping and sequencing, and subunit C3 appears to be the major component based on its peptide signal intensity. As a result, the measured molecular weight (25838.3) matched well to the theoretical molecular weight of subunit C3 (25838.7), considering the N terminus is acetylated (Coux et al., 1996; Lupas and Baumeister, 1998; Voges et al., 1999). However, in this fraction, the molecular weights of the other two subunits, C2 and MECL-1, were not observed because only the dominant component C3 was detected by intact molecular weight measurement due to its abundance. A similar approach was carried out for every fraction. Only five measured molecular weights corresponded well

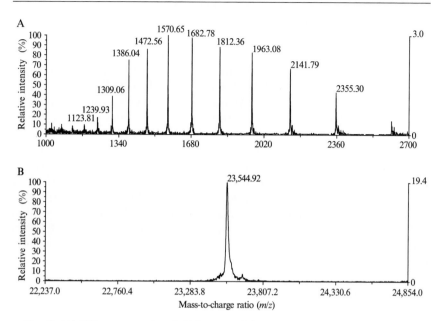

FIG. 5. (A) ESI spectrum of peak 13 in Fig. 3. (B) Transformed spectrum of Fig. 5A.

with the calculated molecular weights based on the protein identified, assuming that the N termini of subunit C3, C9, or C10-II were acetylated or the presequences of subunit X and Z (subunit Z has already been cloned and sequenced) were removed. The rest of the measured molecular weights did not match well with calculated values of identified subunits.

An ESI spectrum of peak 13 during the LC–MS run is illustrated in Fig. 5A, and the transformed ESI spectrum is shown in Fig. 5B. Again, one major molecular weight (MW) was observed and measured as 23546.4, which was several thousand daltons less than the expected MW values of the two subunits identified in this fraction: C5 (26479.4) and IOTA (27399.6). During the peptide analysis of the tryptic digest from this fraction, one particular peptide sequence was determined as FFPYYVY-NII̲GG LDEEGK, identical to human and mouse proteasome subunit C5, but shown as only homologous to the partial sequence of rat proteasome subunit C5 (FFPYYVYNII̲EGLDEEGK) with one amino acid difference (E<->G). Interestingly, this peptide was also found in the tryptic digest from a 2-D spot identified as subunit C5, confirming one sequence error from cDNA-deduced sequence. It is well known that β subunits were usually synthesized as premature forms first and then processed to matured forms for their individual functions in the assembled 20S complex. There-fore, we recalculated the molecular weight of subunit C5 based on cDNA sequence after the removal of the presequence and correction of one amino acid (E→G), and the calculated value (23547.1) proved to corre-spond well to the experimental value (23546.4). This finding suggests that cautions need to be taken when intact protein molecular measurement is used to determine protein modifications due to the presence of different processed forms for one protein and cDNA sequence errors. After se-quence corrections, an additional 10 molecular weights matched well with expected values. However, 11 measured molecular weights still remain unassigned. A total of 17 subunits in all 13 fractions were identified by peptide mass mapping and peptide sequencing, which were the same as identified from 2-D separated proteasome subunits. Even though several subunits were present in several different fractions during the LC–MS run, it is difficult to tell the presence of isoforms because abundant and hydrophobic proteins may elute over a long period of time during the separation. Because each subunit has a different number of forms present in the purified protein complex mixture, the relative quantity of each subunit is difficult to determine.

Part A: Summary and Conclusion

We have successfully identified and characterized rat 20S proteasome subunits separated by 2-D gel electrophoresis using MALDI–TOF and –PSD (Huang *et al.*, 1997, 2000). All the known rat proteasome subunits in the databases, seven α and eight β subunits, have been identified and characterized (Table I). In addition, two new subunits and different iso-forms of most of the proteasome subunits were revealed. The pattern of 20S proteasome from rat liver obtained on 2-D gels reported here is generally similar to that observed by other groups (Ni *et al.*, 1995; Rivett, 1998; Rivett and Sweeney, 1991) and also similar to that of human and mouse 20S proteasomes (Akiyama *et al.*, 1994; Elenich *et al.*, 1999; Groettrup *et al.*, 1995; Kristensen *et al.*, 1994; Nandi *et al.*, 1996; Stohwasser *et al.*, 1997). Furthermore, the pattern of the identified rat proteasome subunits in major spots matched well with that obtained using immunoblotting (Rivett and Sweeney, 1991). Intact molecular weight measurement of individual proteasome subunits from rat liver using LC–MS has revealed not only the heterogeneity of the proteasome subunit composition but also the cDNA sequence errors present in the databases. The cDNA sequence errors can be an obstacle for successfully characterizing proteins because wrong con-clusions may be drawn in terms of chemical modification of proteins based on the differences between experimental molecular weight and calculated molecular weight from cDNA sequences. Extra precautions are needed for the full characterization of proteins.

The rat 20S proteasome subunit composition appears to be more complicated than expected because most of the proteasome subunits in rat occur in more than one form. These different isoforms may reflect multiple forms of the proteinase for diverse functions, possibly from different cell types or subcellular localization, or may show different stages of differentiation (Mason *et al.*, 1996). Morphometric studies have shown that proteasomes are localized in the nucleus and the cytoplasm and are closely associated with the endoplasmic reticulum (ER); they also exist in several different molecular forms (Rivett, 1998). However, the precise distribution and the mechanisms of interconversion between different molecular forms of proteasomes are not yet clear. It has been suggested that the proteasome multisubunit complex acquires functional diversity by a change in molecular composition in response to environmental stimuli (Yuan *et al.*, 1996). The subunit changes of proteasomes in various physiological conditions may be one reason why the proteasome has such a complex composition (Coux *et al.*, 1996). In higher eukaryotes, the subunit composition of the 20S proteasome can vary in a given species and is subject to precise regulation. Furthermore, how the isoforms of proteasome subunits have been processed still remains unknown. Further studies will be focused on clarifying the structural relationships between different isoforms.

The 2-D gel electrophoresis/MS is a powerful tool for investigation of protein complex subunit composition and heterogeneity (Huang *et al.*, 1999, 2000) due to the high sensitivity of MS for protein and peptide analysis and the advantages of 2-D gel electrophoresis (i.e., high resolving power for protein separation, capability of parallel comparison for a number of samples at the same time, direct visualization of protein relative abundance, and the possibility of the characterization of posttranslational modification and isoform identification). However there are some limitations associated with 2-D gel electrophoresis-based techniques because it is only semiquantitative, has limited dynamic range, and often misses proteins with extreme pI as well as molecular weights and membrane proteins. Therefore, direct analysis of protein complexes using LC and MS is developed without any prior separation of proteins (Link *et al.*, 1999; Washburn *et al.*, 2001). Recently, a 26S proteasome complex was purified from yeast and digested, and the entire digested proteasome mixture was directly analyzed by LC–MS/MS (Verma *et al.*, 2000). This single experiment identified all known 26S subunits except for Pre8 and Pup1. 15 of the 33 subunits that were identified yielded only 1 or 2 peptide sequences, and 5 subunits yielded 5 to 10 peptide sequences. Such an approach has the advantage of identifying most subunits in a single run. However, most of the protein sequence coverage is poor since only a few peptides were used for protein identification. Due to the inherent problems of the technique associated with direct analysis of protein mixtures, it is difficult to have a quantitative overview of protein relative abundance, and it is also hard to tell the degree of protein purification, molecular weight, and pI. Most importantly, it is difficult to identify protein isoforms due to their high sequence homology and difficult to find posttranslational modifications due to the complexity of peptide mixtures being analyzed. Even though such an analysis can directly identify the components of protein complexes, if not all, it is not suitable to generate a comprehensive picture of protein complex subunit composition and heterogeneity. For such a purpose, the 2-D gel electrophoresis-based MS technique is preferred.

Part B. MS Analysis of the 20S Proteasome Complex and Its Activator Protein PA26 from T. brucei

T. brucei, a member of the Kinetoplastidae family, is the causative agent of African sleeping sickness. It is generally regarded as a relatively primitive eukaryote further removed from mammals than yeast (Sogin et al., 1989). The 20S proteasome from T. brucei has been identified, isolated, and characterized (Hua et al., 1996). The morphology and dimensions of T. brucei 20S proteasome are similar to those of archaebacterial, yeast, and mammalian 20S proteasomes. However, the diameter of the portal in the T. brucei 20S proteasome is apparently larger than that in the rat 20S proteasome (Hua et al., 1996). In addition, its profile of peptidase activity differs from that of other eukaryotic 20S proteasomes. Instead of the primary chymotrypsin-like activities commonly observed among mammalian 20S proteasomes, it exhibits mainly trypsin-like activity. These observations raise the possibility that the T. brucei proteasome could prove to be a target for a specific antitrypanosomal drug. To help prove this hypothesis, it is important to look for unique structural features of the T. brucei proteasome subunits that may account for the observed differences and, if possible, produce a recombinant 20S proteasome that can be used for more detailed structure–function analyses. Therefore, the 20S proteasome from T. brucei was purified and separated by 2-D gel electrophoresis, and proteins were visualized by silver staining. The major spots were digested and analyzed by MS. Because the protein sequences of the 20S proteasome from T. brucei were not present in any public database at the time of this analysis, novel peptide sequences determined by MS were used for homology searches using newly developed bioinformatics tools to assign unknown protein functions for study of the subunit composition (Huang *et al.*, 1999). The MS sequences for each spot were also used for designing oligonucleotide probes for cDNA sequencing to obtain full-length sequences.

There are at least two different pathways leading to activation of eukaryotic 20S proteasomes. One involves the binding to a 19S protein complex at both ends of the proteasome to form the 26S proteasome, and the other involves binding to a heptamer ring of an activator protein PA28 at each end of the proteasome to yield the activated 20S proteasome (Coux *et al.*, 1996; Peters, 1994; Rechsteiner *et al.*, 1993; Rubin and Finley, 1995). The 26S proteasomes are responsible for ATP-dependent degradation of ubiquitinated proteins and have been identified among all eukaryotes. The activated 20S proteasome is only capable of digesting peptides *in vitro* (Ciechanover and Schwartz, 1998) and has been found only among mammalian cells until the recent identification of an activated 20S proteasome species in *T. brucei* (To and Wang, 1997). The 19S complex is a multisubunit complex that binds to ubiquitinated proteins and hydrolyzes ATP (Ciechanover and Schwartz, 1998; Peters, 1994). This complex contains at least one subunit that binds polyubiquitinated proteins and six homologous subunits that contain ATP binding domains (Coux *et al.*, 1996; Hochstrasser, 1996). PA28 has no hydrolytic activity of its own and has no homologues in yeast (Dubiel *et al.*, 1992; Ma *et al.*, 1992). It is generally conceived to be involved in the major histocompatibility complex class I antigen processing because synthesis of PA28 is strongly induced by interferon γ (Ahn *et al.*, 1995). Three isoforms, PA28: PA28α, PA28β, and PA28γ, share about 50% amino acid sequence identity (Mott *et al.*, 1994). The α and β isoforms form a complex with 20S proteasome in the form of a heptamer ring of three PA28α and four PA28β or three PA28β and four PA28α (Zhang *et al.*, 1998). The crystal structure of recombinant human PA28α has been recently resolved in the form of a self-assembled heptamer ring (Knowlton *et al.*, 1997). It contains a central channel that has an opening of 20 Å diameter at one end and 30 Å diameter at the presumed proteasome-binding surface. Presumably, binding to such a ring structure may cause conformational changes that could open the pore in the proteasome α-ring to allow the passage of peptide substrates.

During the purification and characterization of the 20S proteasome from *T. brucei*, an activated form of the 20S proteasome, similar to the mammalian-activated 20S proteasome, was isolated and identified (To and Wang, 1997). This activated 20S proteasome demonstrated enhanced peptidase activities, up to 100-fold of the original level. It consisted of the 20S proteasome and an extra protein with an estimated molecular mass of 26 kDa, designated as the proteasome activator protein, or PA26. The isolated PA26 was found capable of reconstituting the activated 20S proteasome *in vitro* with purified 20S proteasome. To understand its activation mechanism and study its assembly with 20S proteasome complex, PA26 was separated from 20S proteasome, purified, and analyzed using MS to obtain its sequence and structure.

Part B1: MS Analysis of the 20S Proteasome Activator Proteins PA26 from T. brucei: *Protein Purification*

PA26 was separated on a one-dimensional (1-D) gel as one single band, and the molecular weight was estimated to be around 26 kDa (To and Wang, 1997). After in-gel digestion of the protein band, the tryptic peptide masses were measured by MALDI–TOF. For database searching, 32 peptide masses were submitted into an MS-Fit program (NCBInr, 12/98). No protein entries, matched to more than 30% of peptide masses submitted, were found. No proteasome entries were pulled out in the searching result because it was known that this protein activates 20S proteasome activity, indicating that PA26 might have homologous sequences to other proteasome activator proteins, such as PA28 from mammalian cells. To find the PA26 identity and design of the oligonucleotide probe for cDNA cloning of this unknown protein, the unseparated digests were further separated by RP-HPLC to obtain better mass spectrometric signal and peptide coverage. All the fractions were collected and concentrated, and the tryptic peptide masses in each fraction were further measured using MALDI–TOF (Yao *et al.*, 1999).

De Novo *Sequencing by MALDI–PSD*

MALDI–PSD has been widely used as a powerful tool for peptide partial sequencing to identify gel-separated proteins before the availability of Qqoa-TOF mass spectrometers (Chang *et al.*, 2000; Huang *et al.*, 2001, 2000, 1999, 1997). However, this technique has not been successfully used for *de novo* sequencing due to poor mass accuracy (~1 Da) and mass resolution (<100) of fragment ions. Here, we demonstrate the possibility of using PSD as a *de novo* sequencing technique for obtaining unknown peptide sequences, especially when it is the only tool available for peptide sequencing.

Three types of diagnostic ions featured in a typical PSD spectrum are formed during fragmentation. Immonium ions that present in the low-mass region of the spectrum [mass-to-charge ratio (m/z) 50–300 Da] are indicative of the peptide amino acid composition (Biemann, 1990a,b; Medzihradszky and Burlingame, 1994), which can be used to confirm unknown sequences. Sequence ions are derived from the three possible types of bond cleavages that can occur in the peptide bonds for each residue along the peptide backbone. If the positive charge is localized on the C-terminus, these ions are designated as x_i, y_i, and z_i ions and are numbered starting from the C-terminal residue. Similarly, if the charge is localized on the N terminus, these ions are designated as a_i, b_i, and c_i ions and are numbered starting from the N-terminal residue. In high-energy CID, all the sequence ions are usually shown in the tandem mass spectra (Biemann, 1990b; Medzihradszky and Burlingame, 1994). However, in PSD spectra, the most dominant and commonly observed sequence ions are y and b ions. In addition, internal ions formed during multiple-step fragmentation reactions, which do not contain either end of the molecular chain, can also be observed and are useful for confirmation of proposed sequence. All of these ion types can be accompanied by satellite ions due to the loss of ammonia (-17 u) when an ion contains Arg, Lys, or Gln or due to a loss of water (-18 u) when Ser or Thr are present in the ion. Because of the lack of side chain fragmentation information in PSD, isoleusine and leusine cannot be distinguished in structure analysis.

To determine the peptide partial sequence, manual interpretation of the PSD spectrum was attempted based on the following steps:

1. Examine immonium ions (usually around 50 to 200 Da) to predict possible amino acid composition without knowing the sequence order (Biemann, 1990b; Medzihradszky, 1994). Check dipeptide ions (115~373 Da) to predict possible dipeptide present in the sequence, which is very useful for sequence determination. (The table of dipeptide ion masses can be found at the University of California, San Francisco, CA, Web site at http://prospector.ucsf.edu.)

2. Search for characteristic y_1 ions expected for tryptic peptides with either Arg or Lys at the C-terminus; note that y_1 = residue mass + 19. For peptides ending in Arg, y_1 is 175 u, whereas for peptides ending in Lys, y_1 is 147 u. A 175 ion usually is more abundant and easier to observe than a 147 ion.

3. Search for ion pairs, such as y_i and b_{n-i}. The n is the total number of amino acid residues in the peptide. Choose one dominant ion, and assume it belongs to either the y or b ion series; then use equation (1) to find another type of ion (Medzihradszky, 1994):

$$\text{Mass}(y_i) = MH^+(\text{parent}) - \text{Mass}(b_{n-i}) + 1 \qquad (1)$$

Allow a 1 Da error for this calculation due to the instrument mass accuracy. Note: b_1 (first b ion) = the first amino acid residue mass at N-terminus + 1, which normally does not exist; b_n (last b ion) = $MH^+(\text{parent}) - 18$; y_n (last y ion) = MH^+ (parent).

4. Search for a-type ions to help determine b-type ions in each ion pair obtained from step 3 because $a_i = b_i - 28$.

5. Search for satellite ions (NH_3 and H_2O loss) for a-type, b-type, and y-type ions. In PSD spectra, it is commonly observed that b-type and a-type ions often lose H_2O when ions contain Ser or Thr, whereas y-type ions often have a loss of NH_3 (-17 u) due to the presence of Arg or Lys at C-terminus.

6. Determine b or y ions, and calculate the mass differences between consecutive b ions ($\Delta(b_{i+1} - b_i)$) and consecutive y ions ($\Delta(y_i + 1 - y_i)$) to obtain meaningful masses, which should correspond to amino acid residue masses and allow the determination of a partial peptide sequence. The mass difference between b_{i+1} and b_i should match with the mass difference between $y_{(n-i)+1}$ and y_{n-i}.

7. Evaluate the proposed sequence by matching all the observed ions with the theoretical fragment ions generated using MS-Product (University of California, San Francisco, CA).

As an example, a PSD spectrum of a tryptic peptide with MH^+ at 996.5 from HPLC fraction 9 is given in Fig. 6A. We first examined the immonium ion region, where it indicates that A(44), R(59, 73, 100), R/P(70), R/N(87), I/L(86), T(74), E(102), and H(110) may be present in the sequence. The 175 ion reveals that Arg is at the C-terminus. The ion pairs found are 343.7/654.5, 414.4/583.7, 551.8/446.4, 608.6/389.6, 710.6/288.3, and 823.2/175.3, shown in Table IIA. The dipeptide ions found with the mass under 300 u are 143.1 u (AA), 195.2 u (HG or PP), 201.2 u (EA, I/LS, or CP), 209.1 u (HA), and 214.1 u (RG or NV). The order of each dipeptide ion is not clear. Among the ion pairs, 343.7 u, 414.4 u, 551.8 u, 608.6 u, and 710.6 u all have a-type ions, which show a loss of 28 u from b ions, indicating these ions are b ions. Therefore, their corresponding ion pairs are y ions (i.e., 654.5 u, 583.7, 446.4, 389.6, and 288.3), and most of these ions have a 17 u loss as shown in Table IIA. Since 175.3 u is a y_1 ion, its ion pair 823.2 would be a b ion. The last b ion is 979.7 u, about -18 u of parent ion. Combining $\Delta(b_{i+1} - b_i)$ and $\Delta(y_{i+1} - y_i)$ in Table IIA, the partial sequence can be determined as [343.7]AHGT[I/L]R. Since y ions are missing after y_6, one only can try to use b ions to determine the N-terminal sequence. However, there is no obvious indication of b_1, b_2, and b_3 ions, and one can only guess the ion types by calculating the mass difference between the assumed b ion and known b ion. If one assumes that 214.1 u is a b ion because it is a dominant ion, then the mass difference between 343.7 u and 214.1 u is 129.6 u, suggesting an E; the dipeptide ion 201.2 u supports EA composition. Then, one can assume that 143.0 u is also a b ion, with the mass between 214.1 u and 143.1 u at 71.1 u, indicating an A, and 272.3 u

A

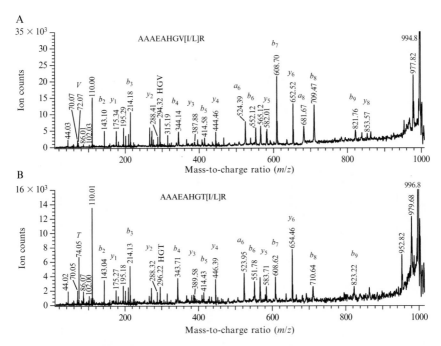

FIG. 6. MALDI–PSD spectra of two tryptic peptides of 20S proteasome activator protein PA26 digest. (A) MH^+ 994.5. (B) MH^+ 996.5.

matches to the internal fragment [AEA]. Furthermore, 143.0 u can only be AA as a dipeptide. Therefore, the full partial sequence is suggested as AAAEAHGT[I/L]R. To evaluate the accuracy of this sequence, it was submitted to MS-Product to obtain theoretical fragment ions that were compared with the rest of the ions (internal ion fragments and satellite ions) unassigned in the spectrum. As a result, the sequence was further confirmed as AAAEAHGT[I/L]R.

Figure 6B illustrates a PSD spectrum of a tryptic peptide with MH^+ at 994.5 from HPLC fraction 12, which gives similar fragmentation as shown in Fig. 6A. Table IIB displays the summary of a, b, and y ions observed in Fig. 6B. Compared with the immonium ion region in Fig. 6A, there is a big 72 u (V) instead of a 74 u (T). Interestingly, b_2, b_3, b_4, b_5, b_6, y_1, and y_2 ions are the same as in both spectra, whereas y_3, y_4, y_5, y_6, y_{10}, b_9, and b_{10} ions in Fig. 6B are 2 u less than those ions in Fig. 6A, except b_8 ions only differ by 1.1 u due to their poor mass resolution of that fragment. Nevertheless, the sequence of this peptide (MH^+ = 994.5) is proposed as AAAEAHGV[I/L]

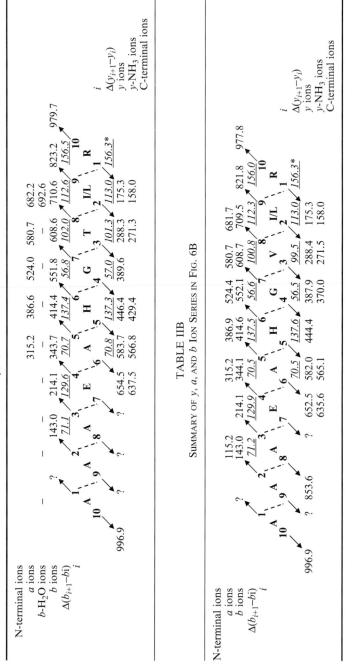

TABLE IIA
SUMMARY OF y, a, AND b ION SERIES IN FIG. 6A

TABLE IIB
SUMMARY OF y, a, AND b ION SERIES IN FIG. 6B

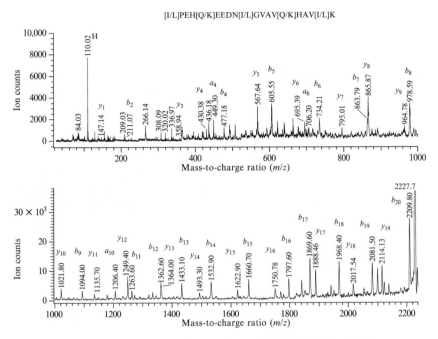

FIG. 7. MALDI–PSD spectrum of a tryptic peptide MH^+ 2226.2 of 20S proteasome activator protein PA26 digest.

R. The only difference between these two peptides is from T to V at position three starting from C terminus, suggesting the presence of dimorphorism of genes in *T. brucei*. Part of the sequences, AAAEAHG, has been used to design probes for cDNA cloning.

It is always easier to solve short peptide partial sequences with MH^+ less than 1000 u due to their simple composition. However, longer peptides ($MH^+ > 1500$ u) are usually preferred because they can provide more sequence information in terms of looking for homologues to predict the protein family and functions. Figure 7 displays the PSD spectrum of a tryptic peptide with MH^+ at 2226.2 from HPLC fraction 20. The y_1 is 147 u, indicating that this peptide ends in K at its C terminus. The immonium ions suggest that R/P(70), V(72), I/L(86), Q/K(84, 101, 129), E(102), and H (110,138, 166) are present in the sequence. As shown in Table III, sixteen ion pairs are calculated as 147.1/2081.4, 358.9/1869.6, 430.4/1797.6, 567.6/ 1660.7, 695.4/1532.9, 795.0/1433.1, 865.9/1362.6, 964.8/1263.6, 1021.8/1206.4, **1135.7/1094.0**, 1249.4/978.6, 1364.0/863.8, 1493.3/734.2, 1622.9/605.6, 1750.8/

477.2, and 2017.5/211.1. Among these ion pairs, the mass error of **1135.7/ 1094.0** is more than 1.5 Da, probably due to poor mass resolution of both ions. Searching for a-type ions and satellite ions reveals that 477.2, 605.6, 734.2, 1094.0, 1206.4, 1362.6, 1433.1, 1532.9, 1797.6, 1869.6, 1968.4, and 2081.4 u are b-type ions. The last b ion is 2209.8 u, which is 18 u less than the parent ion, and 1750.8, 1493.3, 1135.7, 1021.8, 865.9, 795.0, 695.4, 430.4, and 358.9 u are y ions. If 211.1 u is a y ion, then the mass between 211.1 and 147.1 u (y_1) should be an amino acid residue mass. However, the mass difference in-between does not make any sense. Therefore, 211.1 u is a b ion instead of a y ion, which could only be [I/LP]. The possible b ion between b ions 477.2 u and 734.2 u could be either 605.6 or 567.6 u. The sensible mass differences between these ions suggest that only 605.6 could be a b ion. Thus, 1622.9 u of the ion pair with 605.6 u would be a y ion, 567.6 u would be a y ion, and 1660.7 u of the ion pair with 567.6 u would be a b ion. Similarly, the possible b ion between 1206.4 u and 1362.6 u is either 1263.6 u or 1249.4 u. The mass of 1263.6 u is suggested as the b ion since the masses between 1263.6 u and 1206.4 u and between 1263.6 u and 1362.6 u are 57.2 u (G) and 99.0 u (V), respectively, indicating the corresponding ion pair of 964.8 u as a y ion. Therefore, 1249.4 u is a y ion, and its ion pair 978.6 u is a b ion. Since 1362.6 u is a b ion, 1364.0 u can only be a y ion, and its ion pair 863.8 u is a b ion.

After sorting out the y and b ion series, calculating the mass differences between both consecutive b ion and y ion series, and comparing $\Delta(b_{i+1} - b_i)$ and $\Delta(y_{n-i+1} - y_{n-i})$, the partial peptide sequence is suggested as [I/L]PEHE[Q/K]EE**DNI/L**GVAV[Q/K]HAV[I/L]K. The order and accuracy of **DNI/L** is not for certain due to the poor mass accuracy obtained as shown in Table III. To confirm the sequence, MS/MS of this peptide using a Sciex API 300 triple quadrupole mass spectrometer equipped with nanoelectrospray source was also acquired (data not shown). Unfortunately, compared with PSD spectrum, the partial sequences of this ambiguous region still cannot be determined. Even though the three amino acids are questionable, they are the most reasonable choices. Q and K can usually be distinguished using accurate mass measurement of parent ion mass; because the mass accuracy obtained by internal calibration using MALDI–TOF can reach approximately 10 ppm, this mass accuracy is sufficient to differentiate between the element compositions of glutamine and lysine. However, it will be difficult to determine when there is more than one Q or K present in the sequence. Fortunately, a tryptic peptide with MH^+ at 1621.8522 was found in both unseparated and separated digests, which showed a similar fragmentation pattern in PSD spectra as that of 2226.2. The mass matched well with the peptide with sequence of EE**DN[I/L]**GVAV[Q/K]HAV[I/L]K

TABLE III
SUMMARY OF y, a, AND b ION SERIES IN FIG. 7

N-terminal ions	a ions	b ions	$\Delta(b_{i+1}-b_i)$	i		i	$\Delta(y_{i+1}-y_i)$	y ions	C-terminal ions
		?		1	I/L	20		2227.7	
			?				113.6		
		211.1		2	P	19		2114.1	
			?				96.6		
		?		3	E	18		2017.5	
			?				129.0		
	449.3	477.2		4	H	17		1888.5	
			128.4				137.7		
		605.6		5	K	16		1750.8	
			128.6				127.9		
	706.2	734.2		6	E	15		1622.9	
			129.6				129.6		
		863.8		7	E	14		1493.3	
			114.8				129.3		
		978.6		8	D	13		1364.0	
			115.4				114.6		
	1064.8	1094.0		9	N	12		1249.4	
			112.4				113.7		
	1178.7	1206.4		10	I/L	11		1135.7	
			57.2				113.9		
		1263.6		11	G	10		1021.8	
			99.0				57.0		
	1334.4	1362.6		12	V	9		964.8	
			70.5				98.9		
	1405.0	1433.1		13	A	8		865.9	
			99.8				70.9		
	1503.9	1532.9		14	V	7		795.0	
			127.8				99.6		
		1660.7		15	Q/K	6		695.4	
			136.9				127.8		
	1770.1	1797.6		16	H	5		567.6	
			72.0				137.2		
	1841.5	1869.6		17	A	4		430.4	
			98.8				71.5		
	1940.7	1968.4		18	V	3		358.9	
			113.0				?		
	2053.4	2081.4		19	I/L	2		?	
			128.4				128.1*		
		2209.8		20	K	1		147.1	

TABLE IV
SUMMARY OF NOVEL PEPTIDES OF PA26

MALDI peptide masses (Da)[a]	Δ(ppm)[b]	Peptide sequences determined by PSD and/or QTOF MS/MS[c]	Sequences derived after cloning
697.3230	−5.6	NSTYGR	NSTYGR
754.4977	5.0	[Q/K][I/L][I/L][Q/K]PR	KLIQPR
98.5511	4.1	AA[I/L][I/L]QN[I/L]R	AALIQNLR
972.5585	−3.3	[I/L][I/L]DE[I/L]E[I/L]K	IIDELEIK
994.5454	2.0	AAAEAHGV[I/L]R	AAAEAHGVIR
996.5307	8.0	AAAEAHGT[I/L]R	AAAEAHGTIR
1239.7223	13.1	SPE[Q/K][I/L][I/L] GV[I/ L][Q/K]R*	SPEQLLGVLQR
2226.1794	−4.0	[I/L]PEHKEEDN[I/L] GVAVQHAV[I/L]K	IPEHKEEDNLG VAVQHAVLK

[a] Peptide masses were determined using MALDI–TOF.
[b] Mass error between measured and calculated values from cDNA sequence after cloning.
[c] Sequences determined using QTOF tandem mass spectrometry were marked with *; otherwise, the sequences were determined by PSD.

($MH^+ = 1621.8550$), suggesting that this peptide ($MH^+ = 2226.1794$) can be further cleaved under trypsin digestion. Therefore, a K is considered to be located closer to the N terminus. Then, the sequence is proposed as [I/L] PEHEKEE**DN[I/L]**GVAV[Q/K]HAV[I/L]K. To determine the amino acid (Q or K) at the sixth position from C terminus, the mass error between the measured and theoretical masses was calculated. The mass error is −4 ppm if glutamine is present, whereas the mass error is −20 ppm if lysine is in the sequence. The sequence was further evaluated with internal fragment ions and satellite ions present in the spectrum. The final sequence was determined as [I/L]PEHEKEE***DN[I/L]***GVAVQHAV[I/L]K, where the underlined sequence was used for cDNA cloning. All the novel peptides obtained by *de novo* sequencing using MS are summarized in Table IV.

cDNA Cloning and Sequencing. Based on the amino acid sequence of two peptides (AAAEAHG, as in Fig. 6A, and GVAVQHAV, as in Fig. 7) derived from the PA26 activator, degenerate oligonucleotides of both sence and antisence strands were synthesized. The deduced sequence contained 231 amino acid residues. Two cDNA sequences were obtained with one amino acid difference (T← →V) at position 49 (Yao *et al.*, 1999). Sequence alignments of the cloned PA26 with α, β, and γ isoforms of mammalian PA28 show exceedingly low sequence homology. PA26 has a 23 to 24% sequence identity with PA28α, whereas the identity with PA28β and PA28γ falls below 18%. A sample of rabbit polyclonal antibodies against human PA28 was tested against PA26 in immunoblottings, and no immunostaining of PA26 was detectable. These results suggest that PA26 is a protein significantly different from mammalian PA28.

Recombinant PA26 Can Self-Assemble to Form a Heptamer Ring. We tried to monitor whether self-assembly of recombinant *T. brucei* PA26 occurs by first purifying the two isoforms of the tagged recombinant protein, His_6-PA26T and His_6-PA26V expressed in transformed *E. coli* to apparent homogeneity (Yao *et al.*, 1999). Gel filtration chromatography was first used to examine the possibility of recombinant PA26 to self-assemble to form a complex. As a result, the protein was exclusively eluted at an estimated high molecular mass of 170 kDa, which is 6.5 times its monomeric mass, suggesting that His_6-PA26 is polymerized into either a hexamer or a heptamer. This large protein complex was further analyzed under an electron microscope, and the photos indicated the presence of single ring structures. The resolution of the photos, however, did not allow the determination of the stoichometry of the complex. Therefore, ESI–TOF–MS was used to measure the molecular mass of His_6-PA26 ring formed apparently via noncovalent linkages (Yao *et al.*, 1999). The ESI mass spectrum of the His_6-PA26T ring, shown in Fig. 8, has three different charge state distributions with *m/z* around 1800, 6000, and 8000, whereas the *m/z* 6000 represents the most abundant species. Deconvolution from charge scale to mass scale for the *m/z* 1800 species revealed the presence of two components. The minor component was found to have an average molecular mass of 26,525 Da, same as that of the His_6-PA26T monomer measured by LC–ESI and by the calculated mass of His_6-PA26T (26,525 Da). The major component, however, was discovered to have an average molecular mass of 26,823 Da, which is 296 Da greater than the anticipated value for His_6-PA26T. This additional mass was found linked to Cys-83 of the protein by sequencing the corresponding tryptic peptide using MALDI–PSD.

It is likely that a chemical moiety with a molecular mass of 296 Da (determined by peptide mass measurement using MALDI–TOF) became associated with Cys-83 in PA26T, presumably through a disulfide linkage, while the protein was expressed in transformed *E. coli*. Deconvolution for the ions with *m/z* around 6000 gave three molecular masses of 187,201, 187,484, and 187,760 Da, corresponding to the heptamers of His_6-tagged PA26T$_2$PA26(T$_{modified}$)$_5$, PA26T$_1$PA26(T$_{modified}$)$_6$, and PA26(T$_{modified}$)$_7$, respectively. Deconvolution for the ions in the *m/z* range of 8000 showed three hexamers of His_6-tagged PA26T$_2$PA26(T$_{modified}$)$_4$, PA26T$_1$PA26 (T$_{modified}$)$_5$, and PA26(T$_{modified}$)$_6$. The monomer and the hexamers were the minor components and most likely resulted from partial dissociation of

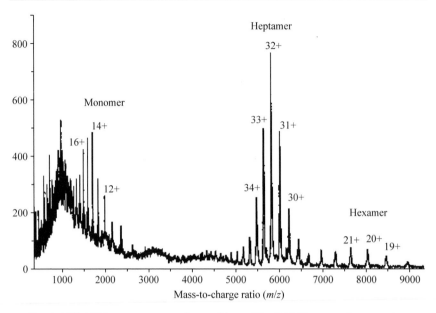

FIG. 8. ESI–TOF mass spectrum of recombinant His_6-PA26T noncovalent complex.

various His_6-PA26T heptamers during ionization under high declustering voltage (250 V). These ions carried fewer charges than those of the heptamer and were not observed at lower declustering voltages. The same results were also observed for the His_6-PA26V isoform. Therefore, the results from MS determined the stoichiometry of the PA26 noncovalent complex and confirmed that it forms a heptamer.

Part B1: Summary and Conclusion

In our study as described in the previous paragraphs, we found two isoforms of PA26, and the only difference we observed was at position 49, T->V. We determined that Thr49 in PA26 may be a potential phosphorylation site because it has a TIR motif in the protein, a specific substrate epitope for threonine phosphorylation by protein kinase C (Hanks *et al.*, 1988). Protein kinase C has been identified in *T. brucei* (Keith *et al.*, 1990). The phosphorylated PA26T could have an effect on heptamer formation, affinity of binding to 20S proteasome, or potency in activating proteasome function, thus placing the proteasome function under at least partial control by one of the potential signal transduction pathways (Grant and Jarvis, 1996). Nevertheless, no direct evidence was observed during our experiment. Either there was no *in vivo* phosphorylation of PA26T, or the phophorylated group(s) may have been removed during protein isolation. Further investigation by including various phosphatase inhibitors during isolation of PA26 will be necessary to clarify this issue.

Both native and recombinant PA26 can by themselves form a ring structure and stimulate the 20S proteasome activity. The molecular weight measurements of both recombinant His_6-PA26T and His_6-PA26V in native condition using MS have demonstrated that the His_6-PA26 forms a heptameric complex (Yao *et al.*, 1999); this finding is similar to that for recombinant human PA28 α, which forms a monomeric heptamer, and that for PA28 α and β which form a hetero-oligomeric heptamer complex as shown by MS (Zhang *et al.*, 1999). MS has also shown that both native and recombinant PA26 can enhance rat and *T. brucei* 20S proteasome peptidase activity, whereas recombinant human PA28 α can only activate rat 20S proteasome activity. Furthermore, PA28α did not change *T. brucei* 20S proteasome activity even when recombinant PA26 was present, suggesting failure in competing with PA26 for binding to *T. brucei* 20S proteasome. Thus, PA26 could be a relatively simple prototype activator protein that may possess a ubiquitous capability of binding to and activating a variety of 20S proteasomes (Yao *et al.*, 1999).

Part B2: MS Analysis of the 20S Proteasome Complex from T. brucei

Subunit Composition of the 20S Proteasome from T. brucei. The 20S proteasomes from *T. brucei* were separated by 2-D gel electrophoresis. Silver staining allowed the visualization of 25 spots, as shown in Fig. 9 (Huang *et al.*, 2001). The pattern of the major spots on 2-D gels is highly reproducible from batch to batch, but the intensity of minor spots varies from different proteasome preparation. Compared with the 2-D gel picture of rat 20S proteasome (Fig. 1) (Huang *et al.*, 2000), the 2-D pattern of 20S proteasome from *T. brucei* is different, and the number of spots is much less. This indicates that the composition of 20S proteasome from *T. brucei* is less complicated than that of rat proteasome. Nevertheless, the number of spots suggests that, most likely, the 20S proteasome complex from *T. brucei* consists of at least 14 subunits as eukaryotic 20S proteasome complexes.

Determination of Partial Peptide Sequences for Each Protein Spot by De Novo *Peptide Sequencing.* Each protein spot was excised and digested by trypsin. The unseparated tryptic digests of each spot were first analyzed by MALDI–TOF–MS. The measured peptide masses were submitted to the MS-Fit program for database searching. At the time when these experiments were performed, the cDNA sequences of 20S proteasome

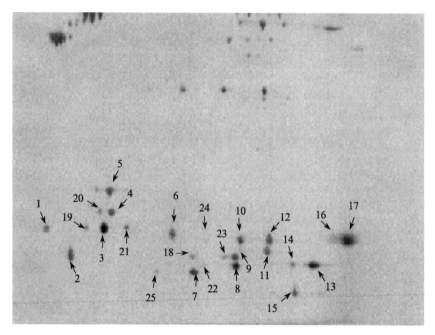

FIG. 9. A 2-D gel picture of the 20S proteasome complex from *T. brucei*.

subunits from *T. brucei* were not present in any public databases. Mass mapping could only find proteasome homologues with conserved tryptic peptides across different species. Not surprisingly, peptide mass mapping did not find any proteasome homologue in the databases. Therefore, more detailed structural information, such as peptide sequence, was obtained for protein identification. *De novo* peptide sequencing was performed to obtain novel partial peptide sequences for each spot using either MALDI–PSD and/or nanoESI tandem mass spectrometry. All the spectra were interpreted manually as discussed earlier. Figure 10 illustrates the nanoESI QSTAR tandem spectrum of a tryptic peptide of MH_2^{2+} 809.5 from spot 7. In this particular spectrum, one dominant immonium ion (102) was observed, indicating the presence of E in the sequence. Sequence ions, such as the y ion series from y_1 to y_{13} and the b ion series from b_2 to b_{11}, were observed. Thus, the sequence was proposed as [259.10]SSYG[I/L] TTFSPSGR. Normally, the b_1 ion is not present due to its instability except when the N terminus is modified. However, b_1(130) was observed in this spectrum, suggesting that the N terminus was modified. Therefore, the sequence was further proposed as [130.06]ESSYG[I/L]TTFSPSGR.

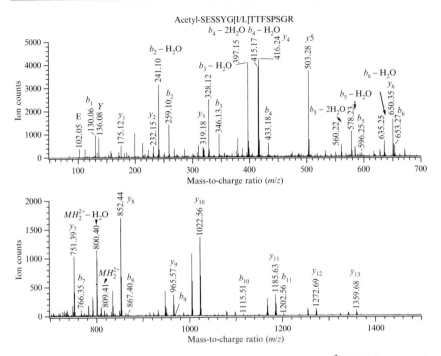

FIG. 10. NanoESI tandem mass spectrum of a tryptic peptide MH_2^{2+} 809.4 from spot 7 in Fig. 9.

Interestingly, the nanoESI QSTAR tandem mass spectrum of another tryptic peptide with MH_2^{2+} 788.5 from spot 7 also showed the same fragmentation pattern, except the b_2 ion was 42 Da less. This finding suggests that the N terminus of this peptide is acetylated. Therefore, the final sequence of the peptide of MH_2^{2+} 809.5 is proposed as Acetyl-SESSYG[I/L] TTFSPSGR, where I and L cannot be distinguished in these spectra.

To solve unknown sequences based on MS analysis data, it is always very straightforward if all the sequence ions (y ion and b ion series) generated from the bond breakage along the peptide backbone are present. However, the type of sequence ions observed in low-energy CID spectra is sequence dependent. In a low-energy CID spectrum obtained using nanoESI QSTAR tandem mass spectrometry, y ions are often observed as the dominant sequence ions compared with the b ion series, especially at the high mass region. Satellite ions (loss of H_2O, NH_3) and internal fragment ions are also frequently observed, which can be used to confirm proposed sequences. Another nanoESI QSTAR tandem mass spectrum of a tryptic

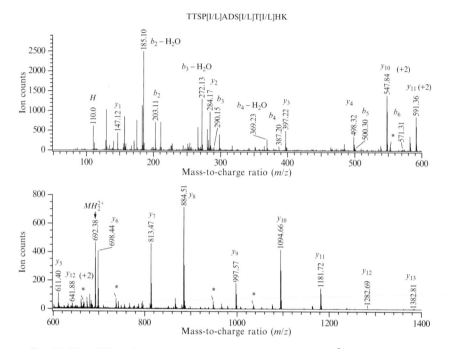

FIG. 11. NanoESI tandem mass spectrum of a tryptic peptide MH_2^{2+} 692.4 from spot 7 in Fig. 9.

peptide with MH_2^{2+} at 692.41 from spot 7 is displayed in Fig. 11. As shown in Fig. 11, the entire y ions from y_1 to y_{13} were observed, whereas b ions were only observed from b_2 to b_6 ions. However, some internal ions at the high mass region labeled with the symbol of star (*) in Fig. 11 helped to confirm the peptide sequence as TTSP[I/L]ADS[I/L]T[I/L]HK. As a result, four novel peptide sequences were obtained for spot 7, as summarized in Table V.

A similar approach was employed to each spot to obtain as many peptide sequences as possible for homology searches for protein identification. Unlike MALDI–PSD, nanoESI QSTAR tandem mass spectrometry has proven to be a superior technique for *de novo* peptide sequencing due to its high resolution and high mass accuracy for both parent and fragment ions. This technique can precisely predict peptide sequences even for the peptide with two different modifications, such as the two oxidized methionines shown Fig. 12. Part of the MALDI–TOF spectrum of tryptic digest of spot 8 is shown in Fig. 12A. The peptide with MH^+ 1384.6 has two

TABLE V
NOVEL PEPTIDE SEQUENCES FOR SPOT 7

Sequence #	Novel peptide sequences	Sequences used for MS-pattern search	Sequences used for gapped blast search
Spot7a0	**TTSP[IIL][IIL]ADS[IIL]T[IIL]HK**	**TTSP[IIL][IIL]ADS[IIL]T[IIL]HK**	**TTSPLLADSLTLHK**
Spot7a1	S-Actyl ESSYG[IIL]TTFSPSGR	SESSYG[IIL]TTFSPSGR	SESSYGLTTFSPSGR
Spot7a2	[AT or DG]DGVV[IIL]AAE[QIK]K	[AT or DG]DGVV[IIL]AAE[QIK]K	AT DGVVLAAEQK
Spot7a3	[IIL]V[QIK][IIL]EYATTAASK	[IIL]V[QIK][IIL]EYATTAASK	LVQLEYATTAASK

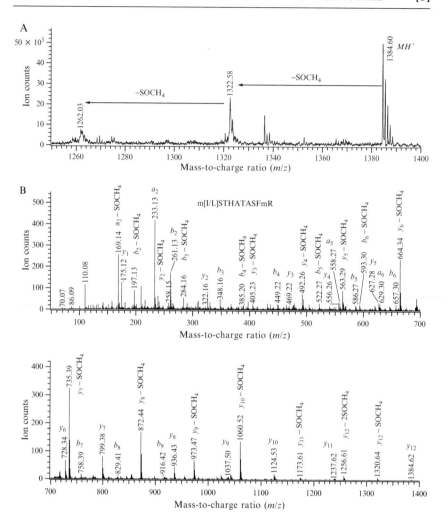

FIG. 12. (A) MALDI–TOF mass spectrum of tryptic digest of spot 8. (B) NanoESI tandem mass spectrum of a tryptic peptide MH_2^{2+} 692.8 from spot 8.

continuous ∼62 Da losses, which is an indication of a peptide containing oxidized methionine. The nanoESI QSTAR tandem mass spectrum of this tryptic peptide with NH_2^{2+} 692.81 from spot 8 is obtained as shown in Fig. 12B. Several immonium ions suggest the presence of R/P(70), I/L (86), and H(110). The characteristic loss of $SOCH_4$ (64 Da) starting from both y_2 and b_2 ions suggests the presence of oxidized methionine at both the C and N termini. Based on the observed y ion series, some b ion series, and their satellite ions (-64 Da), this peptide sequence is proposed as m[I/L]STHATASFmR, where m represents oxidized methionine.

Even though PSD does not provide mass accuracy of fragment ions well enough to distinguish isobaric residues, its fragmentation pattern may be complimentary to that of nanoESI tandem mass spectrometry due to different ionization techniques and energy deposition. Sometimes PSD might be beneficial for determining the peptide sequences when the fragmentation information from both techniques have been considered. A nanoESI tandem mass spectrum and a MALDI–PSD spectrum of a tryptic peptide of MH^+ 1050.6 from spot 13 are illustrated in Fig. 13A and B, respectively. The typical immonium ions in both spectra indicates the presence of R/P(70), I/L(86), and P(126). As shown in a nanoESI tandem mass spectrum (Fig. 13A), all the y ions except y_2 and y_3 were observed, whereas only a b_2 ion was observed. Based on these ions, the sequence can be proposed as [I/L]PPGTT[309]R. The ion pair $y_2(288)/b_8(763)$ observed in PSD helped to determine an I or L present at second position from the C-terminus. All the internal fragment ions shown

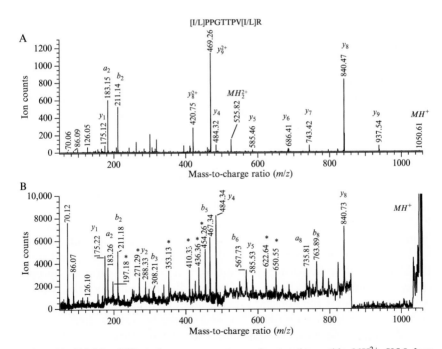

FIG. 13. (A) NanoESI tandem mass spectrum of a tryptic peptide MH_2^{2+} 525.8 from spot 13. (B) MALDI–PSD spectrum of the same peptide MH^+ 1050.6 from spot 13.

in PSD supported the sequence determination, and the final sequence was proposed as [I/L]PPGTTPV[I/L]R.

The peptide sequences from 15 major spots were determined by MAL-DI–PSD and/or nanoESI QSTAR tandem mass spectrometry. The number of sequences obtained for each spot ranged from 2 to 8, and the length of peptide sequences ranged from 5 to 18 residues, averaging ~10 amino acids. The detailed results are summarized in Huang *et al.* (2001).

Strategy for Identifying Unknown Proteins Using Novel Peptide Sequences Obtained by MS. Our strategy for identifying unknown proteins using novel peptide sequences obtained by nanoESI Qq-TOF tandem mass spectrometry and/or PSD is illustrated as a flow chart in Fig. 14, as described in Huang *et al.* (2001). To explain briefly, two approaches were proposed and carried out to identify possibly homologous sequences from databases, using individual sequences and catenated sequences, respectively. Two different searching tools were used for individual sequences: MS-Pattern and Gapped BLAST. MS-Pattern is a simple text string (keyword, accession number, or sequence) search program (University of California, San Francisco, CA) that is used to search for proteins containing similar peptide sequences with a defined number of mismatched amino acids in the static NCBInr nonredundant database. There is no limitation on the length of peptide sequences used for such a search. The peptide sequences with ambiguous amino acids, such as I or L, and the uncertainty of the order of first two amino acids at the N terminus can be submitted into the program at the same time. The number of maximum amino acid substitutions is arbitrary, varying with the length of sequences for searching. The higher the chosen number of substitutions, the more false-positive protein entries will be listed.

Individual sequence was also submitted for homology searches using Gapped BLAST (Washington University, version 2.0a), generating *n* single BLAST search results if *n* sequences were applied. To accommodate the ideology of the peptide sequences determined by MS, we have modified the scoring matrices BLOSUM62 and PAM30 into BLOSUM62MS and PAM30MS, accounting for Ile/Leu and Gln/Lys ambiguities generated during sequence determination by low-energy CID (Huang *et al.*, 2001).

Congruence analyses were carried out on both MS-Pattern and single BLAST search results using MS-Shotgun, a modified form of Shotgun (Pegg and Babbitt, 1999) originally designed to find all superfamily members with a distant relationship (<35% identity) using full-length protein query sequences. This approach is based on the hypothesis that examination of congruence across searches for a set of homologues from the same protein is useful for distinguishing real homologues from unrelated hits even though scores for individual database searches using each of these

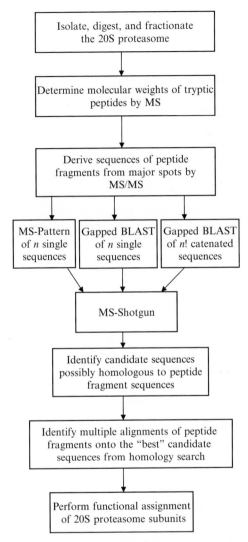

FIG. 14. Flow chart showing the strategy for identification of 20S proteasome from *T. brucei* using different homology searching tools.

peptide sequences may fall below the level of statistical significance. The MS-Shotgun score represented the number of query sequences finding the same homologous proteins. In general, it was found that short peptides do not generate hits with statistical significance (i.e., their probability values are close to one, which means the likelihood of finding the protein hits by chance is high). To amplify statistical significance of BLAST search results and find the correct order of the query sequences, the catenated peptide sequences with all possible permutations and catenations were used as query sequences for Gapped BLAST searches, generating $n!$ multiBLAST search results. Congruence analysis of the search results using MS-Shotgun was also carried out to obtain useful information for determining the true homologues of the unknown proteins. Finally, multiple alignments of peptide sequences and top-scoring protein hits were done to evaluate the protein functional assignment and correctness of its linear ordering of peptide sequences in the same spot. Therefore, we determined that even though all single sequences find the same protein, evaluation using multiple alignments of query sequences against protein hits will be necessary to judge the certainty of the unknown protein identity.

Functional Assignment of 20S Proteasome Subunits of T. brucei *Using Homology Searches.* Four novel peptide sequences in spot 7 were determined by *de novo* sequencing, as shown in Table V. We first used the MS-Pattern to search for any protein containing homologous sequences. For sequences Spot7a0, Spot7a1, and Spot7a3, the MS sequences were directly submitted to the program for searching. Due to the lack of fragmentation information, the first two amino acids of sequence Spot7a2 were not unambiguously assigned and were proposed as either [AT] or [DG]. Therefore, two lists of sequences were generated using AT and DG as N-terminal residues, respectively. The differences in these lists of sequences are substitutions of I and L, Q and K, as well as the different order of the first two amino acids. The MS-Pattern search results from the list of sequences using {AT} as the N terminus are shown in Fig. 15A, whereas the MS-Pattern search results when {DG} was at the N terminus are displayed in Fig. 15B. The top five hits out of several thousands of total hits generated from both lists of sequences (allowing five maximum mismatched amino acids) are proteasome homologues with the least number of mismatched amino acids. Comparing Fig. 15A and B, the best homologous match for sequence Spot7a2 is yeast proteasome C3 with only one mismatch amino acid when the query is ATDGVV-LAAEKK. In addition, the matched homologous sequences from almost all of the 20S proteasome hits have A or AT at their N termini. Thus, we used ATDGVV[I|L]AAE[Q|K]K for the following homology searches. Similarly, sequences Spot7a1 and Spot7a3 also found proteasome C3, with the least number of mismatched amino acids among the search results. Sequence Spot7a0 found 104 proteins containing sequences with a maximum of six mismatched AAs (data not shown); however, no proteasome proteins were found.

The congruence analysis of a single MS-Pattern search is illustrated in Fig. 16, indicating that three out of four sequences matched to 20S proteasome subunit C3, which suggests the subunit in spot 7 is a homologue of proteasome subunit C3. Then we used the single fragments again for Gapped BLAST search. The congruence analysis of the BLAST search results has also revealed that three out of four query sequences have found proteasome C3 from different species, as shown in Fig. 17. However, sequence Spot7a0 still did not find proteasome protein. The probability of finding the protein is close to one, which means that the likelihood of the protein being found by chance is very high. Using MS-Pattern and BLAST searches on individual fragments, proteasome proteins have been found along with other nonproteasome proteins. If one does not know if this unknown protein from spot7 should belong to the proteasome family, it would be hard to make a definite conclusion out of these search results from single fragments. To amplify the real hit among false positives, we used catenated fragments with all possible permutations for the Gapped BLAST search. The amplification can be achieved through finding the right catenation order of the sequences. Catenation of a total of 4 fragments in spot 7 with all possible permutations generates 24 catenated sequences for BLAST searching. Congruence analysis of the Gapped BLAST search results has shown that 24 out of 24 catenated sequences have found various 20S proteasome homologues. To illustrate the MS-Shotgun output more explicitly, we plotted the log of probability of proteins found versus the query sequences, in which can be seen in Fig. 18 (Huang *et al.*, 2001).

As shown in Fig. 18, another important way in which MS-Shotgun with catenation analysis improves the ability to identify function from sequence analysis is its ability to separate correct from incorrect scoring orders. The difference in *p* values obtained for correct and incorrect sort orders is illustrated graphically for spot 7 in Fig. 18; the score for the catenation representing the correct sort order, 1320, and variants of the effective sort order, 132, have the best *p* values and stand out well from the background scores representing other catenations. It is clear that the true protein hit has the lowest probability number, and the query sequence, by which the true hit was found, has the right order of the original sequences. Based on the results, proteasome C3 subunit stands out among other proteasome proteins because this hit has the highest statistical significance with highest MS-Shotgun score, meaning the likelihood of finding this protein by chance is the lowest among other searched hits. The top-scoring hits represent proteasome sequences with the *p* values for the highest scoring proteasome hit improving from statistically insignificant (~1) for all of the BLAST searches; these searches are on single peptide sequences to $2.5e^{-11}$ for the catenated analysis of the correctly ordered permutation, peptide order

MS-Pattern Search Results of Spot7a2A

Sample ID (comment): TBP SPOT7a2A
Database searched: nrStatic
Molecular weight search (1000 – 100,000 Da) selects 342,270 entries.
Full pI range: 359,481 entries.
Combined molecular weight and pI searches select 342,270 entries.
MS-Edman search selects 2196 entries (results displayed for top 200 matches).

Peptide masses:	Peptide	Peptide	Max # AA		List of
Monoisotopic	N terminus:	C terminus:	substitutions:		sequences
Cysteines	Hydrogen (H)	Free acid (OH)	5		ATDGVVLAAEQK
modification:				Search	ATDGVVLAAEQK
Acrylamide				type:	ATDGVVIAAEQK
				List of sequences	ATDGVVIAAEQK
					ATDGVVLAAEKK
					ATDGVVLAAEKK
					ATDGVVIAAEKK
					ATDGVVIAAEKK
					TADGVVLAAEQK
					TADGVVLAAEQK
					TADGVVIAAEQK
					TADGVVLAAEKK
					TADGVVLAAEKK
					TADGVVIAAEKK
					TADGVVIAAEKK

	Number of substitutions	Peptide sequence	Matching sequence	Protein MW (Da)/pI	Species	Accession #	Protein name
1	1	ATDGVVLAAEKK	(K) ATDGVVLATEKK(P)	26404.2/5.00	Schizosacchar omyces pombe	3790251	(AL031966) 20s proteasome component C3
2	2	ATDGVVLAAEKK	(L) AKDGIVLAAEKk(T)	28100.8/5.65	Dictyostelium discoideum	464458	(L22212) proteasome
3	2	ATDGVVIAAEKK	(K) ATNGVVIATEKK(S)	27162.0/5.52	Unreadable_pdb	130880	PDB_ENTRY_NAME_TO O_LONG
4	2	ATDGVVLAAEKK	(K) ATNGVVLATEKK(Q)	25834.6/6.34	Xenopus laevis	130852	(S51111) proteasome subunit XC3
5	2	ATDGVVLAAEKK	(V) AKDGIVLAAEKK(V)	27925.6/5.88	Schizosacchar omyces pombe	1172601	(Z50112) proteosome A-type subunit

MS-Pattern Search Results of Spot7a2A

Sample ID (comment): TBP SPOT7a2B
Database searched: nrStatic
Molecular weight search (1000–100,000 Da) selects 342,270 entries.
Full pI range: 359,481 entries.
Combined molecular weight and pI searches select 342,270 entries.
MS-Edman search selects 2181 entries (results displayed for top 200 matches).

Peptide masses: Monoisotopic	Cysteines modification: Acrylamide	Peptide N terminus: Hydrogen (H)	Peptide C terminus: Free acid (OH)	Max # AA substitutions: 5	Search type: List of sequences	List of sequences
						DGDGVVLAAEQK
						DGDGVVLAAEQK
						DGDGVVIAAEQK
						DGDGVVIAAEQK
						DGDGVVLAAEKK
						DGDGVVLAAEKK
						DGDGVVIAAEKK
						DGDGVVIAAEKK
						GDDGVVLAAEQK
						GDDGVVLAAEQK
						GDDGVVIAAEQK
						GDDGVVLAAEKK
						GDDGVVLAAEKK
						GDDGVVIAAEKK
						GDDGVVIAAEKK

	Number of substitutions	Peptide sequence	Matching sequence	Protein MW (Da)/pI	Species	Accession #	Protein name
1	3	DGDGVVLAAEKK	(L) AKDGIVLAAEKK(T)	28100.8/5.65	Dictyostelium discoideum	464458	(L22212) proteasome
2	3	GDDGVVLAAEKK	(L) AKDGIVLAAEKK(T)	28100.8/5.65	Dictyostelium discoideum	464458	(L22212) proteasome
3	3	GDDGVVLAAEKK	(R) GTDIVVLAVEKK(S)	27337.3/6.86	Arabidopsis thaliana	266839	proteasome:SUBUNIT = alpha
4	3	GDDGVVLAAEKK	(V) AKDGIVLAAEKK(V)	27925.6/5.88	Schizosaccharomyces pombe	1172601	(Z50112) proteosome A-type subunit
5	3	DGDGVVLAAEKK	(V) AKDGIVLAAEKK(V)	27925.6/5.88	Schizosaccharomyces pombe	1172601	(Z50112) proteosome A-type subunit

FIG. 15. (A) Top five hits of MS-Pattern search results using list of sequences of spot7a2A from spot 7. (B) Top five hits of MS-Pattern search results using list of sequences of spot7a2B from spot 7.

Single MS-pattern shotgun results of Spot 7

File name	Original sequence
spot7a0mm5.htm	TTSP[L|I]ADS[L|I]T[L|I]HK
spot7a1mm6.htm	SESSYG[L|I]TTFSPSGR
spot7a2mm5.htm	ATDGVV[L|I]AAE[Q|K]K
spot7a3mm5.htm	[L|I]V[Q|K][L|I]EYATTAASK

Ms-shotgun score	File name	Matching sequence	Protein MW (Da)/pI	Species	Accession #	Protein name
3	spot7a1mm6.htm spot7a2mm5.htm spot7a3mm5.htm	(D)SQYSFSLTTFSPSGK(L) (K)ASNGVVIATEKK(L) (K)LVQIEHALTAVGS(G)	24827.5/5.70	*Arabidopsis thaliana*	3152562	Comes from this gene.
3	spot7a1mm6.htm spot7a2mm5.htm spot7a3mm5.htm	(D)RGYSFSLTTFSPSGK(L) (K)ASNGVVLATEKK(Q) (K)LVQIEYALAAVAA(G)	25877.6/5.99	*Carassius auratus*	3107927	(AB013342) proteasome alpha 2 subunit
3	spot7a1mm6.htm spot7a2mm5.htm spot7a3mm5.htm	(D)SQYSFSLTTFSPSGK(L) (K)ASNGVVIATEKK(L) (K)LVQIEHALTAVGS(G)	25701.4/5.53	*Arabidopsis thaliana*	2511574	(AF043520) 20S proteasome subunit PAB1
3	spot7a1mm6.htm spot7a2mm5.htm spot7a3mm5.htm	(K)RGYSFSLTTFSPSGK(L) (K)AANGVVLATEKK(Q) (K)LVQIEYALAAVAG(G)	25925.8/8.38	*Mus musculus*	1709759	(X70303) proteasome, 25 kDa subunit
3	spot7a1mm6.htm spot7a2mm5.htm spot7a3mm5.htm	(T)ERYSFSLTTFSPSGK(L) (I)ASNGVVIATENK(H) (K)LVQLEYALAAVSG(G)	25906.6/6.21	*Drosophila melanogaster*	730372	(X70304) proteasome, 25 kDa subunit

FIG. 16. Top five hits of MS-Shotgun output from four single MS-Pattern search results of spot 7.

A

Query	BLAST Prob.	MS-Shotgun score	
spot7a1.save	0.850000	3	gi|3152562 - (AC002986) Similar to proteosome com...
spot7a1.save	0.870000	3	gnl|PID|e1179971 - (Y13176) multicatalytic endopeptidas...
spot7a1.save	0.900000	2	gi|2570505 - (AF022735) proteasome component [Ory...
spot7a1.save	0.960000	1	gnl|PID|e327987 - (Y12515) alpha proteasome [Entamoeba...
spot7a2.save	0.997000	3	gnl|PID|e1334470 - (AL031966) 20s proteasome component ...
spot7a1.save	0.999000	3	sp|P40301|PRC3_DROME - PROTEASOME 25 KD SUBUNIT MULTICATAL...
spot7a1.save	0.999000	3	gnl|PID|d1026855 - (AB013342) proteasome alpha 2 subuni...
spot7a1.save	0.999000	3	sp|P49722|PRC3_MOUSE - PROTEASOME COMPONENT C3 (MACROPAIN S...
spot7a1.save	0.999000	3	sp|P24495|PRC3_XENLA - PROTEASOME COMPONENT C3 (MACROPAIN S...
spot7a1.save	0.999000	3	sp|P17220|PRC3_RAT - PROTEASOME COMPONENT C3 (MACROPAIN S...

B

```
gi|3152562 - (AC002986) Similar to proteosome com...
  --Query file--------BLAST score--BLAST prob.--Shotgun score: 3
  spot7a1.save            54             0.85
  spot7a2.save            43             1.0
  spot7a3.save            36             1.0
gnl|PID|e1179971 - (Y13176) Multicatalytic endopeptidas
  --Query file--------BLAST score--BLAST prob.--Shotgun score: 3
  spot7a1.save            54             0.87
  spot7a2.save            43             1.0
  spot7a3.save            36             1.0
gnl|PID|e1334470 - (AL031966) 20s proteasome component
  --Query file--------BLAST score--BLAST prob.--Shotgun score: 3
  spot7a2.save            52             0.997
  spot7a1.save            44             1.0
  spot7a3.save            37             1.0
sp|P25787|PRC3_Human proteasome component C3 (Macropai ...
  --Query file--------BLAST score--BLAST prob.--Shotgun score: 3
  spot7a1.save            50             0.999
  spot7a2.save            42             1.0
  spot7a3.save            37             1.0
```

Fig. 17. (A) Top ten hits of MS-Shotgun output (sorted by p value) from four single BLAST search results of spot 7. (B) The detailed output of the top five hits listed.

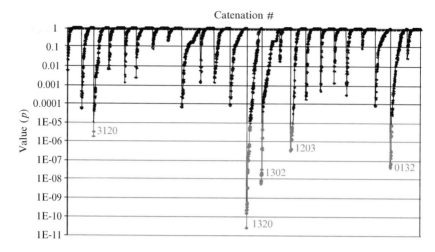

FIG. 18. Plot of the log of the p value (y-axis) for the hits found versus the 24 catenations (x-axis) analyzed for spot 7. Each peak represents a different sequence catenation. Each point on a peak represents the log of the p value for one particular protein hit that was found using each catenated sequence as the query sequence in the database search. Peaks representing the correct sort order, 1320, and its best scoring variants are labeled.

1320. In contrast, the best p value of a nonproteasome hit is six orders of magnitude larger (data not shown). Therefore, the results demonstrate that the unknown protein in spot 7 is a homologue of proteasome C3, which belongs to proteasome α subfamily. To further confirm the identity of this unknown protein, multiple alignment of sequences with correct catenation order (proteasome C3 subfamily from the best homologues and the another type α subunit) was employed for evaluation as described in Huang *et al.* (2001). As a result, all the fragments mapped onto the regions, which are only conserved among homologues of the proteasome C3 subfamily and not in other α subfamilies (data not shown). This finding further supports the identity of the unknown protein from spot 7.

The information provided by both the MS-Pattern and simple Gapped BLAST search results for identifying proteasome homologues using individual sequences was highly variable, which is dependent on whether the subset of peptide sequences that happen to be available include one or more regions that are highly conserved in a homologue in the database. Therefore, the MS-Pattern and simple BLAST searches were only marginally useful for identification of most spots although, in a few cases, highly conserved peptide sequences were available that matched well with homologues in the databases. The success of finding unknown protein homologues using Gapped BLAST search on catenated sequences is mainly dependent on the total number of amino acids contained in all of the sequences, how well these sequences conserved in the homologues, and how far apart these sequences are located in the structure. In general, the number of available peptide fragments and their lengths were not as important as a high degree of conservation of a peptide sequence fragment with a homologue in the database. In some cases, it is difficult to make a definite identification only based on the BLAST search results, even on catenated sequences; this is true simply because no statistically significant hit can be easily found compared to other hits, mainly resulting from the fact that the query sequences happened to be located in the unconserved region of their homologues. By using new informatics approaches developed in this work, we could identify the functions of proteasome proteins represented by 9 of the 15 major spots: spots 1, 2, 4, 6, 7, 11, 12, 15, and 17, as described in Huang *et al.* (2001). Of the remainder, functions for 3, 5, 8, 9, and 13 could not be predicted from sequence information alone, even with the aid of MS-Shotgun with catenation. However, provisional assignments of subunit function could be made using additional information obtained from keyword searches of the MS-Shotgun with catenation results. Because we know this experiment was designed to isolate only proteins from 20S proteasome complex, the appearance of known proteins in our search outputs helped to assign a subunit type for each spot. The summary of subunit assignment is listed in Table VI. As a result, 14 essential subunits for the *T. brucei* 20S proteasome complex were identified, including 7 α and 7 β subunits, the same composition as those of eukaryotic 20S proteasome complexes.

cDNA Sequencing of 14 Major Spots. To investigate the detailed structure and function relationship and its assembly mechanism, all 14 subunits were cloned, and full-length protein sequences were obtained from cDNA sequences. The gapped BLAST searches of the full-length sequences confirmed the protein identities determined by our homology analysis of MS sequences.

Part B2: Summary and Conclusions

We, the authors of this chapter's discussion, have developed a suite of bioinformatics homology searching tools to identify the unknown 20S proteasome subunits from *T. brucei*. The database search tools and protocols commonly used for assignment of function from full-length protein sequences are largely inadequate for dealing with sequence information generally obtained from MS/MS, even when close homologues exist in the databases (Huang *et al.*, 2001). By taking advantage of congruence across

TABLE VI
SUMMARY OF IDENTIFICATION OF THE 20S PROTEASOME SUBUNITS FROM *T. BRUCEI*

| 2-D gel spot# | Suggested subunit assignment from homology analysis of catenated sequences and full-length sequences | |
	Traditional name	Systematic name
1	ZETA	$\alpha 5$
2	C10-II	$\beta 3$
3	C8	$\alpha 7$
4	C2	$\alpha 6$
5	C9	$\alpha 3$
6	N3	$\beta 7$
7	C3	$\alpha 2$
8	C7-I	$\beta 4$
9	C5	$\beta 6$
11	C6-I	$\alpha 4$
12	IOTA	$\alpha 1$
13	Z	$\beta 2$
15	X	$\beta 5$
17	Y	$\beta 1$

database search results for all peptide sequences available for a single protein, we have developed an approach that substantially improves our ability to identify remote homologues (Huang *et al.*, 2001). By allowing all available sequence information from a single spot to be used in consonance, we were able to find matches with homologues at substantially improved statistical significance scores. This approach can be applied to other biological systems for unknown protein identifications. Development of automated homology searches can be useful for high-throughput unknown protein identification, especially when the full-length genomic sequence is not available.

The 20S proteasome in eukaryotic cells is known to consist of 14 distinctive subunits, which have been identified in 20S proteasome in *T. brucei*. However, the presence of 25 identifiable protein spots in the 2-D gel of the 20S proteasome purified from *T. brucei* (9) indicates the subunit composition heterogeneity because the other spots have been recently identified as different isoforms of the various subunits. This seems to be the commonly observed phenomenon since all purified 20S proteasomes from eukaryotic cells have been found to consist of more than 20 subunit proteins on 2-D gels (Elenich *et al.*, 1999; Huang *et al.*, 1999, 2000; Kristensen *et al.*, 1994). The 20S proteasomes from rat were found to have significant composition heterogeneity in the purified population (Huang *et al.*, 2000). The presence of various isoforms of some of the β subunits or even some of the α subunits that has constituted the composition heterogeneity of the 20S proteasomes may be required by the multifunctional proteolytic capability of the proteasome complex. The presence of three nonessential but interferon γ inducible β subunits, MECL-1, LMP2, and LMP7, in mammalian cells is known to be required for generating peptides for MHC class I antigen presentation (Akiyama *et al.*, 1994; Kloetzel, 2001; Stohwasser *et al.*, 1997; Stoltze *et al.*, 2000). These findings provide a good example for the necessity in having a heterogeneous population of the 20S proteasomes to fulfill different biological functions of the mammalian cell. Nevertheless, it remains to be seen whether a similar situation exists also in *T. brucei* to cope with the changing living environments throughout its life cycle.

Acknowledgments

    We thank Professor Burlingame's laboratory members for their help, especially Mr. David Maltby for his help in MS analysis. This work was supported by the National Institutes of Health Grants RR 01614 (to A.L.B.).

References

Ahn, J. Y., Tanahashi, N., Akiyama, K., Hisamatsu, H., Noda, C., Tanaka, K., Chung, C. H., Shibmara, N., Willy, P. J., Mott, J. D., *et al.* (1995). Primary structures of two homologous subunits of PA28, a gamma-interferon-inducible protein activator of the 20S proteasome. *FEBS Lett.* **366,** 37–42.

Akiyama, K., Kagawa, S., Tamura, T., Shimbara, N., Takashina, M., Kristensen, P., Hendil, K. B., Tanaka, K., and Ichihara, A. (1994). Replacement of proteasome subunits X and Y by LMP7 and LMP2 induced by interferon-gamma for acquirement of the functional diversity responsible for antigen processing. *FEBS Lett.* **343,** 85–88.

Biemann, K. (1990a). Appendix 5. Nomenclature for peptide fragment ions (positive ions). Academic Press, San Diego, CA.

Biemann, K. (1990b). Mass spectrometry. *Methods Enz.* **193,** 351–360.

Chang, W. W., Huang, L., Shen, M., Webster, C., Burlingame, A. L., and Roberts, J. K. (2000). Patterns of protein synthesis and tolerance of anoxia in root tips of maize seedlings acclimated to a low-oxygen environment, and identification of proteins by mass spectrometry. *Plant Physiol.* **122,** 295–318.

Ciechanover, A., and Schwartz, A. L. (1998). The ubiquitin-proteasome pathway: The complexity and myriad functions of proteins death. *Proc. Natl. Acad. Sci. USA* **95,** 2727–2730.

Clauser, K. R., Baker, P., and Burlingame, A. L. (1999). Role of accurate mass measurement (+/−10 ppm) in protein identification strategies employing MS or MS/MS and database searching. *Anal. Chem.* **71,** 2871–2882.

Coux, O., Tanaka, K., and Goldberg, A. L. (1996). Structure and functions of the 20S and 26S proteasomes. *Ann. Rev. Biochem.* **65,** 801–847.

Dubiel, W., Pratt, G., Ferrell, K., and Rechsteiner, M. (1992). Purification of an 11S regulator of the multicatalytic protease. *J. Biol. Chem.* **267**, 22369–22377.

Elenich, L. A., Nandi, D., Kent, A. E., McCluskey, T. S., Cruz, M., Iyer, M. N., Woodward, E. C., Conn, C. W., Ochoa, A. L., Ginsburg, D. B., and Monaco, J. J. (1999). The complete primary structure of mouse 20S proteasomes. *Immunogenetics* **49**, 835–842.

Feng, D. F., and Doolittle, R. F. (1987). Progressive sequence alignment as a prerequisite to correct phylogenetic trees. *J. Mol. Evol.* **25**, 351–360.

Fu, H., Doelling, J. H., Arendt, C. S., Hochstrasser, M., and Vierstra, R. D. (1998). Molecular organization of the 20S proteasome gene family from *Arabidopsis thaliana*. *Genetics* **149**, 677–692.

Grant, S., and Jarvis, W. D. (1996). Modulation of drug-induced apoptosis by interruption of the protein kinase C signal transduction pathway: A new therapeutic strategy. *Clin. Cancer Res.* **2**, 1915–1920.

Groettrup, M., Ruppert, T., Kuehn, L., Seeger, M., Standera, S., Koszinowski, U., and Kloetzel, P. M. (1995). The interferon-gamma-inducible 11 S regulator (PA28) and the LMP2/LMP7 subunits govern the peptide production by the 20S proteasome *in vitro*. *J. Biol. Chem.* **270**, 23808–23815.

Groll, M., Ditzel, L., Lowe, J., Stock, D., Bochtler, M., Bartunik, H. D., and Huber, R. (1997). Structure of 20S proteasome from yeast at 2.4-Å resolution. *Nature* **386**, 463–471.

Haass, C., and Kloetzel, P. M. (1989). The *Drosophila* proteasome undergoes changes in its subunit pattern during development. *Exp. Cell Res.* **180**, 243–252.

Hanks, S. K., Quinn, A. M., and Hunter, T. (1988). The protein kinase family: Conserved features and deduced phylogeny of the catalytic domains. *Science* **241**, 42–52.

Hochstrasser, M. (1996). Ubiquitin-dependent protein degradation. *Ann. Rev. Genet.* **30**, 405–439.

Hua, S., To, W. Y., Nguyen, T. T., Wong, M. L., and Wang, C. C. (1996). Purification and characterization of proteasomes from *Trypanosoma brucei*. *Mol. Biochem. Parasitol.* **78**, 33–46.

Huang, L., Jacob, R. J., Pegg, S. C., Baldwin, M. A., Wang, C. C., Burlingame, A. L., and Babbitt, P. C. (2001). Functional assignment of the 20S proteasome from *Trypanosoma brucei* using mass spectrometry and new bioinformatics approaches. *J. Biol. Chem.* **276**, 28327–28339.

Huang, L., Shen, M., Chernushevich, I., Burlingame, A. L., Wang, C. C., and Robertson, C. D. (1999). Identification and isolation of three proteasome subunits and their encoding genes from *Trypanosoma brucei*. *Mol. Biochem. Parasitol.* **102**, 211–223.

Huang, L., To, W. Y., Wang, C. C., and Burlingame, A. L. (1997). Characterization and identification of the 20S proteasome complex from rat liver using mass spectrometry. Proc. 45th ASMS Conf. Mass Spectrom. Allied Top, p. 283. Palm Springs, CA.

Huang, L., Wang, C. C., and Burlingame, A. L. (2000). Investigation of heterogeneous composition of the 20S proteasome complex from rat liver. *In* "Mass Spectrometry in Biology and Medicine" (A. L. Burlingame, S. A. Carr, and M. A. Baldwin, eds.), pp. 217–235. Humana Press, Totowa, NJ.

Keith, K., Hide, G., and Tait, A. (1990). Characterisation of protein kinase C-like activities in *Trypanosoma brucei*. *Mol. Biochem. Parasitol.* **43**, 107–116.

Kloetzel, P. M. (2001). Antigen processing by the proteasome. *Nat. Rev. Mol. Cell. Biol.* **2**, 179–187.

Knowlton, J. R., Johnston, S. C., Whitby, F. G., Realini, C., Zhang, Z., Rechsteiner, M., and Hill, C. P. (1997). Structure of the proteasome activator REGalpha (PA28alpha). *Nature* **390**, 639–643.

Kristensen, P., Johnsen, A. H., Uerkvitz, W., Tanaka, K., and Hendil, K. B. (1994). Human proteasome subunits from 2-dimensional gels identified by partial sequencing. *Biochem. Biophys. Res. Commun.* **205,** 1785–1789.

Link, A. J., Eng, J., Schieltz, D. M., Carmack, E., Mize, G. J., Morris, D. R., Garvik, B. M., and Yates, J. R., 3rd. (1999). Direct analysis of protein complexes using mass spectrometry. *Nat. Biotechnol.* **17,** 676–682.

Lowe, J., Stock, D., Jap, B., Zwickl, P., Baumeister, W., and Huber, R. (1995). Crystal structure of the 20S proteasome from the archaeon *T. acidophilum* at 3.4Å resolution. *Science* **268,** 533–539.

Lupas, A., and Baumeister, W. (1998). Chapter 5: The 20S proteasome. Plenum Press, New York.

Ma, C. P., Slaughter, C. A., and DeMartino, G. N. (1992). Identification, purification, and characterization of a protein activator (PA28) of the 20S proteasome (macropain). *J. Biol. Chem.* **267,** 10515–10523.

Mason, G. G., Hendil, K. B., and Rivett, A. J. (1996). Phosphorylation of proteasomes in mammalian cells. Identification of two phosphorylated subunits and the effect of phosphorylation on activity. *Eur. J. Biochem.* **238,** 453–462.

Medzihradszky, K. F., and Burlingame, A. L. (1994). The advantages and versatility of a high energy collision-induced dissociation-based strategy for the sequence and structural determination of proteins. *Methods* **6,** 284–303.

Mott, J. D., Pramanik, B. C., Moomaw, C. R., Afendis, S. J., DeMartino, G. N., and Slaughter, C. A. (1994). PA28, an activator of the 20S proteasome, is composed of two nonidentical but homologous subunits. *J. Biol. Chem.* **269,** 31466–31471.

Nandi, D., Jiang, H., and Monaco, J. J. (1996). Identification of MECL-1 (LMP-10) as the third IFN-gamma-inducible proteasome subunit. *J. Immunol.* **156,** 2361–2364.

Ni, R., Tomita, Y., Tokunaga, F., Liang, T. J., Noda, C., Ichihara, A., and Tanaka, K. (1995). Molecular cloning of two types of cDNA encoding subunit RC6-I of rat proteasomes. *Biochim. Biophys. Acta* **1264,** 45–52.

Pegg, S. C., and Babbitt, P. C. (1999). Shotgun: Getting more from sequence similarity searches. *Bioinformatics* **15,** 729–740.

Peters, J. M. (1994). Proteasomes: Protein degradation machines of the cell. *Trends Biochem. Sci.* **19,** 377–382.

Rechsteiner, M., Hoffman, L., and Dubiel, W. (1993). The multicatalytic and 26 S proteases. *J. Biol. Chem.* **268,** 6065–6068.

Rivett, A. J. (1998). Intracellular distribution of proteasomes. *Curr. Opin. Immunol.* **10,** 110–114.

Rivett, A. J., and Sweeney, S. T. (1991). Properties of subunits of the multicatalytic proteinase complex revealed by the use of subunit-specific antibodies. *Biochem. J.* **278**(Pt. 1), 171–177.

Rubin, D. M., and Finley, D. (1995). Proteolysis. The proteasome: A protein-degrading organelle? *Curr. Biol.* **5,** 854–858.

Schmidtke, G., Kraft, R., Kostka, S., Henklein, P., Frommel, C., Lowe, J., Huber, R., Kloetzel, P. M., and Schmidt, M. (1996). Analysis of mammalian 20S proteasome biogenesis: The maturation of beta-subunits is an ordered two-step mechanism involving autocatalysis. *EMBO J.* **15,** 6887–6898.

Shevchenko, A., Wilm, M., Vorm, O., and Mann, M. (1996). Mass spectrometric sequencing of proteins silver-stained polyacrylamide gels. *Anal. Chem.* **68,** 850–858.

Sogin, M. L., Gunderson, J. H., Elwood, H. J., Alonso, R. A., and Peattie, D. A. (1989). Phylogenetic meaning of the kingdom concept: An unusual ribosomal RNA from *Giardia lamblia. Science* **243,** 75–77.

Stohwasser, R., Standera, S., Peters, I., Kloetzel, P. M., and Groettrup, M. (1997). Molecular cloning of the mouse proteasome subunits MC14 and MECL-1: Reciprocally regulated tissue expression of interferon-gamma-modulated proteasome subunits. *Eur. J. Immunol.* **27,** 1182–1187.

Stoltze, L., Nussbaum, A. K., Sijts, A., Emmerich, N. P., Kloetzel, P. M., and Schild, H. (2000). The function of the proteasome system in MHC class I antigen processing. *Immunol. Today* **21,** 317–319.

Tanaka, K. (1995). Molecular biology of proteasomes. *Mol. Biol. Rep.* **21,** 21–26.

Thompson, J. D., Higgins, D. G., and Gibson, T. J. (1994). CLUSTAL W: Improving the sensitivity of progressive multiple sequence alignment through sequence weighting, position-specific gap penalties and weight matrix choice. *Nucl. Acids Res.* **22,** 4673–4680.

Thomson, S., and Rivett, A. J. (1996). Processing of N3, a mammalian proteasome beta-type subunit. *Biochem. J.* **315**(Pt. 3), 733–738.

To, W. Y., and Wang, C. C. (1997). Identification and characterization of an activated 20S proteasome in *Trypanosoma brucei. FEBS Lett.* **404,** 253–262.

Verma, R., Chen, S., Feldman, R., Schieltz, D., Yates, J., Dohmen, J., and Deshaies, R. J. (2000). Proteasomal proteomics: Identification of nucleotide-sensitive proteasome-interacting proteins by mass spectrometric analysis of affinity-purified proteasomes. *Mol. Biol. Cell* **11,** 3425–3439.

Voges, D., Zwickl, P., and Baumeister, W. (1999). The 26S proteasome: A molecular machine designed for controlled proteolysis. *Annu. Rev. Biochem.* **68,** 1015–1068.

Washburn, M. P., Wolters, D., and Yates, J. R., 3rd. (2001). Large-scale analysis of the yeast proteasome by multidimensional protein identification technology. *Nat. Biotechnol.* **19,** 242–247.

Yao, Y., Huang, L., Krutchinsky, A., Wong, M. L., Standing, K. G., Burlingame, A. L., and Wang, C. C. (1999). Structural and functional characterizations of the proteasome-activating protein PA26 from *Trypanosoma brucei. J. Biol. Chem.* **274,** 33921–33930.

Yuan, X., Miller, M., and Belote, J. M. (1996). Duplicated proteasome subunit genes in *Drosophila melanogaster* encoding testes-specific isoforms. *Genetics* **144,** 147–157.

Zhang, Z., Krutchinsky, A., Endicott, S., Realini, C., Rechsteiner, M., and Standing, K. G. (1999). Proteasome activator 11S REG or PA28: Recombinant REG alpha/REG beta hetero-oligomers are heptamers. *Biochemistry* **38,** 5651–5658.

Zhang, Z., Realini, C., Clawson, A., Endicott, S., and Rechsteiner, M. (1998). Proteasome activation by REG molecules lacking homolog-specific inserts. *J. Biol. Chem.* **273,** 9501–9509.

[10] The Analysis of Multiprotein Complexes: The Yeast and the Human Spliceosome as Case Studies

By GITTE NEUBAUER

Abstract

The yeast and human spliceosomes represent the first two multiprotein complexes of which protein components were identified solely by mass spectrometry (MS). In this chapter, the different approaches used for the purification of these protein complexes, the MS analysis of the components, and some functional characterization strategies adopted are discussed. Even though from the time of analysis up to 2005 much has been

METHODS IN ENZYMOLOGY, VOL. 405
Copyright 2005, Elsevier Inc. All rights reserved.
0076-6879/05 $35.00
DOI: 10.1016/S0076-6879(05)05010-X

achieved in terms of purification techniques, MS protein analysis and sequence information in public databases, the key knowledge gained from the very early complex analyses still hold true today. The analysis of protein complexes is a powerful approach for understanding the organization of proteins and how they act in units to exert their biological effects. The analysis also creates hypotheses for the role of novel proteins in the context of the cellular function of the protein complex under study. However, the work on the spliceosomes described in this chapter also illustrates the relative ease of protein identification by MS and the difficulty to provide detailed functional information for the vast amount of data generated in such a study.

Introduction

Strategies for the Analysis of Multiprotein Complexes by Mass Spectrometry

The genomes of a large number of organisms have been sequenced in recent years, including the human genome announced in summer 2000 (McPherson *et al.*, 2001; Venter *et al.*, 2001). Even though a vast amount of DNA sequence data information is now available, little is known about the molecular function of the gene products, the proteins, in their often very diverse roles. One way to define more closely the functional context in which proteins act is by the analysis of whole protein assemblies. Most, if not all, proteins do not function in isolation but in close contact with other proteins to exert their biological effects. If a functionally uncharacterized protein is found to be part of a larger protein complex, its putative function will be associated with the function of the whole protein complex, allowing directed investigations to characterize its specific role in this cellular process.

The amount of genome sequence data available, the sensitive analysis of proteins by mass spectrometry (MS), and the availability of search tools for the correlation of MS data to sequences in the databases now allow the rapid and sensitive characterization of multiprotein complexes. To characterize a protein assembly, specific biochemical purification is needed to isolate the complex of interest from, for example, a whole cell extract. Once this is achieved, the protein components are separated by one- or two-dimensional (1- or 2-D) gel electrophoresis, and each of the resulting protein bands or spots are excised and subjected to in-gel digestion with a protease. The resulting peptides are then extracted and subjected to MS and bioinformatic analysis.

In many cases, a peptide mass fingerprint of a given protein acquired with high mass accuracy by MALDI–TOF is sufficient to identify the protein in the database (Jensen *et al.*, 1996; Shevchenko *et al.*, 1996a). If the peptide mass fingerprint does not lead to an unambiguous identification, partial sequence information of the individual peptides has to be generated by tandem mass spectrometric experiments. For cases in which the protein complex was isolated from an organism with a completely sequenced genome (as was the case for the Ul and the [U4/U6.U5] snRNPs from yeast), the MS data will always yield an identification of either a protein or an open reading frame (ORF), given sufficient amount of material. In many cases, the identified protein is already well characterized, and, together with knowledge of the function of the isolated complex, the function of the identified protein is clear and requires no further analysis. For cases in which an ORF (which has not previously been characterized) is identified, a detailed bioinformatic sequence analysis with the complete sequence of the ORF can be performed. The results can lead to a proposal of the function or at least help to design biochemical or genetic functional experiments (Fig. 1). In the case of the human spliceosome, the protein complex was isolated from an organism, and the genome was not fully sequenced at the time of analysis; however, a substantial part (more than 50% [Mann, 1996]) of the expressed genes were represented in public expressed sequence tag (EST) databases. Therefore, in the case of a human protein complex, roughly half of the proteins were found in a protein sequence database, the other half could be identified as ESTs, and some were not found in any of the databases. For the cases in which the protein could not be identified in a protein sequence database, cloning of the full-length gene via an EST had to be performed to further functionally characterize the protein. In the case of an EST "hit," the full insert sequence of the clone and/or assembled ESTs mapping to the same protein may provide valuable homologies to proteins of the same or other organisms with a functional relationship and can help to elucidate the domain structure of the novel protein.

The Spliceosome

The spliceosome is a multiprotein complex consisting of three preformed protein–RNA subunits (Ul, U2, and [U4/U6.U5] snRNPs [small nuclear ribonucleoprotein particles]) and a number of auxiliary proteins that are not stably bound to these subcomplexes. The spliceosome catalyzes the excision of introns (noncoding sequences) and the rejoining of exons (coding sequences) of pre-mRNA to form mature mRNA molecules. The splicing process is a two-step mechanism, requiring the concerted function of 50 to 100 proteins that dynamically assemble and disassemble

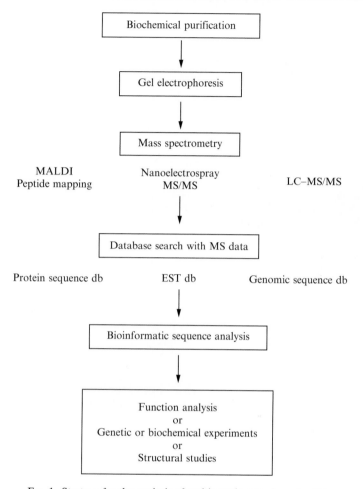

Fig. 1. Strategy for the analysis of multiprotein complexes by MS.

on the pre-mRNA substrate (Fig, 2) to fulfill their task (Kramer, 1996; Will *et al.*, 1995).

For the analysis presented in this chapter, MS was used alone for the first time to characterize the protein components of the *Saccharomyces cerevisiae* U1 and [U4/U6.U5] snRNPs (Gottschalk *et al.*, 1998, 1999; Neubauer *et al.*, 1997) as well as the total human spliceosome (Neubauer *et al.*, 1998). These three complexes act as case studies for a general strategy on the analysis of protein complexes. The approach consists of four different steps: the protein complex purification, the MS analysis and

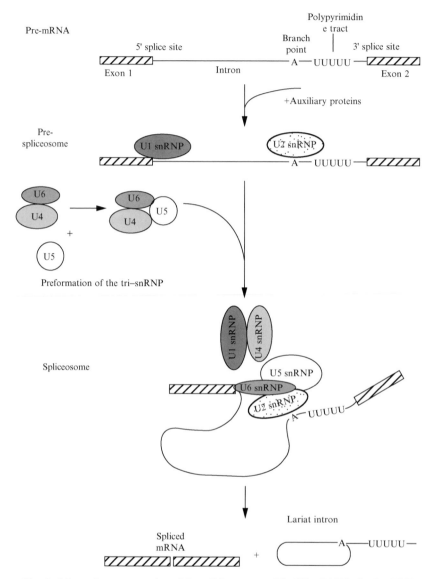

FIG. 2. Schematic representation of the splicing process. The U1 snRNP is the first RNP to contact the pre-mRNA, followed by the recruitment of U2 snRNP to the branch point of the intron to form the prespliceosome. The association of the preformed [U4/U6.U5] tri-snRNP with the prespliceosome leads to the formation of the mature spliceosome. After structural rearrangements take place, the two-step splicing mechanism occurs; finally, the spliced mRNA and the lariat intron are released, and the spliceosome disassembles.

protein identification, the bioinformatic sequence analysis, and, finally, the functional characterization of the identified components. In this chapter, the analyses of the yeast spliceosomal subunits and of the human spliceosome are described and contrasted.

Purification of Protein Complexes or Interaction Partners

The specific purification of a protein assembly is a prerequisite for the successful analysis of its components. Some knowledge of at least one of its components is key to a specific purification of the protein complex. Antibodies against this component can be used to affinity purify the protein and its interacting partners. Alternatively, fusion proteins can be constructed that consist of the target protein and a tag. Here, the tag provides the specific properties that are the basis for the purification. In this case, one can take advantage of strong interactions of peptide or protein tags with their ligands, as, for example, in glutathione-S-transferase (GST)-glutathione or epitope-antibody interactions. In addition, other properties of the protein complex, like its size, can be used for a further purification step (e.g., in a gel filtration or centrifugation step). In many cases, a combination of different methods is used to increase specificity.

In the analysis of the yeast spliceosome subunit UI snRNP and the yeast [U4/U6.U5] tri-snRNP particle, two-step purification procedures were used. In both purifications, the yeast strains expressed a known component of the respective snRNP particle that contained a N-terminal histidine tag. In the first step, antibodies against the m3G-capped RNA molecule, which is a common component of all snRNPs, were used to enrich for snRNPs. To specifically isolate the snRNP of interest, nickel-nitrilotriacetic acid (Ni-NTA) chromatography was employed to purify the histidine-tagged protein and its interaction partners. For the purification of the UI snRNP, the yeast strain constructed for purification expressed the histidine-tagged UI-70k proteins; in the case of the [U4/U6.U5] tri-snRNP isolation, the prp4, a U4/U6-specific component, was histidine tagged (Gottschalk et al., 1999). Since U5 snRNP proteins were identified in the UI snRNP preparation, the [U4/U6.U5] tri-snRNP was subjected to an additional gradient centrifugation step before MS analysis of the protein components of the particle. Nonetheless, in both snRNP isolations, some of the proteins identified in the subsequent MS analyses were found to be part of other snRNP subunits (see following paragraphs).

Purification of the Human Spliceosome

Because homologous recombination is a very efficient process in yeast, tagging a target gene and constructing a strain expressing the tagged protein are very often the steps of choice for the isolation of complexes that are assembling with the target protein. In human cell lines, however, a different approach has to be used since homologous recombination is not as efficient. Therefore, either overexpression of a tagged protein introduced by a vector or the use of other biological handles are more common. In the case of the purification of the human spliceosome from HeLa cells, a radiolabeled, biotinylated pre-mRNA substrate was used as the "bait" on which the spliceosome could assemble and which was used for subsequent purification. The purification of the human spliceosome was performed in the laboratory of A. I. Lamond according to the protocol of Bennett (1992). To briefly explain, nuclear extracts were incubated with a nonspliceable pre-mRNA substrate that was both biotinylated and radiolabeled. In the first purification step, the proteins assembled on the pre-mRNA substrate were fractionated by gel filtration and separated from other RNA–protein complexes. The spliceosomal complexes were then further affinity purified by binding of the biotinylated pre-mRNA to avidin-agarose beads (Bennett *et al.*, 1992). After the RNA was degraded and the complex denatured, the protein components were separated by 2-D electrophoresis and stained with Coomassie.

The Tandem Affinity Purification Method

A novel purification procedure has been developed and termed tandem affinity purification (TAP) (Rigaut *et al.*, 1999). This procedure provides a widely applicable and fast method for protein complex purification with high yield of protein interaction partners under standard purification conditions. The development of this method was prompted by the availability of rapid and sensitive methods for protein identifications by MS and database searching. While the bottleneck for the characterization of multiprotein complexes previously existed in the determination of the full protein sequences of its components, it has become clear that the biochemical purification of the complex is the limiting step. Generic methods were not available, and the purification of low-abundance protein complexes proved to be time-consuming and difficult to achieve. The TAP procedure provides such a generic method for protein complex purification. In this method, the target gene is tagged with the TAP tag, consisting of two IgG binding domains of protein A and a calmodulin binding peptide (CBP) separated by a tobacco etch virus (TEV) protease cleavage site. The first purification step is the binding of the protein A tag to an IgG column. The elution is done by TEV protease cleavage, resulting in the release of the tagged protein and its interaction partners. The TEV protease cleavage allows for very mild elution conditions of the protein complex. In the second purification step, CBP interacts with calmodulin bound to a column.

The removal of calcium from the buffer by ethylene-bis tetraacetic acid (EGTA) allows again a mild elution step of the target protein and its interaction partners. The whole purification procedure starting from a cell lysate takes approximately one day; because of its general applicability, is amenable to automation.

Because homologous recombination in yeast is extremely efficient, the TAP tag can be directly integrated into the genome, providing the expression of the tagged protein under its natural promoter and therefore avoiding overexpression. Using the TAP methodology, the yeast U1 snRNP complex could be purified with only 4 L of yeast cells (Rigaut *et al.*, 1999) instead of the 16 L used in the original procedure (Neubauer *et al.*, 1997).

MS Analysis and Database Searching

Sample Preparation for MS

Individual protein bands are excised from the gel; after extensive washing and reduction/alkylation of the cysteine residues, the proteins are in-gel digested with trypsin as previously described (Shevchenko *et al*, 1996b). Particular care has to be taken to avoid keratin contamination from human skin, woolen cloth, and dust in the laboratory before and during digestion, especially when processing low amounts of protein. Filtering all solutions for preparing the gel, using fresh solvents for the washing and digestion procedure, wearing gloves, and visually inspecting for fibers in solvents and tubes are measures that should be taken to avoid keratin contamination. A clean room with little traffic or a laminar flow hood is also useful to avoid contamination. After digestion, the resulting peptides are extracted in several steps, and the peptide mixture is dried down in a vacuum centrifuge (Shevchenko *et al*, 1996).

Protein Identification by Matrix-Assisted Laser Desorption/Ionization Peptide Mapping

For the analysis of the peptide mixture by matrix-assisted laser desorption/ionization time-of-flight (MALDI–TOF) mass spectrometry, approximately 5% of the supernatant of the digest (before extraction) is sufficient. The peptides can be applied to the MALDI target by mixing with the organic acid matrix and cocrystallization of matrix and analyte on the target (dried droplet method). Alternatively, the peptide mixture can be deposited onto the target after a thin layer of matrix has been dried on it (thin film method). The most common matrix used for the analysis of peptide mixtures by MALDI–UV is a cyanocinnamic acid.

In the analysis of the spliceosomal subunits, the thin film method (Vorm *et al.*, 1994) was applied, with modifications. A saturated solution of α-cyanocinnamic acid in acetone was mixed 4:1 with a 10g/L solution of nitrocellulose in acetone. An amount of 0.5 μL of this solution was deposited on the MALDI target plate. All measurements were carried out on a Bruker REFLEX TOF mass spectrometer. Internal two-point calibration was carried out on matrix and/or trypsin autolysis products of the acquired spectra, leading to a mass accuracy of 30 to 70 ppm. PeptideSearch (Mann *et al.*, 1994), a program that correlates MS data with protein sequence information in databases, was used to search the resulting peptide masses against a nonredundant protein sequence database (NRDB).

MALDI peptide mapping is particularly well suited for the identification of yeast proteins because the genome is fully sequenced and relatively small and because the method can provide for a relatively high-throughput analysis. However, in the analysis of the U1 snRNP and the [U4/U6.U5] tri-snRNP complexes, MALDI peptide mapping was of limited use because many of the bands contained protein mixtures, and small or modified proteins could not be identified with certainty (Figs. 3 and 4). In particular, in the analysis of the [U4/U6.U5] tri-snRNP, almost all bands contained protein mixtures of up to four proteins (see following paragraphs and Table IB). Even though in all cases the most abundant protein was clearly identified, lower abundant proteins could not always be identified with certainty. Likewise, proteins with low molecular weight (below 25 kDa), in particular the Sm proteins of the snRNPs, could not be identified by MALDI because tryptic cleavage yielded less than three peptides in the mass range accessible for accurate mass measurement by MALDI–TOF, too little information for a reliable identification (Table I and Fig. 3).

Nanoelectrospray Tandem MS

For nanoelectrospray MS, a desalting step after digestion of the protein band is necessary. To minimize sample handling and therefore sample loss, a miniaturized capillary column was used for this step, which allowed elution of the peptide mixture directly into the spraying needle of the nanoelectrospray source (Wilm and Mann, 1996). The sorbent used was POROS R2 material, which was filled into the capillary as a slurry in methanol to a final bed volume of approximately 100 nL. The column was then equilibrated with 5% methanol/5% formic acid. The dried-down peptides were dissolved in the same solvent and applied to the column. After washing three times with 3 μL of 5% methanol/5% formic acid, the peptide mixture was eluted in a single step with 50% methanol/5% formic

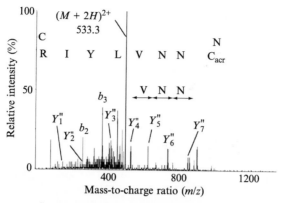

Sequence retrieved from NRDB (gene P9705.4):

MSESSDISAMQPVNPKPFLKGLVNHRVGVKLKFNSTEYRGTLVSTDNYFNLQLNEAEEFVAGVSHGTLGEIFIRCNNVLYIRELPN

FIG. 3. Identification of SmF, a modified protein, by nanoelectrospray MS/MS. The product ion spectrum of the tryptic peptide, which was used for identification in the database, contained an acrylamidated cysteine residue at its C-terminus. The only other peptide that could be detected and sequenced was the N-terminal one, which also was modified: instead of the leading methionine, the first amino acid was found to be an acetylated serine residue. The identification by peptide mass fingerprinting would have been impossible.

acid, directly into the spraying needle of the nanoelectrospray ion source. To minimize the loss of small, hydrophilic peptides, a double column arrangement for the loading and washing steps were used. In this case, the second column contained oligoR3 POROS material, a more hydrophobic resin, aligned with the R2 column. The peptides not bound on the R2 column were then captured by the oligoR3 column. The peptides from each column were individually eluted into the nanoelectrospray needle and analyzed separately by nanoelectrospray tandem mass spectrometry (MS/MS) (Neubauer *et al.*, 1999).

In the case of the spliceosomal proteins, all MS/MS analyses were carried out on an API III triple quadrupole instrument equipped with a nanoelectrospray ion source. In a first experiment, a Ql scan was acquired to identify the peptide ions. In cases where the signal-to-noise ratio was particularly poor due to limited peptide material present, a precursor ion scan for the immonium ion of Ile/Leu (*m/z* 86) was also acquired, which aided the detection of the peptide ions. In the next experiment, the detected peptides were individually selected by the first quadrupole (Ql) and fragmented by the collision with argon in the collision cell (Q2). The resulting fragments were separated by the third quadrupole (Q3) and

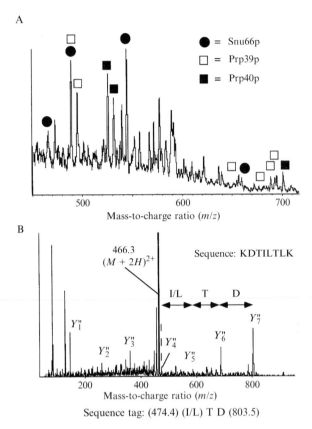

FIG. 4. Analyis of [U4/U6.U5] tri-snRNP proteins by nanoelectrospray MS present in a mixture. (A) The QI spectrum of the peptides produced by digestion of protein band 8 of the [U4/U6.U5] tri-snRNP preparation. Of the marked peptide ions, product ion spectra were acquired and partial sequences determined. Database searches using the sequence tag approach identified three different proteins (see also Table IA). (B) The product ion spectrum of one peptide ion. The sequence tag (474.4) (I/L) T D (803.5) was used for a database search, and Snu66p could be identified.

detected. The collision conditions for MS/MS experiments were chosen to favor dominant Y''-ion formation in the high mass-to-charge ratio (m/z) area of the spectra. This allowed the relative easy determination of a partial Y''-ion series and the construction of a peptide sequence tag (Mann *et al.*, 1994), which also contains—apart from the partial sequence of typically two to four residues—the starting and ending masses of the

TABLE IA
PROTEINS IDENTIFIED IN THE U1SNRNP PREPARATION. IN BOLD:
PROTEINS FIRST IDENTIFIED AS YEAST U1SNRNP PROTEINS IN THIS STUDY

Band #	app. MW	U1 snRNP proteins	U5 snRNP proteins/ contaminants	Bioinformatic sequence analysis comments
1	216		Prp8p	U5 snRNP protein
2	115		Snu114p	U5 snRNP protein
3	77	**Snu71p**		**19 Rs-, RD-, or RE dipeptides, putative nuclear localization signal, highly acidic**
4	69	Prp39P/ Prp40p		Yeast-specific U1-snRNP proteins
5	57	**Nam8p**		**3 RNA binding motifs (RBDs)**
6	55	**Np13p**		**Similar to human hRNP proteins, 2RBDs, one RGG**
7	52	**Snu56p**		**Rich in Ser and Asn, no known functional motifs**
8	42		Aar2	U5 snRNP protein
9	38		Glyceraldehyde dehydrogenase2	contaminant
10	37	Mud1p (U1-A)		Homologue of human U1-A
11	36		Glyceraldehyde dehydrogenase1	contaminant
12	34	Snp1p (U1-70k)		Homologue of human U1-70k
13	32	**U1snRNP-C**		**Homologue of human U1-C**
14	31	U1snRNP-C		Homologue of human U1-C
15	18	SmD1		Homologue of human SmD1
16	15	**SmD2**		**Homologue of human SmD2**
17	12	**SmE**		**Homologue of human SmE**
18	11	SmF		Homologue of human SmF
19	10	SmD3		Homologue of human SmD3
20	9	SmG		Homologue of human SmG

TABLE IB
PROTEINS IDENTIFIED IN THE [U4/U6.U5] TRI-ANRNP PREPARATION.
IN BOLD: NOVEL YEAST TRI-SNRNP PROTEINS

Band #	app. MW	[U4/U6.U5] tri-snRNP proteins	U1 and U2 snRNP proteins	Bioinformatic sequence analysis
1	214	Prp8p		Homologue of human U5 220kDa protein
2	197	Snu246p		Homologue of human U5 200kDa protein; DEXH RNA-unwindase
3	159		Rse1p	U2 snRNP protein
4	115	Snu114p		Homologue of human U5116kDa protein; GTPase domain, EF2-like
5	92	Prp6p		Contains TPR repeats
6	80	**Snu66p (YOR308c)**		-
7	78		Snu71p	U1 snRNP protein
8	69	Snu66p (YOR308c)	Prp39p/Prp40p	U1 snRNP proteins
9	66	Snu66p		
10	56	Prp31p		-
11	54	Prp3p		Homologue of human [U4/U6] 90 kDa protein (hPrp3p)
12	53	Prp4p		WD-repeats, homologue of human [U4/U6] 60 kDa protein (hPrp4p)
13	50		Snu56p	U1 snRNP protein
14	48	Spp381p		Serine-rich, acidic; PEST proteolysis motif
15	42		EF1α	Elongation factor, contaminant,
16	38		U1-A, U1-70k	U1 snRNP proteins
17	37		U1-A	U1 snRNP protein
18	32		Exm2p	U1 snRNP protein
19	31		U1-C	U1 snRNP protein
20	28	Prp38p		Serine-rich, acidic
21	27	**Snu23p (YDL098c)**		**C$_2$H$_2$-Zinc-Finger**
22	26.5	Sm B	Hsh49p	Hsh49p: homologue of human SAP49 (U2 snRNP protein)

(*continued*)

TABLE IB (*continued*)

Band #	app. MW	[U4/U6.U5] tri-snRNP proteins	U1 and U2 snRNP proteins	Bioinformatic sequence analysis
23	25.5	Lsm4p (Uss1p)		U6-specific, Sm-like domains
24	24	SmB, Lsm4p		
25	22	SmB, Lsm4p		
26	21	SmB, Lsm4p		
27	17.5	**Dib1p (YPR082c)** SmD1		**Dib1p: homologue of human U5 15 kDa protein**
28	16.8	**Lsm7p (YNL147w),** SmE		**Lsm7p: U6 specific, Sm-like domains**
29	14	SmD2		
30	12	SmE, Lsm8p (YJR022w), **Snu13p (YEL026w)**, SmF		Lsm8p: U6 specific, Sm-like domains; **Snu13p: homologue of human [U4/U6.U5] 15.5 kDa pro.**
31	10.6	SmF, SmD3		
32	10	SmD3, **Lsm2p (YBL026w)**		**Lsm2p: U6 specific, Sm-like domains**
33	9.1	**Lsm5p (YER146w)**, SmD3		**Lsm5p: U6 specific, Sm-like domains**
34	8.5	Lsm5p (YER146w), SmG, SmD3		
35	8	Lsm6p (YDR378c), SmG		Lsm6p: U6 specific, Sm-like domains

sequence stretch as well as the intact molecular weight of the peptide. This information was used to search the nonredundant protein database. One single peptide is usually sufficient to identify the protein unambiguously in the database, provided the peptide sequence is unique. Since more than one peptide per protein is typically fragmented, redundant and independent information is generated, which confirms the protein's identity. Mixtures of proteins can be determined in this way because more than one peptide is typically fragmented during an experiment (Fig. 4). Small proteins, which yielded too few peptides after proteolysis for MALDI peptide mapping, could be identified this way. This is exemplified in the case of the identification of SmF, one of the core proteins of the spliceosomal snRNPs, (Fig. 3) first identified in the analysis of the U1 snRNP. After proteolytic digestion of this 11-kDa protein, only two peptides were observed in the Ql mass spectrum, both of them modified: the N-terminal peptide of the protein lacked the leading methionine and instead contained an acetylated N-terminal serin, and the second peptide retrieved contained an acrylamidated cystein, a modification that often occurs during gel electrophoresis. This example also demonstrates the power of MS/MS for the determination of peptide modifications.

Results of the MS Analysis of the Yeast Spliceosomal Subunits

U1 snRNP

In the case of the U1 snRNP, 20 different Coomassie-stained protein bands were analyzed, and 21 different proteins could be identified in the database. All of the known yeast U1-specific proteins were identified, namely U1–70k, U1-A, the previously unknown yeast homologue to U1-C, and the yeast-specific proteins Prp39p and Prp40p (Kao *et al.*, 1996; Liao *et al.*, 1993; Lockhart *et al.*, 1994; Smith *et al.*, 1991). In addition, four novel previously unidentified proteins were found to be U1 snRNP specific, namely Snu71p, Nam8p, Np13p, and Snu56p (Gottschalk *et al.*, 1998). Two of these proteins (Snu71p and Snu56p) were identified as proteins of unknown function, and Nam8 and Np13p were previously implicated in mRNA metabolism but were not found to be associated with U1 snRNP. The low molecular weight protein bands ranging in apparent molecular weight from 9 to 18 kDa (Table IA) were all found to be Sm proteins, which are core snRNP proteins found in all spliceosomal snRNPs. The homologues of the human Sm Dl, D2, D3, E, F, and G were identified. D2 and E were described in yeast for the first time in this analysis (Table IA). As was described previously for Sm Dl, D3, F, and G (Hermann *et al.*, 1995; Seraphin, 1995), the Sm D3 and E are evolutionarily highly conserved (with 47 and 58.3% sequence identity between human and yeast, respectively). SmB/B', another known core Sm protein, was not found in the U1 snRNP preparation, which is due to loss during the purification procedure (Gottschalk *et al.*, 1998).

[U4/U6.U5] tri-snRNP

The [U4/U6.U5] tri-snRNP purification resulted in 34 separated Coomassie-stained protein bands on a 1-D SDS polyacrylamide gel. All of the protein bands were excised and individually digested, and the resulting peptides were analyzed both by MALDI peptide mapping and by sequencing with nanoelectrospray MS/MS. Since several of the bands contained mixtures of two or more proteins, both analysis methods were applied to ensure the identification of all the proteins present in one band (Table IB). The analysis of the 34 separate bands yielded the identification of 37 different proteins, of which 29 were already previously characterized. A total of 18 of the 29 proteins were known to be [U4/U6.U5] tri-snRNP proteins. Of these, seven were the Sm core proteins (SmB, Dl, D2, D3, E, F, and G), which associate with Ul, U2, U4, and U5 snRNAs; two were the Sm-like proteins (Lsm4p and Lsm8p), which were known to be associated with U6 and U4/U6 snRNPs; and nine were found to be previously characterized [U4/U6.U5] tri-snRNP-specific proteins (Prp8p, Brr2p, Snu114p, Prp6p, Prp31p, Prp3p, Prp4p, Spp381p, and Prp38p) (Table IB). Of the proteins, which were not previously demonstrated to be associated with the [U4/U6.U5] tri-snRNP, four were found to bear Sm-like domains (Lsm2p, Lsm5p, Lsm6p, and Lsm7p), totalling the number of Sm-like proteins associated with the [U4/U6.U5] tri-snRNP to at least seven. The remaining four proteins were novel [U4/U6.U5] tri-snRNP proteins, termed Snu13p, Diblp, Snu23p, and Snu66p (Table IB). The functional analysis of these new factors is described below.

MS Analysis of the Human Spliceosomal Proteins

On the Coomassie-stained 2-D gel, 69 spots were clearly visible (Fig. 5) and were excised for MS analysis. The in-gel digestion, peptide extraction, and sample preparation procedure was identical to the treatment of the protein bands from the snRNPs described earlier, apart from one change: the more abundant spots were digested in 50% ^{18}O-labeled water for the generation of longer sequence stretches. Since the spliceosome complex was isolated from human cells, it was anticipated that a proportion of the proteins would not be found in protein sequence databases because the human genome was not fully sequenced at the time of analysis and would have to be either identified in EST databases or be sequenced *de novo*. Digestion of the protein in 50% ^{18}O/^{16}O-labeled water yields doublets spaced by two mass units for the Y″ ion series, which allows the distinction of Y″ ions from other fragment ions, especially in the low mass-to-charge ratio area. The ^{18}O-labeling procedure is particularly useful when high resolution and good sensitivity for the fragment ions are achieved (Shevchenko et al., 1997). This is the case when a quadrupole TOF hybrid tandem mass spectrometer is used for generating sequence information (Morris et al., 1996). Since the mass accuracy and the resolution of the quadrupole time of flight instrument are extremely high (mass accuracy is better than 0.1 Da; resolution is 3000–10,000 full-width at half maximum [FWHM]) (Morris et al., 1996; Shevchenko et al., 1997), it is possible to automate data interpretation even for *de novo* sequencing.

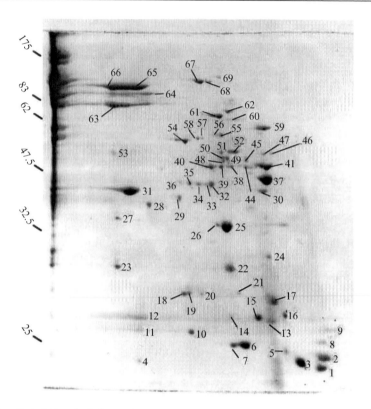

Fig. 5. A 2-D gel of the purified human spliceosomal proteins. A total of 69 distinct protein spots were excised, then in-gel digested, and finally analyzed by nanoelectrospray MS/MS. Figure adapted from Neubauer *et al.* (1998).

Searching EST Databases with MS Data

Expressed sequence tags (ESTs) are cDNA clones that are generated from reverse transcribed mRNA (messenger RNA) of a given tissue or cell line, which is subsequently made double-stranded and cloned into a vector. Typically, a clone contains 1000 to 3000 bases of insert sequence and may represent only part of the coding sequence of a gene. Part of this sequence is then determined by one-pass DNA sequencing to yield an expressed sequence tag. This method allows high throughput with the downsides of producing only short sequences (about 300 bases) with relatively high error rates.

Using the peptide sequence tag approach, MS data can be used to search the EST database. In this case, the peptide sequence tag (or full sequence data) is reverse translated into a degenerate DNA sequence, which has the advantage of DNA sequences in the database of not having to be translated into the six reading frames. Once a peptide sequence matches to an EST, the clone can be ordered; after resequencing the full cDNA insert, normally more peptides can be aligned to the sequence (Mann, 1996).

A total of 39 spots could be identified in the nonredundant database (NRDB) by a search using sequence tags or MALDI peptide maps (Table II). For cases in which the protein could not be identified in NRDB, the sequence data were used for an EST database search. Since ESTs are short and error prone, it is more advantageous to sequence many small peptides rather than fewer but longer ones because the chances are greater that any of the many peptides sequenced match to the region of the protein represented by the sequence of the EST. Even though the EST database searches can be performed with peptide sequence tags, many of the ESTs were found with full peptide sequence because the analysis and the data interpretation was carried out assuming an unknown protein sequence. In some cases, the peptides sequenced could in fact not be identified in the EST database at the time of analysis but only could be identified months later when more EST sequences were available. The database entries for the individual ESTs also contain information about homologous proteins. This information provides a useful hint to the function of the protein represented by the EST. Because the sequences available were only short, however, the information was not always reliable, and more detailed homology analyses had to be carried out once the full-length sequence was obtained. Almost one-third of the spots (21 spots corresponding to 16 different proteins; see Table II) were found via their ESTs in the original analysis (Neubauer et al., 1998).

Analytical Consideration: The Gel Image

The most abundant spots on the gel were almost all characterized previously as spliceosomal components. With few exceptions, the weak spots corresponded either to protein contaminants or unknown proteins with only limited sequence data (e.g., spots 12, 13, and 14 were identified as α-caseins; spot 11 was an unknown protein, respectively). ESTs were found for both, the more and less abundant spots. Another interesting feature of the protein pattern on the gel was the redundancy of proteins; it was found that the 69 spots correspond to only 45 different proteins. One EST

TABLE IIA

Spliceosome-Associated Proteins Identified in Protein Sequence Databases

Spot number	Accession number (genbank)	Protein name
Characterized spliceosomal proteins		
1,2	P14678 SwissProt	SmB/B'
3	P08579 SwissProt	U2-snRNP B''
6, 7	P09661 SwissProt	U2-snRNP A'
16	P09012 SwissProt	U1-snRNP A
17	P09651 SwissProt	hnRNP A1
19, 20	Q01081 SwissProt	U2AF-35kD
21	P22626 SwissProt	hnRNP A2/B1
22, 23	P07910 SwissProt	hnRNP C
25, 26	L35013 trembl	SAP 49
31	U08815 trembl	SAP 61
34, 35, 36	P26368 SwissProt	U2AF-65kD
37	L21990 trembl	SAP 62
38, 39	P08621 SwissProt	U1-snRNP 70kD
45	Y08765/Y08766	SF1 isoforms
48, 49, 50, 51	P52272 SwissProt	hnRNP M
63	X85237 trembl	SAP 114
65, 66	U41371 trembl	SAP 145
Potentially new spliceosome-associated proteins		
8	P12750 SwissProt	40S ribosomal Protein Rps4x[a]
41, 42, 43	U51432 trembl	Skip[b]
43, 44	P17844 SwissProt	p68 RNA helicase
47	P11940 SwissProt	PolyA binding protein (PAB 1)

(*continued*)

TABLE IIA (*continued*)

Spot number	Accession number (genbank)	Protein name
Contaminants		
5	P07219	Phaseolin *(Phaseolus Vulgaris)*
12, 13, 14	P02662	alpha S1 and S2 caseins *(B. taurus)*
	P02663	
	SwissProt	
53	P55081	Microfibrillar protein
	SwissProt	
Spots not identified		
11, 18, 24, 60		

[a] After identification of the EST, the protein was cloned by other researchers.
[b] Co-localization of GFP-fusion protein with snRNPs.

(Gry-rbp) in particular was found in five different spots (40, 46, 52, 55, and 56) that are distributed over a wide pI and molecular weight range, possibly indicating different alternatively spliced products or extensive modification of the same protein (Fig. 5 and Table IIB). Interestingly, patterns that are generally assumed to be caused by phosphorylation of one protein (a row of spots at the same molecular weight but with different pI values) turned out to be in fact a mixture of two proteins (spots 34–36 correspond to various forms of U2AF, a 65 kD protein, but spot numbers 32 and 33 were identified as an EST, later characterized as U4/U6, 60 kD protein). Estimations of the number of spliceosome-associated proteins, which were based on gel images only, may therefore not reflect the true situation. While it is clear that some factors, which were known to play an active part in splicing, are lost during the purification procedure, this study suggests that there are fewer proteins present in the spliceosome than previously assumed.

Even though the proteins were separated in two dimensions, some of the spots still contained protein mixtures (e.g., spot 22 is a mixture of hnRNPC and SPF 38; spot 43 is a mixture of skip and p68). Mixtures of proteins can only be identified when the entire sequence of at least one protein is known. When different peptides from one spot identify different, not overlapping ESTs, mixtures of two proteins can still be present. In many of these cases, however, the different ESTs that were analyzed displayed strong homology to the same protein, indicating that the peptides were derived from one protein. Comparing the intensities of spots of known and novel spliceosome-associated proteins with those from clear contaminants, it seemed likely that most, if not all, of the new factors

TABLE IIB
Novel Spliceosome-Associated Proteins Identified by
Mass Spectrometry as EST Sequences

Spot number	Protein accession number (genbank)	Protein name or homologue	Bioinformatic sequence analysis
4	AF081788	SPF27[b]	–
9	U89876[a]	ALY (mouse)	RRM2 domain RGG motif
10	AF083385	SPF30[b]	Tudor domain
15	AF083190	SPF31[b]	DNA j-like domain
22	AF083383	SPF38[b]	7 WD domains
27	AF026029[a]	Poly A tail binding protein II	RRM2 domain RGG motif
28	AF083384	SPF45[b]	RRM domain G-path domain
29	AA289699 (mouse) (EST)	U23412 (*C. elegans*)	similar to Prp19 in yeast
30	AF044333[a]	PRL-1; phosphatase regulatory subunit[b]	Seven WD domains
32, 33	AF001687[a]	Hprp4	Prp4 like Splicing factor motif 7 WD domains
40, 46, 52, 55, 56	AF037448[a]	Gry-rbp	3 RRM domain RGG motif
54	Z42055 (EST)	IK factor, S74221	IK factor is identical to the central 40% of the sequence
57, 58	AC004475[a]	F23858_1	G-patch domain
59	AF016370[a]	Hrpr3p	Prp3-like PWI domain
61	U86753[a]	Cdc 5 L[b]	DNA binding domains
62	AF026402[a]	U5 snRNP 100kD protein	DEXD-box HELIc domain RNA helicase
64	U94836[a]	Erprot	SWAP domain CTBP domain G-patch
67, 68	AF017789[a]	CA 150	Putative transcription factor; WW domain FF domain
69	AA317471 (EST)	–	–

[a] After identification of the EST, the protein was cloned by other researchers.
[b] Co-localization of GFP-fusion protein with snRNPs.

are actually integral components of the complex. The fact that some of the novel spliceosome-associated proteins were previously characterized in a different biological context does not exclude a role of these proteins in splicing.

U1 snRNP

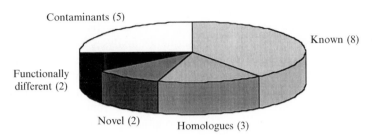

[U4/U6.U5] snRNP

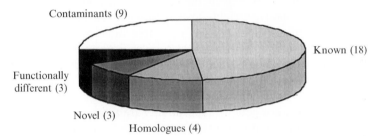

Human spliceosome

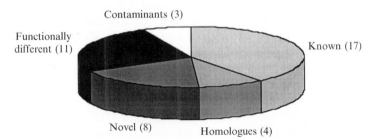

Fig. 6. Schematic representation of the results from the protein identification of the different complexes analyzed. The proteins are grouped into five different categories: (i) the known proteins, which were previously characterized and known to be part of the complex; (ii) the homologues, which were identified in the complex for the first time but have homologous counterparts in other organisms that were already known; (iii) the contaminants, which were shown to associate nonspecifically or were introduced after purification; (iv) the functionally different proteins, which were previously identified in a different cellular context but could play a role in the context of the complex; and (v) the novel proteins, which had no function assigned to them yet. For further studies, the proteins can be grouped into two major

Functional Analysis of the Identified Proteins

After the full-length protein sequence was obtained from either the identification in the protein sequence database or by cloning, homology searches and detailed bioinformatic sequence analyses were performed. As schematically represented in Fig. 6, the proteins fall into four different categories: (i) the protein has already been functionally characterized in the context of the complex under study; (ii) its strong homology to a protein from a different organism allows its functional assignment; (iii) no function and no homologue to other proteins can be inferred; (iv) the protein has previously been implicated in a different cellular mechanism. In the former two cases, no further functional validation of the protein is necessary. In the latter two cases, further analysis is needed for the elucidation of its role in the process under investigation. If the detailed sequence analysis yields insight into known motifs or domains, already the first hints to the protein's possible function might be gained. The protein's implication in a different cellular process does by no means exclude its functional relevant role in the complex of interest. The protein could fulfill a double role and be active in more than one cellular function, or it could indeed provide a functional link between different processes. The first biochemical or genetic experiments are then performed to verify whether the novel protein is truly associated with the complex. Depending on the availability of reagents and methods, this can be coimmunoprecipitation, colocalization, or cosedimentation analysis (examples for each case are given in the following paragraphs). Once the protein's association with the complex has been further proven, several different routes can be taken for functional assignment. In this respect, coimmunoprecipitation experiments can also yield the identification of the protein's closest interaction partners or subcomplexes with novel components, which could not be retrieved by the initial purification method. Biological assays for a given cellular activity might already exist (as is the case for splicing), and an involvement in one particular reaction might already be apparent from this experiment. In yeast, genetic experiments, like mutational analysis and depletion of the gene, are relatively easy to perform and will give clues to whether the gene is essential for cell viability or for proceeding of a certain step in the cellular process under study. Mutational analysis can identify functionally crucial amino acids or sequence stretches in the protein. Finally, the determination of the protein's three-dimensional

---

categories. One group could be the proteins, which have a functional assignment and do not need further functional validation (such as the known proteins, the homologues, and the contaminants); the second group could be the novel proteins, which require further work for full functional assignment (such as the functionally different proteins and the novel proteins).

(3-D) structure provides insight into the detailed molecular action of the novel protein. In the following section, examples of these different approaches are given from all three different projects.

U1 snRNP

In the analysis of the U1 snRNP, nine proteins were found with no previous functional assignment in the U1 snRNP complex. A gradient sedimentation step revealed that of these proteins, four were indeed specific to the U1 snRNP, while five were not cosedimenting with the U1 snRNA (three of these were U5 snRNP proteins, and two were abundant cellular proteins, contaminating the preparation). The four novel U1 snRNP proteins (Nam8p, Snu71p, Snu65p, and Npl3p) were further functionally characterized using coimmunoprecipitation and genetic depletion analysis. By purification of the U1 snRNP from a yeast strain lacking the NAM8 gene, the MS analysis revealed differential assembly of the U1 snRNP proteins: Snu 65p, a protein not identified in the first purification, but in a synthetic lethal screen, was shown to be associated strongly in this preparation. In addition, Snu71p and Snu56p were now barely visible in the mutant strain purification, indicating that Nam8p regulates the U1 snRNP stability. In synthetic lethal screens for functional U1 snRNA-associating genes, Snu56p and Nam8p were found to affect the *in vivo* splicing efficiency of noncanonical introns (Gottschalk *et al.*, 1998).

[U4/U6.U5] tri-snRNP

Of the 37 different proteins identified in the [U4/U6.U5] tri-snRNP preparation, 18 were previously characterized [U4/U6.U5] tri-snRNP proteins, and four showed strong homology to their human counterparts. Of the 15 proteins, which were previously not associated with [U4/U6.U5] tri-snRNP function, 9 were found to copurify with the [U4/U6.U5] tri-snRNP particle without functional significance: 8 were found as copurifying U1 snRNP proteins, and 1 was found to be an elongation factor. The U1 proteins are not surprising because in the U1 preparation, contaminating U5 proteins were also identified. The presence of two U2 snRNP proteins, however, was surprising. The first step for functional analysis of the potentially new [U4/U6.U5] tri-snRNP proteins was coimmunoprecipitation. These experiments showed that Snu23p and Snu66p, the two proteins that had no previous function assigned to them, were indeed part of the [U4/U6.U5] tri-snRNP and could also precipitate at low salt concentration U2 snRNA. A combination of glycerol gradient centrifugation separation and coimmunoprecipitation experiments proved that Snu23p and Snu66p are indeed part of the [U4/U6.U5] tri-snRNP complex and also part of a [U2.U4/U6.U5] tetra-snRNP complex. This finding thus provided a link between the two subcomplexes in the splicing mechanism. Further genetic and biochemical experiments with these two proteins showed that while depletion of Snu66p could inhibit the first splicing step completely in *in vitro* splicing assays, Snu23p depletion only had a mild effect on the splicing efficiency *in vitro*.

The human orthologs of the two other novel proteins, Dip1p and Snu13p, were characterized in parallel in a study of the human [U4/U6.U5] tri-snRNP, and their function in yeast could be inferred from these studies (Nottrott *et al.*, 1999; Reuter *et al.*, 1999). The functional analyses of the yeast proteins are less time-consuming than those for the human proteins because in the case of yeast, genetic manipulations are much simpler. In addition, the full-length protein sequence is readily available after identification in the database, while many of the novel human proteins were identified as ESTs only.

Human Spliceosome

In the analysis of the human spliceosome, 17 proteins were identified as previously characterized spliceosomal proteins, and four additional ones showed homology to already characterized yeast spliceosomal proteins. Almost half the proteins (19) were found to be putative novel spliceosome-associated proteins. Of these 19, 8 were found as ESTs with no homology to a protein with known function, and 7 of the EST hits showed homology to proteins in other organisms with functions other than splicing. Finally, the full-length sequences of four proteins were identified, but their association with the spliceosome was previously not shown. For a first, generally applicable step of testing whether the newly identified proteins were indeed associated with the spliceosome complex, colocalization studies were performed. Fusion proteins of the new factors and the green fluorescent protein (GFP) were constructed and introduced into HeLa cells. If similar subnuclear distributions of the GFP-tagged protein and a known spliceosomal protein (here U1A) were observed, a true spliceosome association of the novel protein would be indicated. For nine putative novel proteins, colocalization with the spliceosome was shown by this method. Interestingly, a previously identified ribosomal protein was found to localize to both the ribosome and the spliceosome complexes, thus providing a link between pre-mRNA splicing and translation (A. I. Lamond, personal communication).

Of the 19 putative novel spliceosomal proteins, further functional characterization was carried out on only three of them in the two research groups initially involved in this project. Only 3 years *after* the completion of the initial MS analysis was the first scientific report published on a detailed characterization for one of these proteins, named CDC5L (Ajuh *et al.*, 2000), thus demonstrating the difficulty in gaining functional insight. On one of the novel proteins, structural studies were carried out by a collaborating group. Interestingly, the further functional characterization of CDC5L involved a new round of MS identifications of proteins copurifying with this novel splicing factor. Indeed, coimmunoprecipitations of CDC5L yielded 13 novel proteins apart from known spliceosomal proteins, and these novel proteins were not identified in the original spliceosome study. CDC5L was found to be part of a multiprotein complex that incorporates into the spliceosome in an ATP-dependent manner and is required for the second catalytic step of splicing. The CDC5L complex contains at least five additional proteins and is a non-snRNA subcomplex of the spliceosome (Ajuh *et al.*, 2000).

Conclusions

The study of the splicesomal complexes were the first complexes that were characterized by MS analysis alone. Even though this chapter recounts an almost historical perspective of the state-of-the-art equipment and processes at the time of analysis, the conclusions that can be drawn from the chapter still hold true. The studies illustrate well the relative ease of protein identification by MS and the difficulty to provide detailed functional information on the vast amount of protein data generated in such a study. It also becomes clear that specific and efficient protein complex purification procedures are key to a successful study. On the one hand, the more specific the purification procedure, the less false-positive hits have to be pursued in the time-consuming functional characterization. On the other hand, the more complete the picture of the complex components, the easier it becomes to gain insight into the full cellular picture. While generic protein complex purification techniques, MS protein identifications, and bioinformatic sequence analyses are now very robust, fast, and efficient tools, the detailed functional characterization of the protein components is still time-consuming and awaits more generic procedures. However, with the growing information in databases and the tools described previously, proteome wide mapping of interacting proteins has been possible to perform (Butland *et al.*, 2005; Gavin *et al.*, 2002).

References

Ajuh, P., Kuster, B., Panov, K., Zomerdijk, J. C., Mann, M., and Lamond, A. I. (2000). Functional analysis of the human CDC5L complex and identification of its components by mass spectrometry. *EMBO J.* **19,** 6569–6581.

Bennett, M., Michaud, S., Kingston, J., and Reed, R. (1992). Protein components specifically associated with prespliceosome and spliceosome complexes. *Genes Dev.* **6**, 1986–2000.

Butland, G., Peregrin-Alvarez, J. M., Li, J., Yang, W., Yang, X., Canadien, V., Starostine, A., Richards, D., Beattie, B., Krogan, N., Davey, M., Parkinson, J., Greenblatt, J., and Emili, A. (2005). Interaction network containing conserved and essential protein complexes in *Escherichia coli. Nature* **433**, 531–537.

Gavin, A. C., Bosche, M., Krause, R., Grandi, P., Marzioch, M., Bauer, A., Schultz, J., Rick, J. M., Michon, A. M., Cruciat, C. M., Remor, M., Hofert, C., Schelder, M., Brajenovic, M., Ruffner, H., Merino, A., Klein, K., Hudak, M., Dickson, D., Rudi, T., Gnau, V., Bauch, A., Bastuck, S., Huhse, B., Leutwein, C., Heurtier, M. A., Copley, R. R., Edelmann, A., Querfurth, E., Rybin, V., Drewes, G., Raida, M., Bouwmeester, T., Bork, P., Seraphin, B., Kuster, B., Neubauer, G., and Superti-Furga, G. (2002). Functional organization of the yeast proteome by systematic analysis of protein complexes. *Nature* **415**, 141–147.

Gottschalk, A., Neubauer, G., Banroques, J., Mann, M., Luhrmann, R., and Fabrizio, P. (1999). Identification by mass spectrometry and functional analysis of the yeast [U4/U6.U5] tri-snRNP. *EMBO J.* **18**, 4535–4548.

Gottschalk, A., Tang, J., Puig, O., Salgado, J., Neubauer, G., Colot, H., Mann, M., Seraphin, B., Rosbash, M., Luhrmann, R., and Fabrizio, P. (1998). A comprehensive biochemical and genetic analysis of the yeast U1 snRNP reveals five novel proteins. *RNA* **4**, 374–393.

Hermann, H., Fabrizio, P., Raker, V. A., Foulaki, K., Hornig, H., Brahms, H., and Luhrmann, R. (1995). SnRNP Sm proteins share two evolutionarily conserved sequence motifs which are involved in Sm protein–protein interactions. *EMBO J.* **14**, 2076–2088.

Jensen, O. N., Podtelejnikov, A., and Mann, M. (1996). Delayed extraction improves specificity in database searches by MALDI peptide maps. *Rapid Commun. Mass Spectrom.* **10**, 1371–1378.

Kao, H. Y., and Siliciano, P. G. (1996). Identification of Prp40, a novel essential yeast splicing factor associated with the U1 small nuclear ribonucleoprotein particle. *Mol. Cell. Biol.* **16**, 960–967.

Kramer, A. (1996). The structure and function of proteins involved in mammalian pre-mRNA splicing. *Annu. Rev. Biochem.* **65**, 367–409.

Liao, X. C., Tang, J., and Rosbash, M. (1993). An enhancer screen identifies a gene that encodes the yeast U1 snRNP A protein: Implications for snRNP protein function in pre-mRNA splicing. *Genes Dev.* **7**, 419–428.

Lockhart, S. R., and Rymond, B. C. (1994). Commitment of yeast pre-mRNA to the splicing pathway requires a novel U1 small nuclear ribonucleoprotein polypeptide, Prp39p. *Mol. Cell. Biol.* **14**, 532–539.

Mann, M. (1996). A shortcut to interesting human genes: Peptide sequence tags, ESTs, and computers. *TiBS* **21**, 494–495.

Mann, M., and Wilm, M. (1994). Error-tolerant identification of peptides in sequence databases by peptide sequence tags. *Anal. Chem.* **66**, 4390–4399.

McPherson, J. D., *et al.* (2001). A physical map of the human genome. *Nature* **409**, 934–941.

Morris, H. R., Paxton, T., Dell, A., Langhom, J., Berg, M., Bordoli, R. S., Hoyes, J., and Bateman, R. H. (1996). High-sensitivity collisionally-activated decomposition tandem mass spectrometry on a novel quadrupole/orthogonal-acceleration time-of-flight mass spectrometer. *Rapid Commun. Mass Spectrom.* **10**, 889–896.

Neubauer, G., and Mann, M. (1999). Mapping of phosphorylation sites of gel-isolated proteins by nanoelectrospray tandem mass spectrometry: Potentials and limitations. *Anal. Chem.* **71**, 235–242.

Neubauer, G., Gottschalk, A., Fabrizio, P., Seraphin, B., Luhrmann, R., and Mann, M. (1997). Identification of the proteins of the yeast U1 small nuclear ribonucleoprotein complex by mass spectrometry. *PNAS (USA)* **94**, 385–390.

Neubauer, G., King, A., Rappsilber, J., Calvio, C., Watson, M., Ajuh, P., Sleeman, J., Lamond, A., and Mann, M. (1998). Mass spectrometry and EST-database searching allows characterization of the multi-protein spliceosome complex. *Nat. Genet.* **20**, 46–50.

Nottrott, S., Hartmuth, K., Fabrizio, P., Urlaub, H., Vidovic, I., Ficner, R., and Luhrmann, R. (1999). Functional interaction of a novel 15.5 kD [U4/U6.U5] tri-snRNP protein with the 5′ stem-loop of U4 snRNA. *EMBO J.* **18**, 6119–6133.

Reuter, K., Nottrott, S., Fabrizio, P., Luhrmann, R., and Ficner, R. (1999). Identification, characterization, and crystal structure analysis of the human spliceosomal U5 snRNP-specific 15 kD protein. *J. Mol. Biol.* **294**, 515–525.

Rigaut, G., Shevchenko, A., Rutz, B., Wilm, M., Mann, M., and Seraphin, B. (1999). A generic protein purification method for protein complex characterization and proteome exploration. *Nat. Biotech.* **17**, 1030–1032.

Seraphin, B. (1995). Sm and Sm-like proteins belong to a large family: Identification of proteins of the U6 as well as the U1, U2, U4, and U5 snRNPs. *EMBO J.* **14**, 2089–2098.

Shevchenko, A., Chernushevich, I., Ens, W., Standing, K. G., Thomson, B., Wilm, M., and Mann, M. (1997). Rapid *de novo* peptide sequencing by a combination of nanoelectrospray, isotopic labeling, and a quadrupole/time-of-flight mass spectrometer. *Rapid Commun. Mass Spectrom.* **11**, 1015–1024.

Shevchenko, A, Jensen, O. N., Podtelejnikov, A., Sagliocco, F., Wilm, M., Vorm, O., Mortensen, P., Shevchenko, A., Boucherie, H., and Mann, M. (1996a). Linking genome and proteome by mass spectrometry: Large-scale identification of yeast proteins from two-dimensional gels. *PNAS (USA)* **93**, 14440–14445.

Shevchenko, A., Wilm, M., Vorm, O., and Mann, M. (1996b). Mass spectrometeric sequencing of proteins from silver-stained polyacrylamide gels. *Anal. Chem.* **68**, 850–858.

Smith, V., and Barrell, B. G. (1991). Cloning of a yeast U1 snRNP 70k protein homologue: Functional conservation of an RNA-binding domain between humans and yeast. *EMBO J.* **10**, 2627–2634.

Venter, J. C., *et al.* (2001). The sequence of the human genome. *Science* **291**, 1304–1351.

Vorm, O., Roepstorff, P., and Mann, M. (1994). Matrix surfaces made by fast evaporation yield improved resolution and very high sensitivity in MALDI–TOF. *Anal. Chem.* **66**, 3281–3287.

Will, C. L., Fabrizio, P., and Luhrmann, R. (1995). Nuclear pre-mRNA splicing. *In* "Nucleic Acids and Molecular Biology" (F. F. Eckstein and D. M. J. Lilley, eds.) Vol. 9. Springer Verlag, Berlin, Germany.

Wilm, M., and Mann, M. (1996). Analytical properties of the nanoelectrospray ion source. *Anal. Chem.* **68**, 1–8.

[11] The Use of Mass Spectrometry to Identify Antigens from Proteasome Processing

By ODILE BURLET-SCHILTZ, STÉPHANE CLAVEROL, JEAN EDOUARD GAIRIN, and BERNARD MONSARRAT

Abstract

Mass spectrometry (MS) is a powerful tool for the characterization of antigenic peptides that play a major role in the immune system. Most of the major histocompatibility complex (MHC) class I peptides are generated during the degradation of intracellular proteins by the proteasome, a catalytic complex present in all eukaryotic cells. This chapter focuses on the contribution of MS to the understanding of the mechanisms of antigen processing by the proteasome. This knowledge may be valuable for the design of specific inhibitors of proteasome, which has recently been recognized as a therapeutic target in cancer therapies and for the development of efficient peptidic vaccines in immunotherapies. Examples from the literature have been chosen to illustrate how MS data can contribute first to the understanding of the mechanisms of proteasomal processing and, second, to the understanding of the crucial role of proteasome in cytotoxic T lymphocytes (CTL) activation. The general strategy based on MS analyses used in these studies is also described.

Introduction

The immune system exerts continuous control on viral infections and cancers by monitoring whether foreign or mutated proteins are synthesized within the cell. Peptides derived from intracellularly expressed proteins play a pivotal role in this surveillance process. Bound to major histocompatibility complex (MHC) class I molecules, peptides are displayed at the cell surface where they act as molecular signals recognized by $CD8^+$ cytotoxic T lymphocytes (CTL), the major effector cells of the immune system to control tumor growth or viral infections. Under "normal" physiological conditions, only MHC class I-presented "self"-peptides deriving from normal autologous protein sequences (Harris, 1994) are displayed, against which the immune system remains, in principle, nonreactive. Under pathophysiological conditions, such as viral infections or malignant transformations, "foreign" peptides derived from viral proteins (Oldstone *et al.*, 1995) or from mutated or overexpressed endogenous proteins, like tumor

METHODS IN ENZYMOLOGY, VOL. 405
0076-6879/05 $35.00
DOI: 10.1016/S0076-6879(05)05011-1

antigens (Boon *et al.*, 1994; Van den Eynde and van der Bruggen, 1997), may also bind to MHC class I molecules and then be displayed at the cell surface where they serve as targets to CTL. If immunogenic, these antigenic peptides induce CTL activation, leading to the destruction of the pathogenic antigen presenting cell.

Processing and Presentation Pathway of MHC Class I-Restricted Antigenic Peptides

Antigen processing and presentation by MHC class I molecules have been studied extensively during the past decade (Androlewicz, 2001; Nandi *et al.*, 1998; Pamer and Cresswell, 1998; Rock and Goldberg, 1999; Yewdell and Bennink, 1999; York *et al.*, 1999). Most of the MHC class I peptides presented to CTL are generated in the cytoplasm during degradation of cytosolic ubiquitinated proteins of diverse origins by the proteasome. Peptides generated from proteasome processing include peptides possessing the correct C-terminus of antigenic peptide sequences. These peptides are then translocated via the transporter associated with antigen processing (TAP) into the endoplasmic reticulum (ER), where they assemble with the MHC class I molecule. When N-terminally extended, antigenic peptide sequences are produced by proteasome; enzymes like aminopeptidases may also be involved in generating the proper antigenic sequence to bind to the MHC class I molecule (Beninga *et al.*, 1998; Craiu *et al.*, 1997; Mo *et al.*, 1999). The peptide–MHC complex is then transported to the cell surface via the Golgi apparatus to be presented to CTL. This antigen processing and presentation pathway is illustrated in Fig. 1. The sequences of MHC class I antigenic peptides are 8 to 11 amino acids long and display an allele-specific MHC binding motif (Falk *et al.*, 1991; Rammensee, 1995). Despite these restrictions, peptide sequences can vary greatly, and several thousands of peptide sequences are estimated to be presented at the cell surface.

Cellular Function of the Proteasome and its Implication in Antigen Processing

The proteasome is a ubiquitous multicatalytic complex present in the cytoplasm and in the nucleus of all eukaryotic cells (Coux *et al.*, 1996; Rock and Goldberg, 1999); the proteasome represents the main protein degradation machinery in the cell. Evidence has accumulated to show that the proteasome is involved in crucial cellular processes, including the cell cycle, apoptosis, morphogenesis, stress response, regulation of intracellular proteins content, removal of abnormal proteins, and MHC class I antigen processing (Ciechanover, 1998). The role of the proteasome in the generation of antigenic peptides has been demonstrated by the use of proteasome

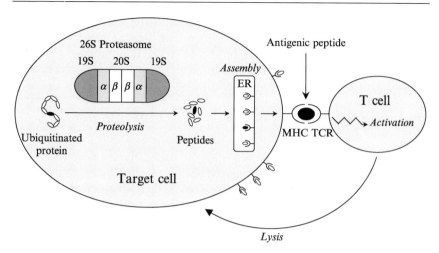

FIG. 1. Schematic representation of the different key steps of the processing and presentation pathway of MHC class I-restricted antigenic peptides. Ubiquitinated proteins are degraded by the 26S proteasome, a ubiquitous multicatalytic complex present in the cytoplasm and the nucleus of all eukaryotic cells. Generated peptides are translocated via the TAP transporter into the endoplasmic reticulum (ER), where they assemble with the MHC class I molecule. The peptide–MHC complex is then transported to the cell surface. If the cytotoxic T cell receptor (TCR) recognizes the appropriate peptide–MHC complex, then the T cell may become activated to kill the infected target cell.

inhibitors, such as lactacystin or peptide aldehydes, that block generation and presentation of MHC class I-restricted CTL epitopes (Cerundolo *et al.*, 1997; Rock *et al.*, 1994); this role has also been confirmed in purified proteasome's ability to produce *in vitro* antigenic peptides from polypeptide precursors (Dick *et al.*, 1994; Niedermann *et al.*, 1996). The active form of the proteasome, the 26S proteasome (Coux *et al.*, 1996; Voges *et al.*, 1999), is a 2500-kDa complex that is composed of a barrel-shaped proteolytic 20S core complex (approximately 750 kDa) flanked by two 19S regulatory complexes. The 19S complexes containing ATPases are necessary to recognize ubiquitinated proteins and to unfold them, allowing their access to the proteolytic 20S core complex.

Structure and Function of the 20S Proteasome

Structural studies of 20S proteasomes have contributed majorly to the understanding of proteasome function (Bochtler *et al.*, 1999; Coux *et al.*, 1996; Tanaka, 1998). Initial electron microscopy analyses (Hegerl *et al.*, 1991; Puhler *et al.*, 1992) and further crystal structures of *Thermoplasma acidophilum* (Lowe *et al.*, 1995) and yeast *Saccharomyces cerevisiae* (Groll

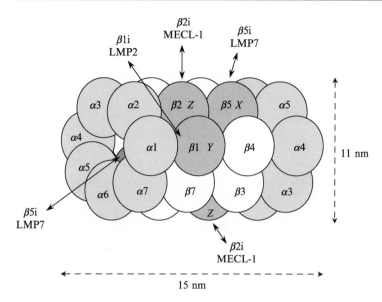

FIG. 2. Schematic representation of mammalian 20S proteasome and immunoproteasome subunit composition. The 20S proteasome is a barrel-shaped core proteolytic complex (a cylindrical structure of 15 nm in length and 11 nm in diameter). Two outer rings are composed of seven different catalytically inactive α subunits, and two inner rings are composed of seven different β subunits forming a central chamber in which proteolysis occurs. In the immunoproteasome structure, the three IFNγ-inducible β subunits, β 1i (or LMP2), β2i (or MECL-1), and β5i (or LMP7), replace their respective constitutive homologues β1 (or Y), β2 (or Z), and β5 (or X).

et al., 1997) 20S proteasomes showed a conserved cylindrical structure (15 nm in length and 11 nm in diameter) made of four stacked rings, each ring containing seven subunits. The composition of eukaryotic 20S proteasomes is more complex. As illustrated in Fig. 2, each of the two outer rings is made of seven different α subunits (numbered α1 to α7), and each of the two inner rings contains seven different β subunits (numbered β1 to β7), forming a central cavity in which proteolysis occurs. Only the three subunits of β1 (or Y), β2 (or Z), and β5 (or X) are catalytically active and lead to three major peptidase activities described for the proteasome: one is a trypsin-like activity (cleavage after basic residues), one is a chymotrypsin-like activity (cleavage after hydrophobic residues), and the latter one is a peptidylglutamyl-peptide hydrolyzing (PGPH) activity (cleavage after acidic residues). Two additional activities, the "branched chain amino acid preferring" (BrAAP) and "small neutral amino acid preferring" (SNAAP) activities, have also been observed (Orlowski *et al.*, 1993). Catalytic sites of

20S proteasome involve a threonine residue located at the N-terminus of the each active β subunit (Lowe et al., 1995), thus classifying the proteasome in a new class of proteases: the N-terminal nucleophile hydrolases (Orlowski and Wilk, 2000).

In the mammalian 20S proteasome, the three catalytic subunits $\beta 1$ (or Y), $\beta 2$ (or Z), and $\beta 5$ (or X) can be replaced by three interferon gamma (IFNγ)-inducible subunits of $\beta 1$i (or LMP2), $\beta 2$i (or MECL-1), and $\beta 5$i (or LMP7) to form what is called the immunoproteasome (Fig. 2). The presence of the IFNγ-inducible β subunits in the 20S proteasome structure modifies its catalytic activities (Rock and Goldberg, 1999). The cleavage activity after acidic residues is decreased, and the cleavage activities after hydrophobic and basic amino acids are enhanced with a number of antigens. Because the C-terminal residue of MHC class I antigenic peptides is predominantly a hydrophobic or a basic residue, the immunoproteasome is expected to process such antigenic peptides more efficiently than the standard proteasome. However, there is now evidence that the immunoproteasome can be detrimental to the generation of antigenic peptides (Morel et al., 2000). The presence of IFNγ also induces the PA28 activator of 20S proteasome, which can replace the 19S regulatory complex. Despite numerous biochemical and structural studies on PA28, the role of this activator in class I antigen processing remains to be elucidated (Kloetzel et al., 1999; Rock and Goldenberg, 1999; Rechsteiner et al., 2000; Tanaka, 1998).

Mass Spectrometry

Mass spectrometry (MS) has become a valuable tool in the field of immunology (Downard, 2000), in particular in the characterization of MHC-bound antigenic peptide sequences (Bazemore et al., 1998; de Jong, 1998). The pioneering work of Hunt and coworkers (Cox et al., 1994; Hunt et al., 1992) on the identification of peptides bound to MHC class I molecules has demonstrated how the sensitivity of electrospray mass spectrometry (ESI–MS) and its ability to analyze complex mixtures when coupled to capillary liquid chromatography (LC) could be efficiently used to identify individual peptides from a mixture of several thousand sequences presented at the cell surface. This approach allowed the detection of peptides at 30 to 600 fmole per 10^8 cells corresponding to a number of copies per cell between 100 to 3000. Since these early studies, strategies have been designed to improve detection limits to access to less abundant peptide sequences. The use of two-dimensional (2-D) chromatography consisting of reversed-phase high-performance liquid chromatography (RP-HPLC) and on-line membrane preconcentration-capillary electrophoresis/MS was shown to enable the sequencing of less than 60 fmole of peptide in a dilute sample obtained from 3×10^9 cells (van der Heeft et al., 1998). Micro-capillary column-switching HPLC (Naylor et al., 1998; Tomlinson et al., 1996) and sheathless preconcentration–capillary zone electrophoresis (Barroso and de Jong, 1999) have also been reported to efficiently concentrate samples before MS analysis of MHC class I peptides. The combination of nanoflow HPLC with electrospray ionization on a Fourier transform MS has allowed researchers to reach detection limits in the attomole range for the detection of a male-specific minor histocompatibility antigen present at 30 copies per cell (Pierce et al., 1999). These strategies are usually conducted in conjunction with CTL activity measurements. New strategies based on MS detection of MHC-bound peptides and on avoiding the need of CTL assays were recently proposed (Brockman et al., 1999; Schirle et al., 2000).

This chapter focuses on the use of MS in the identification of antigenic peptides from proteasome processing (i.e., on the contribution of MS to the understanding of the initial step of the MHC class I antigen presentation pathway). The ultimate goal of these studies, which represents a major challenge, would be to control antigen processing by proteasome. The proteasome could serve as a pharmacological target of specific inhibitors, which would modify antigen formation and consequently CTL activation (Groettrup and Schmidtke, 1999). Understanding the mechanisms of antigen processing by the proteasome would also help in the design of efficient peptidic vaccines and immunotherapy strategies. Examples from the literature, including our own results, meaning the results from the research performed by the authors of this chapter, have been chosen to illustrate how MS data contribute to the understanding of the mechanisms of proteasomal processing and of the crucial role of proteasome in CTL activation. The general strategy based on MS analyses used in these studies will first be described.

General Strategy

The use of MS in the study of antigen processing by 20S proteasome contributes to the identification and relative quantitation of peptides generated by in vitro 20S proteasome digestions of precursor peptides. The general strategy is shown in Fig. 3. The first step involves the digestion of synthetic precursor peptides by purified 20S proteasomes. Precursor peptides are chosen to include the epitope sequence of interest. Typically, 15 to 40 residue-long peptides are considered to be representative models. In particular cases, denatured proteins (up to hundreds of amino acids) can be used as substrates. The digestion is performed at 37° for different periods of time (typically from 30 min to 8 h). Short digestion times (10 to 30 min) may allow the characterization of initial cleavage sites and intermediates of

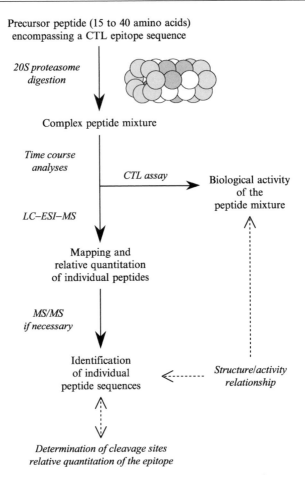

Precursor peptide (15 to 40 amino acids)
encompassing a CTL epitope sequence

20S proteasome
digestion

Complex peptide mixture

Time course
analyses

CTL assay → Biological activity
of the
peptide mixture

LC–ESI–MS

Mapping and
relative quantitation
of individual peptides

MS/MS
if necessary

Identification
of individual
peptide sequences

←------- *Structure/activity*
relationship

Determination of cleavage sites
relative quantitation of the epitope

FIG. 3. General strategy for the study of peptides generated from the digestion of antigen precursor peptides by 20S proteasome *in vitro.*

reaction. Prolonged digestion times (up to 48 h) may allow accumulation of final products and facilitate the identification of epitopes generated in low abundance. However, the relevance of products formed after such prolonged digestion times to support biological observations may be questionable.

The second step corresponds to the molecular mass determination of individual peptides generated during the digestion. Due to the complexity of most peptide digests obtained, initial separation of peptides by HPLC is generally required. MS analyses are then performed on-line using an electrospray source and various possible analyzers (e.g., quadrupoles, ion traps, and hybrid quadrupole time-of-flight, or QTOF). On-line analyses should offer the most comprehensive view of the peptide mixture content because every peptide eluted from the HPLC column is sent to the mass spectrometer. Off-line analyses of collected fractions can also be performed for peptide identification, usually using a matrix-assisted laser desorption/ionization time-of-flight (MALDI–TOF) analyzer. With the sequence of the precursor peptide being known, the molecular mass determination is often sufficient to identify the sequences of generated peptides. If ambiguities arising from isobaric sequences remain, then further MS/MS analyses may be required to obtain sequence information. These analyses can be performed either on-line with instruments having MS/MS capabilities (e.g., triple quadrupole, ion trap, QTOF) or off-line using MALDI post-source decay (PSD) analyses or Edman sequencing. Relative quantitation of peptides can also be extracted from these analyses, either from individual UV absorptions or from the intensity of specific MS signals. This strategy leads to the determination of preferential cleavage sites and to the identification and relative quantitation of epitope sequences. To establish structure/activity relationships at different digestion times, CTL activity measurements can be performed showing the presence of peptides with antigenic potency in the peptide mixture digests.

Purification of 20S Proteasome

Various protocols are described in the literature for the purification of 20S proteasomes (Beyette *et al.*, 2001). They are based on different protein purification strategies, including immunoaffinity chromatography, ion exchange chromatography, gel filtration chromatography, ultracentrifugation, and precipitation. Many protocols use a combination of these techniques. The simplest and most rapid method is based on affinity chromatography using a monoclonal antibody immobilized on CnBr-activated Sepharose and directed against the $\alpha2$ subunit of 20S proteasome (Hendil and Uerkvitz, 1991; Valmori *et al.*, 1999). Using this protocol to purify 20S proteasome from human erythrocytes, 1.5 to 3 mg of purified proteasome may be expected from 100 ml of blood (circa 5×10^{11} cells). Methods including a final step consisting of ion exchange chromatography may allow researchers to obtain more homogeneous 20S proteasome solutions (Dahlmann *et al.*, 2000). The homogeneity of the purified 20S proteasome is controlled by monodimensional or bidimensional gel electrophoresis, and proteasomal activities are assayed with fluorogenic peptides (Beyette *et al.*, 2001; Valmori *et al.*, 1999).

Peptide Digestion

Synthetic 15- to 40-mer peptides (1–4 nmoles) are incubated with purified 20S proteasome (4–20 μg) at 37° (or 60° for archaeal proteasomes) in 10 to 20 mM Tris or Hepes buffers at a pH ranging from 7 to 8 (Morel *et al.*, 2000). More complex digestion buffers including the addition of 2 mM MgAc$_2$ or MgCl$_2$ and 1 mM DTT (Dahlmann *et al.*, 2000; Sijts *et al.*, 2000a), and 1 mM EGTA, 0.5 mM EDTA (Niedermann *et al.*, 1997) may also be used. Digestion of proteins by 20S proteasome requires that the protein be denatured before digestion. For this purpose, reaction buffers include 0.01% SDS (Nussbaum *et al.*, 1998). Reactions are stopped at various time points by either adding trifluoroacetic acid (TFA) or freezing the reaction mixture.

Peptide Separation by HPLC

Standard conditions for peptide separation include the use of reversed-phase C18 columns with eluents consisting of water/acetonitrile/TFA. Considering that the amount of starting material used for the digestion is in the nanomole range and that most of the generated products will then be in the 1 to 100 pmole range, peptide separation is often performed with narrow-bore (2.1 and 1.0 mm internal diameter) columns. The flow rates generally used with these columns (around 200 and 50 μl per minute) and the eluent composition allow a direct coupling to electrospray MS. Developments in the area of peptide separation have consisted in miniaturizing systems to work with lower quantities of material. Reversed-phase columns of 300 down to 75 μm internal diameters are commercially available. The use of these columns allows the concentration of samples into small volumes and enables researchers to work with peptide amounts in the picomole down to subpicomole range. The flow rates used are a few microliters per minute down to 150 nl per minute. Because of advances in the interfaces of capillary HPLC to MS, the smallest columns can be coupled to nanoelectrospray sources for on-line MS analyses. Another improvement for on-line capillary HPLC–MS coupling comes from the development of reversed-phase PepMap columns (LC Packings, Amsterdam, The Netherlands), which can be used with mobile phases containing formic acid or acetic acid instead of TFA and which achieve an equivalent separation efficiency as standard reversed-phase columns (Van Soest *et al.*, 1998). The replacement of TFA by either formic acid or acetic acid increases the sensitivity of electrospray MS analyses. The presence of TFA, however, is fully compatible with analyses by MALDI–MS. The use of these approaches for the identification of peptides generated from proteasome digestions may favor the detection of epitope sequences that may be

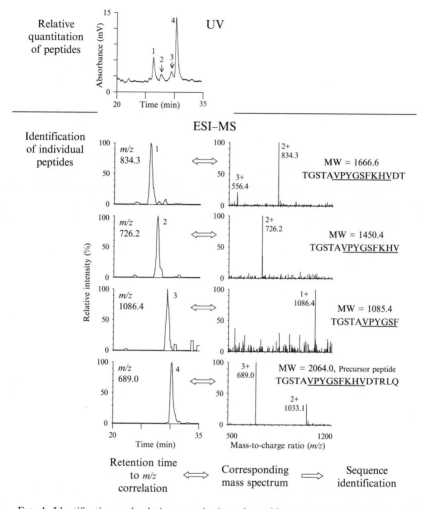

FIG. 4. Identification and relative quantitation of peptides generated from the *in vitro* digestion of the renal ubiquitous protein precursor peptide ($RU1_{29-47}$) by 20S proteasome purified from renal carcinoma cells using on-line LC/ESI MS analyses. Major peptide fragments obtained after 1 h of digestion were separated on a Brownlee Aquapore (Perkin Elmer, Boston, MA) C18 column (2.1 mm × 10 cm) with eluents consisting of water/acetonitrile/TFA and then were analyzed by on-line UV detection (214 nm) and electrospray ionization ESI–MS. Relative quantitation of generated peptides fragments (UV peaks 1, 2, and 3) and of the precursor peptide (UV peak 4) was obtained from the UV chromatogram by measuring corresponding peak areas. Identification of individual peptide fragments was achieved from ESI–MS data at the retention time corresponding to each UV peak. Knowing the precursor peptide sequence, each generated peptide fragment was identified by its averaged or monoisotopic molecular weight calculated from singly (1+), doubly (2+), or triply (3+) charged ions on the corresponding mass spectrum. Amino acids belonging to the epitope sequence are underlined in the identified peptide fragment sequences; MW represents molecular weight.

minor products of the digestion. Their use would also allow working with lower amounts of starting material, including the amount of purified 20S proteasome.

Peptide Identification by MS

The aim of MS analyses may be either to identify all possible fragments formed during 20S proteasome digestions of precursor peptides (to understand the mechanisms of degradation) or to look specifically for one epitope sequence, which may be present as a minor product.

Peptide identification from an on-line LC–MS analysis is illustrated in Fig. 4, with the *in vitro* digestion of the renal ubiquitous protein epitope precursor peptide, $RU1_{29-47}$, by 20S proteasome purified from renal carcinoma cells. After peptide separation by RP–HPLC, a UV absorption profile is acquired at 214 nm, corresponding to the absorption of the peptide bond. If enough resolution is obtained to separate all peptides from the mixture, then UV absorption can be used to estimate the relative amount of peptides generated during the digestion. An equivalent profile is then recorded by ESI–MS, leading to the total ion chromatogram, from which significant MS information is extracted. For example, at the retention time of UV peak 1, a major signal at mass-to-charge ratio (m/z) 834.3 is observed. The intensity of this signal can be used, similar to UV absorption measurements, to estimate the amount of peptide formed during the digestion. When co-eluting peptides are present, only the MS signal will be usable to obtain quantitative information. The full mass spectrum recorded at the retention time of the m/z 834.3 ion indicates an additional ion at m/z 556.4. From the mass spectrum, the molecular mass of the peptide is deduced (1666.6), the observed ions being the doubly charged and triply charged ions of the peptide.

Knowing the precursor peptide sequence, TGSTAVPYGSFKHV DTRLQ, the calculated molecular mass can only be attributed to the indicated peptide fragment sequence, TGSTAVPYGSFKHVDT. It is noticed that this sequence contains two basic amino acid residues, which is in agreement with the observed charge state distribution on the ESI mass spectrum. The number of basic residues in a sequence may be useful to confirm peptide sequences. If isobaric sequences are expected, then further MS/MS analyses or Edman sequencing are required to obtain sequence information. Due to the increased sensitivity in MS/MS mode of the new generation of MS analyzers, such as ion traps or hybrid QTOF, it is expected that these mass spectrometers will provide sequence information on minor products of digestion.

Even though on-line LC–MS analyses are expected to give the most comprehensive view of a peptide digest, MALDI–TOF analyses can be performed either on collected HPLC fractions or directly on a peptide digest if the latter is not too complex. A comparison of an on-line LC–ESI–MS analysis and a MALDI–TOF analysis of a peptide digest is illustrated in Fig. 5. As expected, there is a fairly good correlation between the relative intensities of peptides observed on the UV chromatogram and on the ESI total ion chromatogram but not on the MALDI–TOF mass spectrum. It is well known that ionization efficiencies of peptides by MALDI differ from one peptide to another and that the more complex a peptide mixture, the more suppression effects are observed. As a result, in the case presented in Fig. 5, six peptides (numbered 1, 2, 4, 8, 10, and 13) identified during the LC–ESI–MS analysis were not detected by MALDI–TOF analysis. An explanation may be that peptides with masses below 700 Da may not ionize well and may have to compete with matrix ions. The use of MALDI–TOF analyses may then be recommended after initial HPLC separation to minimize mixture complexity and to further desalt the sample. However, these analyses become more time-consuming than on-line LC–MS experiments. If further MS–MS analyses are required, acquisition of MALDI–PSD spectra may be performed. These analyses, however, are also more time-consuming than on-line MS/MS experiments acquired on a triple quadrupole, an ion trap, or a QTOF analyzer.

Measurement of CTL Activity

In parallel to the structural identification by HPLC–MS of the molecular species generated during antigen processing, biological samples are tested for their ability to activate CTL. The most commonly used experimental procedures for assessment of CTL activation are the measurements of the cytolytic activity or the cytokine production by CTL. Cytolytic activity is classically measured in ^{51}Cr release assays. Basically, target cells (either pulsed with synthetic peptides or with biological samples naturally expressing the antigenic peptide) are first radiolabeled with ^{51}Cr and then incubated for 4 to 5 h in presence of the effector CTL. Radioactivity released in the culture medium reflects the cytolytic activity and is then counted. In cytokine production assays, the cytokine commonly analyzed is IFNγ or tumor necrosis factor alpha (TNFα). The amount of IFNγ secreted by CTL upon interaction with target cells is measured either by enzyme-linked immunosorbent assay (ELISA) (for extracellular content) or by intracellular staining using flow cytometry (for intracellular content). The content of TNFα secreted by CTL in contact with target cells in the culture

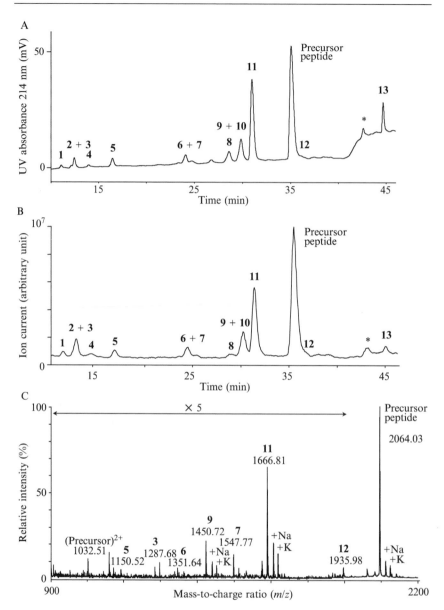

FIG. 5. Comparison of on-line LC–ESI–MS and MALDI–TOF analyses for the identification of peptides generated from the *in vitro* digestion of the renal ubiquitous protein precursor peptide (RU1$_{29-47}$) by 20S proteasome purified from human erythrocytes. (A) UV chromatogram. (B) ESI total ion chromatogram. (C) MALDI–TOF analysis. The digestion of RU1$_{29-47}$ precursor peptide (4 μg) with 20S proteasome (1 μg) was carried out in

medium is determined by testing the cytotoxic effect of the supernatant on TNFα-sensitive target cells (such as WEHI-164 cell line). Both ^{51}Cr release and cytokine production assays are very sensitive assays with a detection level of antigenic peptide in the low femtomole range. The ^{51}Cr release assays can be run in a short period of time (less than 6 h) but require usage of radioactivity. Cytokine production assays do not require radiolabeling of target cells but are run over a longer period of time (24 to 48 h).

Mechanisms of Polypeptide Degradation by Proteasomes

Information about the size and nature of peptides formed during proteasome degradation is essential for the understanding of the mechanisms involved in proteasome degradation at the molecular level and for defining more precisely the proteasome cleavage specificity.

Determination of Product Length Distribution

To confirm the potential role of proteasome in antigen processing, it was important to verify that the length of peptides generated during proteasomal degradation of proteins and/or polypeptides was consistent with antigen length.

Early work by Wenzel *et al.* (1994) showed the use of LC–MS for the analysis of degradation products from two proteins, bovine insulin B-chain

---

10 μl of 10 mM tris-HCl buffer (pH 8) at 37° for 1 h. The reaction was stopped by freezing. On-line LC–ESI–MS analysis of an aliquot of the digestion mixture was performed on a reversed-phase C18 PepMap column (3μm, 1.0 mm × 10 cm) (LC Packings, Amsterdam, The Netherlands, coupled to a TSQ 700 triple quadrupole mass spectrometer (Finnigan Mat, San Jose, CA) equipped with an electrospray source. Peptide separation was achieved with a gradient elution from 5 to 35% B in 35 min (A is 0.05% formic acid in water; B is 0.05% formic acid in acetonitrile) at a flow rate of 40 μl per minute. The UV chromatogram was recorded at 214 nm. Another aliquot of the same digestion mixture was analyzed with a Voyager DE STR MALDI–TOF mass spectrometer (PerSeptive Biosystems, Framingham, MA) in the reflector mode. The peptide mixture was desalted using a ZipTip$_{C18}$ (Millipore, Bedford, MA, http://www.millipore.com) before performing the mass mapping using α-cyano-4-hydroxycinnamic acid as a matrix. Peptides are numbered in the order of increasing retention times from the LC–MS analysis. Peptide sequences and monoisotopic molecular masses are the following: 1 VDTRLQ (730.39), 2 SFHVDTRLQ (1229.64), 3 GSFKHVDTRLQ (1286.66), 4 VPYGSFKH (933.66), 5 PYGSFKHVDT (1149.53), 6 TGSTAVPYGSFKH (1350.65), 7 PYGSFKHVDTRLQ (1546.78), 8 TGSTAVPY (794.37), 9 TGSTAVPYGSFKHV (1449.71), 10 TGSTAVPYGSFKHVD (1564.74), 11 TGSTAV-PYGSFKHVDT (1665.79), 12 TGSTAVPYGSFKHVDTRL (1934.97), 13 TGSTAVPYGSF (1085.49), and the precursor peptide TGSTAVPYGSFKHVDTRLQ (2063.03), in which the epitope sequence is underlined. The asterisk in Figs. 5A and 5B indicates a peak present in a blank LC–MS analysis.

(30 amino acids) and human hemoglobin A (141 amino acids), by *Thermoplasma acidophilum* proteasome, which contains 14 identical catalytically active β subunits. These analyses indicated a surprisingly narrow size range of generated peptides, 45% of peptides being 7-, 8-, or 9-mers. The observed size distribution included peptides from 3 to 15 residues long and led the authors to suggest that a "molecular ruler" would exist, which would determine the length of generated peptides. The ruler would be defined by the distance between multiple active sites, which corresponds to the length of a 7- or 8-mer peptide in an extended conformation.

Similar experiments were then conducted with murine 20S proteasome to explore the length of fragments generated by the more complex mammalian proteasomes composed of three distinct catalytic β subunits among the seven different β subunits (Niedermann *et al.*, 1996). LC–MS analyses of the digestion of 22-mer and 44-mer peptides from ovalbumin showed that the distribution of peptide length obtained with murine 20S proteasome ranged also from 3- to 15-mer and was centered on 9-mer, with 45% of peptides being 8- to 11- mers. This size distribution was again compatible with the molecular ruler hypothesis and with the length of peptides binding to MHC class I molecules. Moreover, in this case, the immunodominant H-$2K^b$ ovalbumin epitope was identified in the digest, showing that it can be directly generated by 20S proteasome.

More recent studies have focused on the degradation of larger proteins by proteasomes from various origins and allowed a more statistical view of generated peptide fragments (Emmerich *et al.*, 2000; Kisselev *et al.*, 1998, 1999; Niedermann *et al.*, 1997; Nussbaum *et al.*, 1998). Interestingly, the length distribution observed for the digestion of a 123 amino acid protein by either mouse or yeast 20S proteasomes led again to similar peptide length distribution ranging from 5- to 11-mer or 5- to 13-mer, respectively, with more than half of the peptides falling into the size range of 8- to 10-mer. A comprehensive study based on the analyses of more than 400 products generated from the digestion of enolase 1 (436 amino acids) by 20S yeast proteasomes allowed progress in the understanding of the degradation mechanism (Nussbaum *et al.*, 1998). Product analyses were performed by HPLC separation and MALDI–TOF analyses to obtain peptide masses. Sequence information was obtained by electrospray MS/MS (on a QTOF instrument) of HPLC fractions and by Edman degradation. HPLC profiles showed a stable peak pattern over time, and no large enolase 1 fragment was detected even at early time points of the digestion corresponding to the consumption of 5% of substrate. These observations were consistent with a processive substrate degradation mechanism, as proposed for the *T. acidophilum* proteasome (Akopian *et al.*, 1997). The peptide length distribution observed with 20S yeast proteasome (Nussbaum *et al.*, 1998) was ranging from 3- mer to -25 mer peptides, with an average peptide length corresponding to 7- to -8 mer. Surprisingly, this length distribution was very similar for enolase 1 digests performed with mutant yeast 20S proteasomes carrying reduced numbers of active β subunits. These studies showed that the fragment length was not influenced by the number of active sites in the proteasome and thus strongly argued against the distance between the active sites influencing the length of digestion products and ruled out the molecular ruler hypothesis. Studies on the degradation of various length proteins by archeal and mammalian proteasomes led to similar conclusions (i.e., that the molecular ruler hypothesis could not be valid and proposed that proteolysis occurs in a processive manner, degrading one protein molecule at a time until products are small enough to diffuse out of the proteasome) (Kisselev *et al.*, 1998, 1999). Thus, proteasomes digest proteins in a fundamentally different manner from other proteases, which generally dissociate from their substrate after each cleavage (Kisselev *et al.*, 1998).

Different results were obtained from the study of the size distribution of products formed during the digestion of reduced carboxamidomethylated lysozyme (129 amino acids) by bovine 20S proteasomes (Wang *et al.*, 1999). MALDI–TOF analyses of HPLC fractions showed the presence of reaction intermediates with masses ranging from 3500 to 10,200 Da after 10 min of digestion. It was then concluded that the mechanism of mammalian 20S proteasome was nonprocessive and that intermediates dissociate from the proteasome before undergoing further degradation. More studies will thus be needed to unravel the digestion mechanism of proteasome and to determine whether it may depend on the nature of the substrates, reaction conditions, or other factors.

Proteasome Cleavage Specificity

The cleavage specificity of proteasome has been initially determined by cleavages of small fluorogenic peptides. These studies led to the description of three distinct proteolytic activities designed as trypsin-like, chymotrypsin-like, and PGPH cleaving after basic, hydrophobic, and acidic residues, respectively. Two additional activities were reported as BrAAP and SNAAP activities (Orlowski and Wilk, 2000; Orlowski *et al.*, 1993).

A study based on the degradation of various substrates with wild-type and mutant yeast 20S proteasomes containing one known inactive β subunit allowed the definition of each active β subunit's contribution to the various proteolytic activities (Dick *et al.*, 1998). Digests were analyzed by LC–MS and MS/MS on a triple quadrupole mass spectrometer. Detailed analyses of the fragments generated led to the following conclusions.

Cleavages after basic and acidic residues strictly correlate with $\beta2$ and $\beta1$ subunits, respectively. The BrAAP activity can be attributed to the $\beta1$ subunit as well. In contrast, cleavages after hydrophobic residues can be the result of $\beta5$ alone or with contributions of $\beta2$ or $\beta1$. Likewise, cleavage after small residues appears not to be restricted to a single subunit. These results are of major importance for the design of subunit-specific inhibitors to drive the proteasomal degradation toward the formation or the destruction of a given epitope in a protein sequence without affecting the whole degradation mechanism, which could lead to toxic effects for the cell (Groettrup and Schmidtke, 1999).

Detailed analysis of the digestion of enolase 1 (436 amino acids) by the wild-type and mutant yeast 20S proteasomes allowed a more statistical view and thus a more precise description of cleavage motifs (Nussbaum et al., 1998). The previously attributed activities remained. However, the influences of the nature of the residue at the cleavage site and of flanking residues on the efficiency of cleavage were also observed. For example, analyses of cleavages performed by $\beta2$ revealed a preference for arginine over lysine residues at the cleavage site as well as an enrichment of lysine residues at position P3 (three residues away from the cleavage site on the N-terminal side). The enrichment of proline residues at position P4 was also observed when all products were considered. The effect of a proline residue within an epitope sequence on its proteasomal formation has been studied (Miconnet et al., 2000; Shimbara et al., 1998). The replacement of the proline residue at position P4 within the epitope sequence by an alanine residue resulted in a marked reduction of the epitope production from a precursor peptide. It has been proposed that proline residues within epitope sequences may protect the sequence from random cleavage by proteasome (Shimbara et al., 1998).

The detailed analyses of all products formed during the digestion of proteins by proteasomes leading to the determination of cleavage motifs represent valuable information for developing algorithms for the prediction of proteasomal cleavages. Moreover, since it appears that proteasomes from various eukaryotic origins lead to similar digestion patterns, all results obtained should be usable to predict cleavage sites for human proteasome. Such algorithms were described very recently (Holzhutter et al., 1999; Kuttler et al., 2000a,b) and were able to reproduce the experimental data with 93 to 100% accuracy. In combination with existing predictive algorithms for MHC binding, these algorithms for the prediction of proteasome cleavages should allow the refinement of the identification of CTL epitopes within a protein sequence.

These algorithms are so far based on results obtained from digestions with 20S proteasomes. Few comparisons of in vitro digestions performed with 20S and 26S proteasomes have been described (Emmerich *et al.*, 2000; Kisselev *et al.*, 1999). These comparisons of *in vitro* digestions of denatured, nonubiquitinated proteins by mammalian 20S and 26S proteasomes have shown overlapping but different sets of peptides despite numerous identical cleavage sites. Thus, more results are awaited to better understand the mechanisms of *in vitro* digestions of proteins by 26S proteasomes as well as the role of the 19S regulatory complexes associated to 20S proteasomes to form the 26S proteasome. Similarly, the role of the IFNγ-inducible PA28 activator of 20S proteasome in antigen processing remains to be elucidated. Comparisons of 20S proteasome digestion profiles with and without PA28 led to different results for the generation of CTL epitopes, some being favored and some being unaffected (Dick *et al.*, 1996; Shimbara *et al.*, 1998).

Relationship Between *In Vitro* Antigen Processing by 20S Proteasome and Biological Activities

MS identification of all peptides generated by the *in vitro* digestion of an antigen precursor by 20S proteasome allows progress in the understanding of the molecular mechanisms involved in proteasome proteolysis, as described earlier. Combined with biological data, this approach sets molecular grounds to explain the observed biological activities. Examples from the literature will illustrate how these MS strategies are integrated to understand biological phenomena.

Determination of Epitope Hierarchies

Each protein that can be degraded by proteasome in the cell contains a number of potential epitope sequences. These sequences can be predicted by algorithms based on defined MHC binding criteria. Among the predicted sequences, some will elicit strong (immunodominant) cytotoxic T cell responses, and others will elicit weak (subdominant) or no (cryptic) responses (Yewdell and Bennink, 2001). A number of factors can contribute to the immunodominance of an epitope. First, the epitope has to be generated and transported to the MHC molecule. Then, it has to bind to the MHC molecule with an affinity constant, K_d, greater than 500 n*M*. Finally, it has to activate a $CD8^+$ T cell from the repertoire. As a result, only 1 out of 2000 of randomly generated peptides achieve immunodominant status with a given class I allele (Yewdell and Bennink, 2001). Among peptides that can bind an MHC molecule with a biologically significant affinity and that can trigger a T cell response as a synthetic peptide, four out of five remain subdominant or cryptic epitopes due to the low quantities of peptide available for MHC presentation. The amount of peptide available depends on the efficiency and cleavage specificity of proteasome degradation and also on the efficiency of transport to the endoplasmic reticulum by the TAP transporter. Thus, antigen processing by proteasome may play a role in determining peptide immunodominance.

MS has been used by several groups to understand the role of antigen processing by proteasome in determining the immunodominance and more interestingly the subdominance of some epitopes. The work described by Dick *et al.* in 1994 was the first illustration of the use of MS to analyze and identify peptides resulting from the digestion of proteins by purified 20S proteasome. These analyses allowed direct evidence for the production of the immunodominant epitope of ovalbumin, Ova257–264, SIINFEKL. The subdominant epitope of ovalbumin, Ova55–62, KVVRFDKL, however, was not detected. These early experiments were performed using HPLC separation followed by fast atom bombardment mass spectrometry (FAB–MS) analysis of the active fractions as determined by specific CTL assays. The identity of the peptide was ascertained by its coelution with its synthetic analogue, the determination of its molecular mass by FAB–MS, and the analysis of its sequence by Edman degradation.

Niedermann *et al.* (1995, 1996) studied in more detail the processing by 20S proteasome of ovalbumin epitopes to understand the role of proteasome processing in epitope hierarchies. Indeed, in the case of ovalbumin, the K_d value of Ova55–62 has been reported to be around 50 nM, which is in the range expected for immunodominant epitopes and may thus not be sufficient to explain the subdominant status of Ova55–62. Moreover, there is no evidence that the low immunogenicity is caused by a defect in the CTL repertoire (Niedermann *et al.*, 1995). Niedermann *et al.* (1996) analyzed the digestion profiles of synthetic precursor peptides (22- to 44-mer) corresponding to partial sequences of ovalbumin and encompassing the epitope sequences. The complete identification of peptides generated in the digests was rendered possible by the use of on-line coupling of HPLC to electrospray MS. Sequence information was achieved by MS/MS on a triple quadrupole analyzer. These analyses clearly showed that, independently of the precursor length, major cleavage sites occurred at the N- and C-termini of the immunodominant epitope sequence, thus liberating large quantities of the epitope. In contrast, a major cleavage site within the subdominant epitope sequence was observed, thus destroying most of this epitope. The sensitivity of the LC–MS analyses compared to the initially performed FAB–MS analyses (and also to pool Edman sequencing also performed) of the digest allowed the detection of the subdominant epitope sequence in trace amounts. The detection of the epitope sequence may then explain the low but detectable presentation of Ova55–62 at the cell surface and lead to a perfect correlation between mass spectrometric data and CTL activity *in vivo* obtained with the same precursors introduced into the cell with liposomes (Niedermann *et al.*, 1995). It was then concluded that the subdominance of Ova55–62 was due to an inefficient processing by proteasome because of a major cleavage occurring within the epitope sequence.

The work of Ossendorp *et al.* (1996) illustrates the complementarity of the various approaches, including MS, that can be used to understand the structural evolution of an epitope during each step of the antigen presentation pathway. The immunodominant epitope sequence, KSPWFTTL, of the endogenous type of murine leukemia virus differs from the exogenous Friend/Moloney/Rauscher (FMR) virus type of epitope sequence, RSPWFTTL, by one amino acid residue. This single-point mutation does not affect peptide binding to MHC class I K^b molecules, and specific CTL against each epitope can be raised. However, cells expressing the FMR virus type were not recognized by specific CTL. Comparative LC–MS analyses of peptides generated from the digestion of KSPWFTTL and RSPWFTTL precursors by purified 20S proteasome clearly showed a major cleavage after the arginine residue within the epitope sequence, RSPWFTTL, leading to the destruction of this epitope. This result may explain the lack of CTL activity observed with cells expressing the FMR virus type. For the immunodominant epitope, KSPWFTTL, the exact epitope sequence was not detected. However, major N-terminally extended epitope sequences were identified. Ossendorp *et al.* (1996) then showed that, unlike the exact 8-mer epitope sequence, the extended 10-mer peptide was efficiently transported by TAP. Moreover, the 8-mer epitope sequence was identified to be presented at the cell surface by screening HPLC fractions of peptides eluted from the cell surface for CTL activity and by coelution with synthetic peptides. Thus, the proteasome can generate N-terminally extended epitope sequences, which can be efficiently transported by TAP and subsequently trimmed to the proper epitope sequence before binding to the MHC class I molecule for presentation at the cell surface.

Effect of a Single Amino Acid Substitution Flanking an Epitope Sequence

Evidence for the importance of the nature of amino acid sequences flanking an epitope sequence on its generation by proteasome was described by several groups (reviewed in Niedermann *et al.*, 1999). Two examples show extreme cases in which a single-point mutation flanking an epitope sequence can affect proteasomal cleavages and lead to the abrogation of an epitope generation as evidenced by LC–MS analyses of 20S proteasome digestion of precursor peptides (Beekman *et al.*, 2000; Theobald *et al.*, 1998).

The first example (Beekman *et al.*, 2000) is given by an arginine to histidine mutation on the p53 tumor suppressor protein at position 273 flanking an epitope sequence, LLGRNSFEV, spanning from position 264 to position 272. Mutations on this protein are found in a majority of human cancers. The mutation described was shown to promote tumor growth *in vivo* due to an escape of recognition by CTL specific of the epitope. Analyses of the digestion products of 27-mer peptides containing the epitope sequence followed by either an arginine (wild type) or an histidine (mutant) residue by human 20S proteasome revealed that the cleavage after the C-terminal valine residue of the epitope sequence was markedly reduced in the mutant digest. Moreover, observed cleavages led to N-terminally extended peptides, and the exact sequence of the epitope was not detected. These extended peptides can form epitope precursors that might be subsequently trimmed to the size of the optimal CTL epitope. The exact epitope sequence may still be formed in the digest in quantities too low to be detected under the conditions used for MS analyses. This hypothesis was tested by using five times more concentrated HPLC fractions for measuring the CTL activity. Lysis of target cells was observed with only one HPLC fraction of the wild-type precursor peptide digest, which contained the 9-mer epitope sequence as identified by MS/MS. Thus, the exact epitope sequence is only detected in the wild-type precursor digest. The sequence presented at the cell surface was also identified as the 9-mer peptide based on HPLC coelution with synthetic analogues. These analyses thus show that a single-point mutation just following an epitope sequence can lead to an abrogation of epitope generation by proteasome. In the present case, altered proteasomal processing prevents the formation of the natural CTL epitope and leads to tumor escape.

The second example is given by the single-point mutation from an asparagine to an aspartic acid residue C-terminally flanking the viral epitope sequence, SSWDFITV, from the Moloney murine leukemia virus (Beekman *et al.*, 2000). In this study, analyses were performed by injection of the peptide digest (after on-line desalting through a precolumn and a gradient elution) into a nanoelectrospray source with a QTOF analyzer, allowing MS and MS/MS experiments. This represents one of the first studies in this field using on-line separation and nanoelectrospray ionization, which should provide an increased sensitivity over electrospray for the identification of minor peptides. Analyses of *in vitro* digestion products of 26-mer precursor peptides with murine 20S proteasomes showed that a major cleavage site was located after the C-terminal valine residue of the epitope sequence when an asparagine residue followed; however, this cleavage site was shifted after the aspartic acid residue when this residue followed the epitope sequence. No detectable cleavage after the C-terminal valine residue of the epitope sequence was detected in this case. The authors (Beekman *et al.*, 2000) proposed that the abolition of precise epitope C-terminal cleavage by a single-point mutation can be a highly efficient mechanism of immune escape.

20S Standard Proteasome Versus Immunoproteasome in Antigen Processing

Immunoproteasomes can either be constitutive of some cells or tissues (Cardozo *et al.*, 1995; Eleuteri *et al.*, 1997, 2000) or induced by IFNγ treatment of other cells. Immunoproteasomes can be characterized by the presence of the three specific catalytic β subunits: LMP2, LMP7, and MECL-1. The presence of LMP2 and LMP7 can be determined by Western blot analysis with corresponding antibodies or by 2-D gel electrophoresis with identification based on the isoelectric point and molecular weight of the subunits. One study (Dahlmann *et al.*, 2000) has shown that cells possess a heterogeneous distribution of proteasome, each proteasome subtype exhibiting different enzymatic properties with fluorogenic peptides. Thus, the purified 20S proteasome of a particular cell line or tissue represents the average of the various proteasome subtypes. The incorporation of the immunosubunits into the proteasome structure induces changes in its catalytic activities (Rock and Goldberg, 1999). In particular, the PGPH activity, corresponding to cleavages after acidic residues, of the immunoproteasome as measured *in vitro* with fluorogenic peptides is decreased. In agreement with this observation, MS analyses of *in vitro* digestions of ovalbumin and antiproliferative B cell translocation gene (BTG1) epitope precursors showed that cleavages after acidic residues were less efficient with immunoproteasome than with standard proteasome (Niedermann *et al.*, 1997). Digestions of proteins or antigen precursor peptides by the immunoproteasome are expected to be different from digestions by the standard proteasome. The consequences on the generation of particular epitopes remain, however, difficult to predict. Immunoproteasomes often favor epitope generation (Gileadi *et al.*, 1999; Schwarz *et al.*, 2000; Sewell *et al.*, 1999; Sijts *et al.*, 2000a,b) but not always (Morel *et al.*, 2000). Thus, immunoproteasomes contribute to antigen processing, but the extent of their role is unclear.

A recent example showing that immunoproteasome is required for efficient antigen generation is given by the study of the viral epitope of hepatic B virus (HBV) core antigen, HBcAg141–151, STLPETTVVRR

(Sijts *et al.*, 2000a). LC–MS analyses of the digestion of a 32-mer precursor encompassing the epitope sequence with purified 20S standard proteasome and immunoproteasome clearly indicated quantitative differences in the generation of peptides. The exact sequence of the epitope was produced at a constant rate in the digestion with immunoproteasome, reaching amounts of the epitope 8 to 10 times larger than with standard proteasome after 8 h of digestion. Interestingly, digestions performed with proteasomes from either mouse or human produced similar qualitative and quantitative differences in the generation of peptides from standard proteasome and immunoproteasome digestions. This is in agreement with a previous study showing highly conserved cleavage patterns of polypeptides obtained with proteasomes from various eukaryotic origins (Niedermann *et al.*, 1997). The results obtained by Sijts *et al.* (2000a) indicate that cells expressing standard proteasome will not process the epitope, whereas cells expressing immunoproteasome will. The consequences for the epitope studied may be that the epitope will not be generated by HBV-infected hepatocytes expressing standard proteasome unless a high level of lymphokines is present to convert the standard proteasome into immunoproteasome. This would then explain why HBV may persist in chronically infected patients (associated with low levels of lymphokines) as opposed to patients with acute hepatitis B (associated with high levels of lymphokines).

The study of the immunodominant epitope of lymphocytic choriomeningitis virus (LCMV), NP118, RPQASGVYM, provides another example of the efficient generation of a viral epitope by the immunoproteasome (Schwarz *et al.*, 2000). In this case, Schwarz *et al.* (2000) showed that the presence of immunosubunits enhanced the formation of N-terminally extended epitope sequences. The research group used B8 cells as a source of 20S standard proteasome and an LMP2/LMP7/MECL-1 triple transfectant as a source of immunoproteasome to compare the digestion products generated from a 25-mer precursor peptide including the NP118 sequence. Analyses of the digests were performed using HPLC separation followed by MALDI–TOF analyses of the collected fractions for peptide identification. Peptide quantitation was obtained from Edman sequencing analyses. Again, the data showed quantitative differences in the generation of peptides from the digests. The amounts of peptides corresponding to the epitope sequence extended on its N-terminus by two and three amino acid residues were threefold and sixfold higher with immunoproteasome than with standard proteasome, respectively, after 48 h of digestion. In this case, the peptides of interest were minor products of the digestion, which could be quantitated only after 48 h of digestion to accumulate enough material. It was expected that the use of on-line microLC/MS strategies with nanoelectrospray ionization may help in increasing the detection limits of such minor products and thus enable shorter digestion times. Nevertheless, data indicated that extended peptides are favored with immunoproteasome. These peptides are better transported by TAP than the exact epitope sequence due to the proline residue in position two of the epitope sequence, and the generation of these extended peptides can be correlated to the amount of epitope available for presentation at the cell surface after N-terminal trimming.

We have shown for the first time that the immunoproteasome can be detrimental to the generation of epitopes (Morel et al., 2000). This is the case of the renal ubiquitous protein epitope, $RU1_{34-42}$, VPYGSFKHV. The source of 20S standard proteasome was a renal carcinoma cell line derived from the primary renal cell carcinoma of a patient. Immunoproteasome was purified from the same renal carcinoma cell line treated with IFNγ. Comparative LC–MS analyses of the digestion of a 19-mer precursor peptide encompassing the epitope sequence with standard proteasome and immunoproteasome showed that the cleavage after the C-terminal valine residue of the epitope sequence was only detected with the standard proteasome. These results correlated perfectly with the CTL assays, which only displayed a specific CTL activity with the standard proteasome digest. Similar results were obtained for the major antigenic peptide of melanoma differentiation antigen Melan-A.

The possibility that some antigens may not be produced by immunoproteasomes may have critical implications. One implication is that these antigens will not be presented by dendritic cells, which have been shown to express the immunoproteasome (Morel et al., 2000). Dendritic cells are considered to be major antigen-presenting cells, a subset of which is involved in the presentation of self-antigens to thymocytes for negative selection. Thus, RU1-specific CTL would escape negative selection in the thymus. Another implication is that these antigens will not be presented by cells in an IFNγ-rich environment, which will favor the presence of immunoproteasome. This, in the case of tumor antigens, could explain tumor escape mechanisms after IFNγ treatment, in which tumor antigens would not be presented and tumor cells would escape CTL attack.

Perspectives

MS entered the field of immunology and in particular of the characterization of antigenic peptides involved in the immune response with the pioneering work of Hunt et al. (1992); the group used very sensitive micro-HPLC coupling to electrospray MS to identify peptides bound to MHC class I molecules and eluted from the surface of cells (Cox et al.,

1994; Hunt *et al.*, 1992). The resolving power and sensitivity of LC–MS techniques were then rapidly used to identify peptides in mixtures generated from the degradation of proteins by purified 20S proteasome. This technique remains a valuable tool for the study of antigen processing by purified proteasome (Wenzel *et al.*, 1994). The new generation of analyzers (ion trap, QTOF) associated to electrospray and also to nanoelectrospray sources can be directly coupled to micro- and nano-HPLC systems and allows researchers to obtain peptide mass and sequence information simultaneously on tens of femtomoles of peptides. These approaches should allow the identification of minor components of complex peptide mixtures, which is often the case of CTL epitope sequences. These strategies will certainly continue to be used to accumulate data on the degradation of proteins by proteasomes to help the refinement of predictive proteasome cleavages algorithms and to correlate antigen processing by proteasome at the molecular level with biological activities. MS analyses of proteasome degradation patterns have also been proposed to be incorporated in epitope prediction procedures (Kessler *et al.*, 2001). By identifying the correct C-terminal cleavage of putative epitope sequences, these analyses limit the number of peptides to be assayed for MHC binding. This novel strategy allowed the identification of four human tumor epitopes from the tumor-associated preferentially expressed antigen from melanoma (PRAME) protein and demonstrated the usefulness of proteasome processing criteria for the identification of CTL epitopes.

Complementary to the analyses of degradation patterns by proteasome, the study of the structure/activity relationship of proteasome itself may be of importance to the understanding of antigen processing by the proteasome. Indeed, it has been shown that proteasomes correspond to a heterogeneous population of molecules in cells, each population displaying different catalytic activities (Dahlmann *et al.*, 2000). The detailed analysis of proteasome subunit composition may then be of particular interest to characterize these different proteasome populations. Proteomics-type approaches based on initial separation of proteasome subunits by 2-D gel electrophoresis followed by MS analyses can be used to characterize proteasome subunits (Huang *et al.*, 1999, 2000). These analyses performed on rat liver 20S proteasomes showed the existence of more proteins than the expected different α and β subunits, suggesting a high degree of structural complexity and diversity (Huang *et al.*, 2000). The combination of these data with analyses of proteasome digestion patterns and with biological activities should help to shed more light on the role of proteasomes in antigen processing.

MS is being introduced into strategies to characterize proteins involved in several steps of the antigen presentation pathway. For example, proteins upregulated by IFNα and IFNγ were identified as proteins involved (upstream from proteasome processing) in the ubiquitin pathway. It is proposed that their upregulation may increase the rate of ubiquitin conjugation and, consequently, may contribute to enhanced antigen presentation (Nyman *et al.*, 2000). Proteins directly associated to yeast 26S proteasome have also been characterized, including a new subunit of the 19S regulatory complex (Verma *et al.*, 2000). The role of these associated proteins remains to be elucidated. Downstream from proteasome processing in the antigen presentation pathway, aminopeptidases able to trim the N-terminus of peptides to reach the correct CTL epitope sequence were identified in the cytosol (Stoltze *et al.*, 2000). The role of these enzymes in antigen processing is still debated (Lucchiari-Hartz *et al.*, 2000). All of these studies on protein identification required the use of proteomics-type approaches based on the combination of separative (2-D gel electrophoresis or multidimensional chromatography) and MS techniques (MALDI–TOF for mass mapping and nanoelectrospray ion trap or QTOF for sequencing). It can be foreseen that these approaches will be very useful for identifying yet other proteins participating in the antigen presentation pathway and for characterizing putative parallel systems to proteasome processing for the generation of class I antigenic peptides (Geier *et al.*, 1999; Glas *et al.*, 1998).

Addendum

This subsection is intended to cover work published in early twenty-first century literature that illustrates the main progress made in the different fields discussed in this chapter since it was first written.

In the last few years, approximately 2002 through 2005, interest in proteasome machinery has increased dramatically, and numerous studies have been conducted to better understand the pivotal role of this macromolecular complex in the regulation of protein degradation. Some of the major advances in proteasome research include the determination of the crystal structure of mammalian 20S proteasome from bovine liver (Unno *et al.*, 2002a), which also allowed the prediction of a model for the immunoproteasome (Unno *et al.*, 2002b). Some progress in the characterization and in the understanding of the role of 20S proteasome regulators, including the IFNγ-inducible PA28 regulator, have been made, but regulation of proteasome activity appears to be complex and is still not fully understood (Bajorek and Glickman, 2004; Rechsteiner and Hill, 2005; Sijts *et al.*, 2002). Studies using proteasome inhibitors (Groll and Huber, 2004) have contributed greatly to the understanding of the role of proteasome in biological processes, and one of them, the *bortezomib*, now opens new perspectives for proteasome as a therapeutic target in cancer therapies, as this inhibitor was approved in 2004 for the treatment of multiple myeloma (Adams, 2004).

Antigenic Peptide Processing for MHC Class I Presentation

The study of antigenic peptide processing has allowed progress in the understanding of the various steps in the antigen presentation pathway and in the role of immunoproteasomes (Goldberg *et al.*, 2002; Kloetzel, 2004b; Paulsson, 2004; Van den Eynde and Morel, 2001). One major issue has concerned the characterization of enzymes acting downstream of the proteasome that are involved in N-terminal peptide trimming as well as play a critical role in the generation of correct antigenic peptide sequences for presentation (Kloetzel, 2004a; Kloetzel and Ossendorp, 2004; Rock *et al.*, 2004). Several of these enzymes have been found to act in the cytosol before the transport of antigenic peptides to the ER, and others have been characterized in the ER. It is now thought that the extent of antigen presentation depends on the balance and interplay between several proteolytic processes that may generate or destroy epitopes.

MS for the Characterization of Antigenic Peptides and Their Identification from Proteasome Processing

The challenge in the characterization of antigenic peptide sequences constituting the peptide repertoire lies in the sample complexity and the very low quantity of some epitopes. The gain in sensitivity of the newest instruments and their capacity to handle high-throughput MS/MS analyses allow a more comprehensive analysis of the peptide repertoire using smaller amounts of starting material (Admon *et al.*, 2003; Engelhard *et al.*, 2002; Purcell and Gorman, 2004). Taking advantage of the strategies developed in proteomics, researchers have proposed new objectives, such as the identification of overpresented MHC ligands when comparing the peptide repertoire of two different samples (Lemmel *et al.*, 2004). The differential quantitative approach described is based on stable isotope labeling followed by MS analysis, and the results show its usefulness for the identification of potential tumor-associated antigens. Another strategy recently published has been developed to detect defined peptides presented with a particular MHC class I molecule with a very high sensitivity prior to CTL assays (Hogan *et al.*, 2005). It is based on the combination of several chromatographic steps and the use of selected reaction monitoring MS and shows that it can detect one peptide present in a quantity equivalent to as low as one copy per cell.

The identification of peptide sequences resulting from the *in vitro* digestion of epitope precursor peptides by proteasomes has in the recent years mainly been applied to the comparison of the cleavage specificities of standard 20S proteasome and immunoproteasome. These studies contributed to a better understanding of the role of immunoproteasomes and of the proteolytic mechanisms at the molecular level. Experimentally, newer generations of mass spectrometers, like QTOF, ion traps, Q-traps, and TOF/TOF, offering improved sensitivity and acquisition of quality MS/MS data and leading to valuable peptide sequence information have been available. For example, these studies have shown that the immunoproteasome could be required for the generation of a tumor-specific antigen, a new MAGE-3 antigen presented by HLA-B40 (Schultz *et al.*, 2002) and could determine the subdominance of epitopes derived from the glycoprotein of the lymphocytic choriomeningitis virus (Basler *et al.*, 2004). The quantitative analysis of the peptides generated from the *in vitro* digestion of enolase-1 by the standard 20S proteasome and the immunoproteasome has led to define specific cleavage motifs that can explain the differential processing of some antigens (Toes *et al.*, 2001). These data can also be used to refine proteasomal cleavage prediction algorithms (Nussbaum *et al.*, 2003).

Analysis of the peptides generated from the *in vitro* digestion of the prion protein has confirmed the different cleavage specificities of standard 20S proteasome and immunoproteasome; this analysis has also evidenced differences in the overall degradation efficiency upon changes in the amino acid sequence of the protein, which may be related to its pathogenicity (Tenzer *et al.*, 2004). In the case of the decamer Melan-A antigenic peptide, the mutation of one residue in the epitope sequence, leading to a higher binding affinity to the MHC molecule and to a stronger CTL induction, does not alter significantly the *in vitro* proteasome digestion pattern of corresponding precursors (Chapatte *et al.*, 2004). Finally, the study of the peptides generated from the *in vitro* proteasome digestion of a series of precursor peptides presenting various deletions in their sequence allowed the validation of the occurrence of peptide splicing in the proteasome for the generation of one epitope of the melanocytic protein gp100 (Vigneron *et al.*, 2004). These results revealed a novel and unanticipated mechanism that may be involved in the generation of antigenic peptides by the proteasome.

MS for the Study of Proteasome Complexes

The continuous improvement in the development of proteomic approaches based on MS has prompted the study of the subunit composition of proteasome complexes present in different cells. The subunit patterns of several proteasome complexes obtained by 2-D gel electrophoresis separation and MS identification have been published for different species. (Claverol *et al.*, 2002; Iwafune *et al.*, 2002; Kurucz *et al.*, 2002; Yang *et al.*, 2004). MS has also contributed to the identification of sites of proteasome subunit posttranslational modifications like acetylations and phosphorylations (Claverol *et al.*, 2002; Iwafune *et al.*, 2002, 2004) and has been used in conjunction with immunological detection and lectin-binding experiments to identify glycosylated proteasome subunits (Sumegi *et al.*, 2003). The role of these modifications is still under investigation. The phosphorylation of the $\alpha7$ subunit of the 20S core complex has been proposed to be involved in 26S proteasome stability (Bose *et al.*, 2004), and the glycosylation of the 19S regulatory complex has been shown to inhibit proteasome activity (Zhang *et al.*, 2003). Another interesting and promising MS analysis of proteasome complexes has been demonstrated using a QTOF instrument and allowed the determination of the molecular mass of the whole noncovalent 20S core complex (Chernushevich and Thomson, 2004). This analysis may represent a new way of studying proteasome heterogeneity and maybe assembly. Recently, a quantitative proteomic approach has been proposed for the study of human 20S proteasome heterogeneity (Froment, 2005). It is based on stable isotope labeling of the subunits followed by 2-D gel electrophoresis separation and MS analyses. This method allows the quantification of the variations in subunit composition of proteasome complexes from different samples. These data may then be related to differences in proteasome activity.

Acknowledgments

We thank Dr. Sandra Morel, Dr. Jacques Chapiro, and Dr. Benoît Van den Eynde (Ludwig Institute for Cancer Research, Brussels) as well as Dr. Frédéric Lévy (Ludwig Institute for Cancer Research, Lausanne) for their fruitful collaborations. We are grateful to Professor Simon J. Gaskell for careful reading of the manuscript.

References

Adams, J. (2004). The proteasome: A suitable antineoplastic target. *Nat. Rev. Cancer* **4,** 349–360.
Admon, A., Barnea, E., and Ziv, T. (2003). Tumor antigens and proteomics from the point of view of the major histocompatibility complex peptides. *Mol. Cell Proteomics* **2,** 1388–398.
Akopian, T. N., Kisselev, A. F., and Goldberg, A. L. (1997). Processive degradation of proteins and other catalytic properties of the proteasome from *Thermoplasma acidophilum. J. Biol. Chem.* **272,** 1791–1798.

Androlewicz, M. J. (2001). Peptide generation in the major histocompatibility complex class I antigen processing and presentation pathway. [In process citation]. *Curr. Opin. Hematol.* **8,** 12–16.

Bajorek, M., and Glickman, M. H. (2004). Keepers at the final gates: Regulatory complexes and gating of the proteasome channel. *Cell. Mol. Life Sci.* **61,** 1579–1588.

Barroso, M. B., and de Jong, A. P. (1999). Sheathless preconcentration-capillary zone electrophoresis-mass spectrometry applied to peptide analysis. *J. Am. Soc. Mass Spectrom.* **10,** 1271–1278.

Basler, M., Youhnovski, N., Van den Broek, M., Przybylski, M., and Groettrup, M. (2004). Immunoproteasomes down-regulate presentation of a subdominant T cell epitope from lymphocytic choriomeningitis virus. *J. Immunol.* **173,** 3925–3934.

Bazemore Walker, C. R., Sherman, N. E., Shabanowitz, J., and Hunt, D. F. (1998). Mass spectrometry in immunology: Identification of a minor histocompatibility antigen. *In* "Mass Spectrometry of Biological Materials" (B. S. Larsen and C. N. McEwen, eds.). Marcel Dekker, New York.

Beekman, N. J., van Veelen, P. A., van Hall, T., Neisig, A., Sijts, A., Camps, M., Kloetzel, P. M., Neefjes, J. J., Melief, C. J., and Ossendorp, F. (2000). Abrogation of CTL epitope processing by single amino acid substitution flanking the C-terminal proteasome cleavage site. *J. Immunol.* **164,** 1898–1905.

Beninga, J., Rock, K. L., and Goldberg, A. L. (1998). Interferon-gamma can stimulate post-proteasomal trimming of the N terminus of an antigenic peptide by inducing leucine aminopeptidase. *J. Biol. Chem.* **273,** 18734–18742.

Beyette, J. R., Hubbell, T., and Monaco, J. J. (2001). Purification of 20S proteasomes. [In process citation]. *Methods Mol. Biol.* **156,** 1–16.

Bochtler, M., Ditzel, L., Groll, M., Hartmann, C., and Huber, R. (1999). The proteasome. *Annu. Rev. Biophys. Biomol. Struct.* **28,** 295–317.

Boon, T., Cerottini, J. C., Van den Eynde, B., van der Bruggen, P., and Van Pel, A. (1994). Tumor antigens recognized by T lymphocytes. *Annu. Rev. Immunol.* **12,** 337–365.

Bose, S., Stratford, F. L., Broadfoot, K. I., Mason, G. G., and Rivett, A. J. (2004). Phosphorylation of 20S proteasome alpha subunit C8 (alpha7) stabilizes the 26S proteasome and plays a role in the regulation of proteasome complexes by gamma-interferon. *Biochem. J.* **378,** 177–184.

Brockman, A. H., Orlando, R., and Tarleton, R. L. (1999). A new liquid chromatography/tandem mass spectrometric approach for the identification of class I major histocompatibility complex associated peptides that eliminates the need for bioassays. *Rapid Commun. Mass Spectrom.* **13,** 1024–1030.

Cardozo, C., Eleuteri, A. M., and Orlowski, M. (1995). Differences in catalytic activities and subunit pattern of multicatalytic proteinase complexes (proteasomes) isolated from bovine pituitary, lung, and liver. Changes in LMP7 and the component necessary for expression of the chymotrypsin-like activity. *J. Biol. Chem.* **270,** 22645–22651.

Cerundolo, V., Benham, A., Braud, V., Mukherjee, S., Gould, K., Macino, B., Neefjes, J., and Townsend, A. (1997). The proteasome-specific inhibitor lactacystin blocks presentation of cytotoxic T lymphocyte epitopes in human and murine cells. *Eur. J. Immunol.* **27,** 336–341.

Chapatte, L., Servis, C., Valmori, D., Burlet-Schiltz, O., Dayer, J., Monsarrat, B., Romero, P., and Levy, F. (2004). Final antigenic Melan-A peptides produced directly by the proteasomes are preferentially selected for presentation by HLA-A*0201 in melanoma cells. *J. Immunol.* **173,** 6033–6040.

Chernushevich, I. V., and Thomson, B. A. (2004). Collisional cooling of large ions in electrospray mass spectrometry. *Anal. Chem.* **76,** 1754–1760.

Ciechanover, A. (1998). The ubiquitin-proteasome pathway: On protein death and cell life. *EMBO J.* **17,** 7151–7160.

Claverol, S., Burlet-Schiltz, O., Girbal-Neuhauser, E., Gairin, J. E., and Monsarrat, B. (2002). Mapping and structural dissection of human 20S proteasome using proteomic approaches. *Mol. Cell. Proteomics* **1,** 567–578.

Cox, A. L., Skipper, J., Chen, Y., Henderson, R. A., Darrow, T. L., Shabanowitz, J., Engelhard, V. H., Hunt, D. F., and Slingluff, C. L., Jr. (1994). Identification of a peptide recognized by five melanoma-specific human cytotoxic T cell lines. *Science* **264,** 716–719.

Coux, O., Tanaka, K., and Goldberg, A. L. (1996). Structure and functions of the 20S and 26S proteasomes. *Annu. Rev. Biochem.* **65,** 801–847.

Craiu, A., Akopian, T., Goldberg, A., and Rock, K. L. (1997). Two distinct proteolytic processes in the generation of a major histocompatibility complex class I-presented peptide. *Proc. Natl. Acad. Sci. USA* **94,** 10850–10855.

Dahlmann, B., Ruppert, T., Kuehn, L., Merforth, S., and Kloetzel, P. M. (2000). Different proteasome subtypes in a single tissue exhibit different enzymatic properties. *J. Mol. Biol.* **303,** 643–653.

de Jong, A. (1998). Contribution of mass spectrometry to contemporary immunology. *Mass Spectrom. Rev.* **17,** 311–335.

Dick, L. R., Aldrich, C., Jameson, S. C., Moomaw, C. R., Pramanik, B. C., Doyle, C. K., DeMartino, G. N., Bevan, M. J., Forman, J. M., and Slaughter, C. A. (1994). Proteolytic processing of ovalbumin and beta-galactosidase by the proteasome to a yield antigenic peptides. *J. Immunol.* **152,** 3884–3894.

Dick, T. P., Nussbaum, A. K., Deeg, M., Heinemeyer, W., Groll, M., Schirle, M., Keilholz, W., Stevanovic, S., Wolf, D. H., Huber, R., Rammensee, H. G., and Schild, H. (1998). Contribution of proteasomal beta-subunits to the cleavage of peptide substrates analyzed with yeast mutants. *J. Biol. Chem.* **273,** 25637–25646.

Dick, T. P., Ruppert, T., Groettrup, M., Kloetzel, P. M., Kuehn, L., Koszinowski, U. H., Stevanovic, S., Schild, H., and Rammensee, H. G. (1996). Coordinated dual cleavages induced by the proteasome regulator PA28 lead to dominant MHC ligands. *Cell* **86,** 253–262.

Downard, K. M. (2000). Contributions of mass spectrometry to structural immunology. *J. Mass Spectrom.* **35,** 493–503.

Eleuteri, A. M., Angeletti, M., Lupidi, G., Tacconi, R., Bini, L., and Fioretti, E. (2000). Isolation and characterization of bovine thymus multicatalytic proteinase complex. *Protein Expr. Purif.* **18,** 160–168.

Eleuteri, A. M., Kohanski, R. A., Cardozo, C., and Orlowski, M. (1997). Bovine spleen multicatalytic proteinase complex (proteasome). Replacement of *X*, *Y*, and *Z* subunits by LMP7, LMP2, and MECL1 and changes in properties and specificity. *J. Biol. Chem.* **272,** 11824–11831.

Emmerich, N. P., Nussbaum, A. K., Stevanovic, S., Priemer, M., Toes, R. E., Rammensee, H. G., and Schild, H. (2000). The human 26S and 20S proteasomes generate overlapping but different sets of peptide fragments from a model protein substrate. *J. Biol. Chem.* **275,** 21140–21148.

Engelhard, V. H., Brickner, A. G., and Zarling, A. L. (2002). Insights into antigen processing gained by direct analysis of the naturally processed class I MHC associated peptide repertoire. *Mol. Immunol.* **39,** 127–137.

Falk, K., Rotzschke, O., Stevanovic, S., Jung, G., and Rammensee, H. G. (1991). Allele-specific motifs revealed by sequencing of self-peptides eluted from MHC molecules. *Nature* **351,** 290–296.

Froment, C., Uttenweiler, J. S., Bousquet-Dubouch, M. P., Matondo, M., Borges, J. P., Esmenjaud, C., Lacroix, C., Monsarrat, B., and Burlet-Schiltz, O. (2005). A quantitative proteomic approach using 2-D gel electrophoresis and isotope-coded affinity tag labeling for studying human 20S proteasome heterogeneity. *Proteomics* **5**, 2351–2363.

Geier, E., Pfeifer, G., Wilm, M., Lucchiari-Hartz, M., Baumeister, W., Eichmann, K., and Niedermann, G. (1999). A giant protease with potential to substitute for some functions of the proteasome. *Science* **283**, 978–981.

Gileadi, U., Moins-Teisserenc, H. T., Correa, I., Booth, B. L., Jr., Dunbar, P. R., Sewell, A. K., Trowsdale, J., Phillips, R. E., and Cerundolo, V. (1999). Generation of an immunodominant CTL epitope is affected by proteasome subunit composition and stability of the antigenic protein. *J. Immunol.* **163**, 6045–6052.

Glas, R., Bogyo, M., McMaster, J. S., Gaczynska, M., and Ploegh, H. L. (1998). A proteolytic system that compensates for loss of proteasome function. *Nature* **392**, 618–622.

Goldberg, A. L., Cascio, P., Saric, T., and Rock, K. L. (2002). The importance of the proteasome and subsequent proteolytic steps in the generation of antigenic peptides. *Mol. Immunol.* **39**, 147–164.

Groettrup, M., and Schmidtke, G. (1999). Selective proteasome inhibitors: Modulators of antigen presentation? *Drug Discov. Today* **4**, 63–71.

Groll, M., Ditzel, L., Lowe, J., Stock, D., Bochtler, M., Bartunik, H. D., and Huber, R. (1997). Structure of 20S proteasome from yeast at 2.4 A resolution. *Nature* **386**, 463–471.

Groll, M., and Huber, R. (2004). Inhibitors of the eukaryotic 20S proteasome core particle: A structural approach. *Biochim. Biophys. Acta* **1695**, 33–44.

Harris, P. E. (1994). Self-peptides bound to HLA molecules. *Crit. Rev. Immunol.* **14**, 61–87.

Hegerl, R., Pfeifer, G., Puhler, G., Dahlmann, B., and Baumeister, W. (1991). The three-dimensional structure of proteasomes from *Thermoplasma acidophilum* as determined by electron microscopy using random conical tilting. *FEBS Lett.* **283**, 117–121.

Hendil, K. B., and Uerkvitz, W. (1991). The human multicatalytic proteinase: Affinity purification using a monoclonal antibody. *J. Biochem. Biophys. Methods* **22**, 159–165.

Hogan, K. T., Sutton, J. N., Chu, K. U., Busby, J. A., Shabanowitz, J., Hunt, D. F., and Slingluff, C. L., Jr. (2005). Use of selected reaction monitoring mass spectrometry for the detection of specific MHC class I peptide antigens on A3 supertype family members. *Cancer Immunol. Immunother.* **54**, 359–371.

Holzhutter, H. G., Frommel, C., and Kloetzel, P. M. (1999). A theoretical approach towards the identification of cleavage-determining amino acid motifs of the 20S proteasome. *J. Mol. Biol.* **286**, 1251–1265.

Huang, L., Shen, M., Chernushevich, I., Burlingame, A. L., Wang, C. C., and Robertson, C. D. (1999). Identification and isolation of three proteasome subunits and their encoding genes from *Trypanosoma brucei*. *Mol. Biochem. Parasitol.* **102**, 211–223.

Huang, L., Wang, C. C., and Burlingame, A. L. (2000). Investigation of intact subunit polypeptide composition of the 20S proteasome complex from rat liver using mass spectrometry. *In* "Mass Spectrometry in Biology and Medicine" (A. L. Burlingame, S. A. Carr, and M. A. Baldwin, eds.). Humana Press, Totowa, NJ.

Hunt, D. F., Henderson, R. A., Shabanowitz, J., Sakaguchi, K., Michel, H., Sevilir, N., Cox, A. L., Appella, E., and Engelhard, V. H. (1992). Characterization of peptides bound to the class I MHC molecule HLA-A2.1 by mass spectrometry. *Science* **255**, 1261–1266.

Iwafune, Y., Kawasaki, H., and Hirano, H. (2002). Electrophoretic analysis of phosphorylation of the yeast 20S proteasome. *Electrophoresis* **23**, 329–338.

Iwafune, Y., Kawasaki, H., and Hirano, H. (2004). Identification of three phosphorylation sites in the alpha7 subunit of the yeast 20S proteasome *in vivo* using mass spectrometry. *Arch. Biochem. Biophys.* **431**, 9–15.

Kessler, J. H., Beekman, N. J., Bres-Vloemans, S. A., Verdijk, P., van Veelen, P. A., Kloosterman-Joosten, A. M., Vissers, D. C., ten Bosch, G. J., Kester, M. G., Sijts, A., Drijfhout, J. W., Ossendorp, F., Offringa, R., and Melief, C. J. (2001). Efficient identification of novel HLA-A*0201-presented cytotoxic T lymphocyte epitopes in the widely expressed tumor antigen PRAME by proteasome-mediated digestion analysis. *J. Exp. Med.* **193**, 73–88.

Kisselev, A. F., Akopian, T. N., and Goldberg, A. L. (1998). Range of sizes of peptide products generated during degradation of different proteins by archaeal proteasomes. *J. Biol. Chem.* **273**, 1982–1989.

Kisselev, A. F., Akopian, T. N., Woo, K. M., and Goldberg, A. L. (1999). The sizes of peptides generated from protein by mammalian 26 and 20S proteasomes. Implications for understanding the degradative mechanism and antigen presentation. *J. Biol. Chem.* **274**, 3363–3371.

Kloetzel, P. M. (2004a). Generation of major histocompatibility complex class I antigens: Functional interplay between proteasomes and TPPII. *Nat. Immunol.* **5**, 661–669.

Kloetzel, P. M. (2004b). The proteasome and MHC class I antigen processing. *Biochim. Biophys. Acta* **1695**, 225–233.

Kloetzel, P. M., and Ossendorp, F. (2004). Proteasome and peptidase function in MHC-class-I-mediated antigen presentation. *Curr. Opin. Immunol.* **16**, 76–81.

Kloetzel, P. M., Soza, A., and Stohwasser, R. (1999). The role of the proteasome system and the proteasome activator PA28 complex in the cellular immune response. *Biol. Chem.* **380**, 293–297.

Kurucz, E., Ando, I., Sumegi, M., Holzl, H., Kapelari, B., Baumeister, W., and Udvardy, A. (2002). Assembly of the *Drosophila* 26S proteasome is accompanied by extensive subunit rearrangements. *Biochem. J.* **365**, 527–536.

Kuttler, C., Nussbaum, A. K., Dick, T. P., Rammensee, H. G., Schild, H., and Hadeler, K. P. (2000a). An algorithm for the prediction of proteasomal cleavages. *J. Mol. Biol.* **298**, 417–429.

Kuttler, C., Nussbaum, A. K., Dick, T. P., Rammensee, H. G., Schild, H., and Hadeler, K. P. (2000b). Erratum. *J. Mol. Biol.* **301**(1), 229.

Lemmel, C., Weik, S., Eberle, U., Dengjel, J., Kratt, T., Becker, H. D., Rammensee, H. G., and Stevanovic, S. (2004). Differential quantitative analysis of MHC ligands by mass spectrometry using stable isotope labeling. *Nat. Biotechnol.* **22**, 450–454.

Lowe, J., Stock, D., Jap, B., Zwickl, P., Baumeister, W., and Huber, R. (1995). Crystal structure of the 20S proteasome from the archaeon *T. acidophilum* at 3.4 Å resolution. *Science* **268**, 533–539.

Lucchiari-Hartz, M., van Endert, P. M., Lauvau, G., Maier, R., Meyerhans, A., Mann, D., Eichmann, K., and Niedermann, G. (2000). Cytotoxic T lymphocyte epitopes of HIV-1 Nef: Generation of multiple definitive major histocompatibility complex class I ligands by proteasomes. *J. Exp. Med.* **191**, 239–252.

Miconnet, I., Servis, C., Cerottini, J. C., Romero, P., and Levy, F. (2000). Amino acid identity and/or position determines the proteasomal cleavage of the HLA-A*0201-restricted peptide tumor antigen MAGE-3271–279. *J. Biol. Chem.* **275**, 26892–26897.

Mo, X. Y., Cascio, P., Lemerise, K., Goldberg, A. L., and Rock, K. (1999). Distinct proteolytic processes generate the C and N termini of MHC class I-binding peptides. *J. Immunol.* **163**, 5851–5859.

Morel, S., Levy, F., Burlet-Schiltz, O., Brasseur, F., Probst-Kepper, M., Peitrequin, A. L., Monsarrat, B., van Velthoven, R., Cerottini, J. C., Boon, T., Gairin, J. E., and Van den Eynde, B. J. (2000). Processing of some antigens by the standard proteasome but not by the immunoproteasome results in poor presentation by dendritic cells. *Immunity* 12, 107–117.

Nandi, D., Marusina, K., and Monaco, J. J. (1998). How do endogenous proteins become peptides and reach the endoplasmic reticulum. *Curr. Top. Microbiol. Immunol.* 232, 15–47.

Naylor, S., Ji, Q., Johnson, K. L., Tomlinson, A. J., Kieper, W. C., and Jameson, S. C. (1998). Enhanced sensitivity for sequence determination of major histocompatibility complex class I peptides by membrane preconcentration-capillary electrophoresis–microspray–tandem mass spectrometry. *Electrophoresis* 19, 2207–2212.

Nussbaum, A. K., Kuttler, C., Tenzer, S., and Schild, H. (2003). Using the World Wide Web for predicting CTL epitopes. *Curr. Opin. Immunol.* 15, 69–74.

Niedermann, G., Butz, S., Ihlenfeldt, H. G., Grimm, R., Lucchiari, M., Hoschutzky, H., Jung, G., Maier, B., and Eichmann, K. (1995). Contribution of proteasome-mediated proteolysis to the hierarchy of epitopes presented by major histocompatibility complex class I molecules. *Immunity* 2, 289–299.

Niedermann, G., Geier, E., Lucchiari-Hartz, M., Hitziger, N., Ramsperger, A., and Eichmann, K. (1999). The specificity of proteasomes: Impact on MHC class I processing and presentation of antigens. *Immunol. Rev.* 172, 29–48.

Niedermann, G., Grimm, R., Geier, E., Maurer, M., Realini, C., Gartmann, C., Soll, J., Omura, S., Rechsteiner, M. C., Baumeister, W., and Eichmann, K. (1997). Potential immunocompetence of proteolytic fragments produced by proteasomes before evolution of the vertebrate immune system. *J. Exp. Med.* 186, 209–220.

Niedermann, G., King, G., Butz, S., Birsner, U., Grimm, R., Shabanowitz, J., Hunt, D. F., and Eichmann, K. (1996). The proteolytic fragments generated by vertebrate proteasomes: Structural relationships to major histocompatibility complex class I binding peptides. *Proc. Natl. Acad. Sci. USA* 93, 8572–8577.

Nussbaum, A. K., Dick, T. P., Keilholz, W., Schirle, M., Stevanovic, S., Dietz, K., Heinemeyer, W., Groll, M., Wolf, D. H., Huber, R., Rammensee, H. G., and Schild, H. (1998). Cleavage motifs of the yeast 20S proteasome beta subunits deduced from digests of enolase 1. *Proc. Natl. Acad. Sci. USA* 95, 12504–12509.

Nyman, T. A., Matikainen, S., Sareneva, T., Julkunen, I., and Kalkkinen, N. (2000). Proteome analysis reveals ubiquitin-conjugating enzymes to be a new family of interferon-alpha-regulated genes. *Eur. J. Biochem.* 267, 4011–4019.

Oldstone, M. B., Lewicki, H., Borrow, P., Hudrisier, D., and Gairin, J. E. (1995). Discriminated selection among viral peptides with the appropriate anchor residues: Implications for the size of the cytotoxic T-lymphocyte repertoire and control of viral infection. *J. Virol.* 69, 7423–7429.

Orlowski, M., Cardozo, C., and Michaud, C. (1993). Evidence for the presence of five distinct proteolytic components in the pituitary multicatalytic proteinase complex. Properties of two components cleaving bonds on the carboxyl side of branched chain and small neutral amino acids. *Biochemistry* 32, 1563–1572.

Orlowski, M., and Wilk, S. (2000). Catalytic activities of the 20S proteasome, a multicatalytic proteinase complex. *Arch. Biochem. Biophys.* 383, 1–16.

Ossendorp, F., Eggers, M., Neisig, A., Ruppert, T., Groettrup, M., Sijts, A., Mengede, E., Kloetzel, P. M., Neefjes, J., Koszinowski, U., and Melief, C. (1996). A single residue exchange within a viral CTL epitope alters proteasome-mediated degradation resulting in lack of antigen presentation. *Immunity* 5, 115–124.

Pamer, E., and Cresswell, P. (1998). Mechanisms of MHC class I-restricted antigen processing. *Annu. Rev. Immunol.* **16,** 323–358.

Paulsson, K. M. (2004). Evolutionary and functional perspectives of the major histocompatibility complex class I antigen-processing machinery. *Cell. Mol. Life Sci.* **61,** 2446–2460.

Pierce, R. A., Field, E. D., den Haan, J. M., Caldwell, J. A., White, F. M., Marto, J. A., Wang, W., Frost, L. M., Blokland, E., Reinhardus, C., Shabanowitz, J., Hunt, D. F., Goulmy, E., and Engelhard, V. H. (1999). Cutting edge: The HLA-A*0101-restricted HY minor histocompatibility antigen originates from DFFRY and contains a cysteinylated cysteine residue as identified by a novel mass spectrometric technique. *J. Immunol.* **163,** 6360–6364.

Puhler, G., Weinkauf, S., Bachmann, L., Muller, S., Engel, A., Hegerl, R., and Baumeister, W. (1992). Subunit stoichiometry and three-dimensional arrangement in proteasomes from *Thermoplasma acidophilum. EMBO J.* **11,** 1607–1616.

Purcell, A. W., and Gorman, J. J. (2004). Immunoproteomics: Mass spectrometry-based methods to study the targets of the immune response. *Mol. Cell. Proteomics* **3,** 193–208.

Rammensee, H. G. (1995). Chemistry of peptides associated with MHC class I and class II molecules. *Curr. Opin. Immunol.* **7,** 85–96.

Rechsteiner, M., and Hill, C. P. (2005). Mobilizing the proteolytic machine: Cell biological roles of proteasome activators and inhibitors. *Trends Cell. Biol.* **15,** 27–33.

Rechsteiner, M., Realini, C., and Ustrell, V. (2000). The proteasome activator 11S REG (PA28) and class I antigen presentation. *Biochem. J.* **345**(Part 1), 1–15.

Rock, K. L., and Goldberg, A. L. (1999). Degradation of cell proteins and the generation of MHC class I- presented peptides. *Annu. Rev. Immunol.* **17,** 739–779.

Rock, K. L., Gramm, C., Rothstein, L., Clark, K., Stein, R., Dick, L., Hwang, D., and Goldberg, A. L. (1994). Inhibitors of the proteasome block the degradation of most cell proteins and the generation of peptides presented on MHC class I molecules. *Cell* **78,** 761–771.

Rock, K. L., York, I. A., and Goldberg, A. L. (2004). Post-proteasomal antigen processing for major histocompatibility complex class I presentation. *Nat. Immunol.* **5,** 670–677.

Schirle, M., Keilholz, W., Weber, B., Gouttefangeas, C., Dumrese, T., Becker, H. D., Stevanovic, S., and Rammensee, H. G. (2000). Identification of tumor-associated MHC class I ligands by a novel T cell-independent approach. *Eur. J. Immunol.* **30,** 2216–2225.

Schultz, E. S., Chapiro, J., Lurquin, C., Claverol, S., Burlet-Schiltz, O., Warnier, G., Russo, V., Morel, S., Levy, F., Boon, T., Van den Eynde, B. J., and van der Bruggen, P. (2002). The production of a new MAGE-3 peptide presented to cytolytic T lymphocytes by HLA-B40 requires the immunoproteasome. *J. Exp. Med.* **195,** 391–399.

Schwarz, K., van den Broek, M., Kostka, S., Kraft, R., Soza, A., Schmidtke, G., Kloetzel, P. M., and Groettrup, M. (2000). Overexpression of the proteasome subunits LMP2, LMP7, and MECL-1, but not PA28 alpha/beta, enhances the presentation of an immunodominant lymphocytic choriomeningitis virus T cell epitope. *J. Immunol.* **165,** 768–778.

Sewell, A. K., Price, D. A., Teisserenc, H., Booth, B. L., Jr., Gileadi, U., Flavin, F. M., Trowsdale, J., Phillips, R. E., and Cerundolo, V. (1999). IFN-gamma exposes a cryptic cytotoxic T lymphocyte epitope in HIV-1 reverse transcriptase. *J. Immunol.* **162,** 7075–7079.

Shimbara, N., Ogawa, K., Hidaka, Y., Nakajima, H., Yamasaki, N., Niwa, S., Tanahashi, N., and Tanaka, K. (1998). Contribution of proline residue for efficient production of MHC class I ligands by proteasomes. *J. Biol. Chem.* **273,** 23062–23071.

Sijts, A., Sun, Y., Janek, K., Kral, S., Paschen, A., Schadendorf, D., and Kloetzel, P. (2002). The role of the proteasome activator PA28 in MHC class I antigen processing. *Mol. Immunol.* **39,** 165.

Sijts, A. J., Ruppert, T., Rehermann, B., Schmidt, M., Koszinowski, U., and Kloetzel, P. M. (2000a). Efficient generation of a hepatitis B virus cytotoxic T lymphocyte epitope requires the structural features of immunoproteasomes. *J. Exp. Med.* **191**, 503–514.

Sijts, A. J., Standera, S., Toes, R. E., Ruppert, T., Beekman, N. J., van Veelen, P. A., Ossendorp, F. A., Melief, C. J., and Kloetzel, P. M. (2000b). MHC class I antigen processing of an adenovirus CTL epitope is linked to the levels of immunoproteasomes in infected cells. *J. Immunol.* **164**, 4500–4506.

Stoltze, L., Schirle, M., Schwarz, G., Schröter, C., Thompson, M. W., Hersh, L. B., Kalbacher, H., Stevanovic, S., Rammensee, H. G., and Schild, H. (2000). Two new proteases in the MHC class I pathway. *Nature Immunol.* **1**, 413–418.

Sumegi, M., Hunyadi-Gulyas, E., Medzihradszky, K. F., and Udvardy, A. (2003). 26S proteasome subunits are *O*-linked *N*-acetylglucosamine-modified in *Drosophila melanogaster*. *Biochem. Biophys. Res. Commun.* **312**, 1284–1289.

Tanaka, K. (1998). Proteasomes: Structure and biology. *J. Biochem. (Tokyo)* **123**, 195–204.

Tenzer, S., Stoltze, L., Schonfisch, B., Dengjel, J., Muller, M., Stevanovic, S., Rammensee, H. G., and Schild, H. (2004). Quantitative analysis of prion–protein degradation by constitutive and immuno-20S proteasomes indicates differences correlated with disease susceptibility. *J. Immunol.* **172**, 1083–1091.

Theobald, M., Ruppert, T., Kuckelkorn, U., Hernandez, J., Haussler, A., Ferreira, E. A., Liewer, U., Biggs, J., Levine, A. J., Huber, C., Koszinowski, U. H., Kloetzel, P. M., and Sherman, L. A. (1998). The sequence alteration associated with a mutational hotspot in p53 protects cells from lysis by cytotoxic T lymphocytes specific for a flanking peptide epitope. *J. Exp. Med.* **188**, 1017–1028.

Toes, R. E., Nussbaum, A. K., Degermann, S., Schirle, M., Emmerich, N. P., Kraft, M., Laplace, C., Zwinderman, A., Dick, T. P., Muller, J., Schonfisch, B., Schmid, C., Fehling, H. J., Stevanovic, S., Rammensee, H. G., and Schild, H. (2001). Discrete cleavage motifs of constitutive and immunoproteasomes revealed by quantitative analysis of cleavage products. *J. Exp. Med.* **194**, 1–12.

Tomlinson, A. J., Jameson, S., and Naylor, S. (1996). Strategy for isolating and sequencing biologically derived MHC class I peptides. *J. Chromatogr. A* **744**, 273–278.

Unno, M., Mizushima, T., Morimoto, Y., Tomisugi, Y., Tanaka, K., Yasuoka, N., and Tsukihara, T. (2002a). Structure determination of the constitutive 20S proteasome from bovine liver at 2.75 Å resolution. *J. Biochem. (Tokyo)* **131**, 171–173.

Unno, M., Mizushima, T., Morimoto, Y., Tomisugi, Y., Tanaka, K., Yasuoka, N., and Tsukihara, T. (2002b). The structure of the mammalian 20S proteasome at 2.75 Å resolution. *Structure (Camb.)* **10**, 609–618.

Valmori, D., Gileadi, U., Servis, C., Dunbar, P. R., Cerottini, J. C., Romero, P., Cerundolo, V., and Levy, F. (1999). Modulation of proteasomal activity required for the generation of a cytotoxic T lymphocyte-defined peptide derived from the tumor antigen MAGE-3. *J. Exp. Med.* **189**, 895–906.

Van den Eynde, B. J., and Morel, S. (2001). Differential processing of class-I-restricted epitopes by the standard proteasome and the immunoproteasome. *Curr. Opin. Immunol.* **13**, 147–153.

Van den Eynde, B. J., and van der Bruggen, P. (1997). T cell defined tumor antigens. *Curr. Opin. Immunol.* **9**, 684–693.

van der Heeft, E., ten Hove, G. J., Herberts, C. A., Meiring, H. D., van Els, C. A., and de Jong, A. P. (1998). A microcapillary column switching HPLC-electrospray ionization MS system for the direct identification of peptides presented by major histocompatibility complex class I molecules. *Anal. Chem.* **70**, 3742–3751.

Van Soest, R. E. J., Salzmann, J. P., van Marle, M., Ursem, M., Vissers, J. P. C., and Chervet, J. P. (1998). New stationary phase for protein/peptide mapping in LC/ESI–MS. *The 46th ASMS Conf. Mass Spectrom. Allied Top.* Orlando, Florida.

Verma, R., Chen, S., Feldman, R., Schieltz, D., Yates, J., Dohmen, J., and Deshaies, R. J. (2000). Proteasomal proteomics: Identification of nucleotide-sensitive proteasome-interacting proteins by mass spectrometric analysis of affinity-purified proteasomes [In process citation]. *Mol. Biol. Cell* **11**, 3425–3439.

Vigneron, N., Stroobant, V., Chapiro, J., Ooms, A., Degiovanni, G., Morel, S., van der Bruggen, P., Boon, T., and Van den Eynde, B. J. (2004). An antigenic peptide produced by peptide splicing in the proteasome. *Science* **304**, 587–590.

Voges, D., Zwickl, P., and Baumeister, W. (1999). The 26S proteasome: A molecular machine designed for controlled proteolysis. *Annu. Rev. Biochem.* **68**, 1015–1068.

Wang, R., Chait, B. T., Wolf, I., Kohanski, R. A., and Cardozo, C. (1999). Lysozyme degradation by the bovine multicatalytic proteinase complex (proteasome): Evidence for a nonprocessive mode of degradation. *Biochemistry* **38**, 14573–14581.

Wenzel, T., Eckerskorn, C., Lottspeich, F., and Baumeister, W. (1994). Existence of a molecular ruler in proteasomes suggested by analysis of degradation products. *FEBS Lett.* **349**, 205–209.

Yang, P., Fu, H., Walker, J., Papa, C. M., Smalle, J., Ju, Y. M., and Vierstra, R. D. (2004). Purification of the *Arabidopsis* 26S proteasome: Biochemical and molecular analyses revealed the presence of multiple isoforms. *J. Biol. Chem.* **279**, 6401–6413.

Yewdell, J. W., and Bennink, J. R. (1999). Immunodominance in major histocompatibility complex class I-restricted T lymphocyte responses. *Annu. Rev. Immunol.* **17**, 51–88.

Yewdell, J. W., and Bennink, J. R. (2001). Cut and trim: Generating MHC class I peptide ligands. *Curr. Opin. Immunol.* **13**, 13–18.

York, I. A., Goldberg, A. L., Mo, X. Y., and Rock, K. L. (1999). Proteolysis and class I major histocompatibility complex antigen presentation. *Immunol. Rev.* **172**, 49–66.

Zhang, F., Su, K., Yang, X., Bowe, D. B., Paterson, A. J., and Kudlow, J. E. (2003). O-GlcNAc modification is an endogenous inhibitor of the proteasome. *Cell* **115**, 715–725.

[12] Glycosphingolipid Structural Analysis and Glycosphingolipidomics

By Steven B. Levery

Abstract

Sphingosines, or sphingoids, are a family of naturally occurring long-chain hydrocarbon derivatives sharing a common 1,3-dihydroxy-2-amino-backbone motif. The majority of sphingolipids, as their derivatives are collectively known, can be found in cell membranes in the form of amphiphilic conjugates, each composed of a polar head group attached to an *N*-acylated sphingoid, or ceramide. Glycosphingolipids (GSLs), which are the glycosides of either

METHODS IN ENZYMOLOGY, VOL. 405
0076-6879/05 $35.00
DOI: 10.1016/S0076-6879(05)05012-3

ceramide or *myo*-inositol-(1-*O*)-phosphoryl-(*O*-1)-ceramide, are a structurally and functionally diverse sphingolipid subclass; GSLs are ubiquitously distributed among all eukaryotic species and are found in some bacteria. Since GSLs are secondary metabolites, direct and comprehensive analysis (metabolomics) must be considered an essential complement to genomic and proteomic approaches for establishing the structural repertoire within an organism and deducing its possible functional roles. The glycosphingolipidome clearly comprises an important and extensive subset of both the glycome and the lipidome, but the complexities of GSL structure, biosynthesis, and function form the outlines of a considerable analytical problem, especially since their structural diversity confers by extension an enormous variability with respect to physicochemical properties.

This chapter covers selected developments and applications of techniques in mass spectrometric (MS) that have contributed to GSL structural analysis and glycosphingolipidomics since 1990. Sections are included on basic characteristics of ionization and fragmentation of permethylated GSLs and of lithium-adducted nonderivatized GSLs under positive-ion electrospray ionization mass spectrometry (ESI-MS) and collision-induced mass spectrometry (CID-MS) conditions; on the analysis of sulfatides, mainly using negative-ion techniques; and on selected applications of ESI–MS and matrix-assisted laser desorption/ionization mass spectrometry (MALDI–MS) to emerging GSL structural, functional, and analytical issues. The latter section includes a particular focus on evolving techniques for analysis of gangliosides, GSLs containing sialic acid, as well as on characterizations of GSLs from selected nonmammalian eukaryotes, such as dipterans, nematodes, cestodes, and fungi. Additional sections focus on the issue of whether it is better to leave GSLs intact or remove the ceramide; on development and uses of thin-layer chromatography (TLC) blotting and TLC–MS techniques; and on emerging issues of high-throughput analysis, including the use of flow injection, liquid chromatography mass spectrometry (LC–MS), and capillary electrophoresis mass spectrometry (CE–MS).

Introduction

Sphingosines, or sphingoids, are a family of naturally occurring long-chain hydrocarbon derivatives sharing a common 1,3-dihydroxy-2-amino-backbone motif (Kanfer and Hakomori, 1983; Merrill *et al.*, 1997). This defining structural motif derives from a biosynthetic pathway that is highly conserved among eukaryotes. In general, the concentrations of free sphingoids, their biosynthetic intermediates, and their simple derivatives, such as

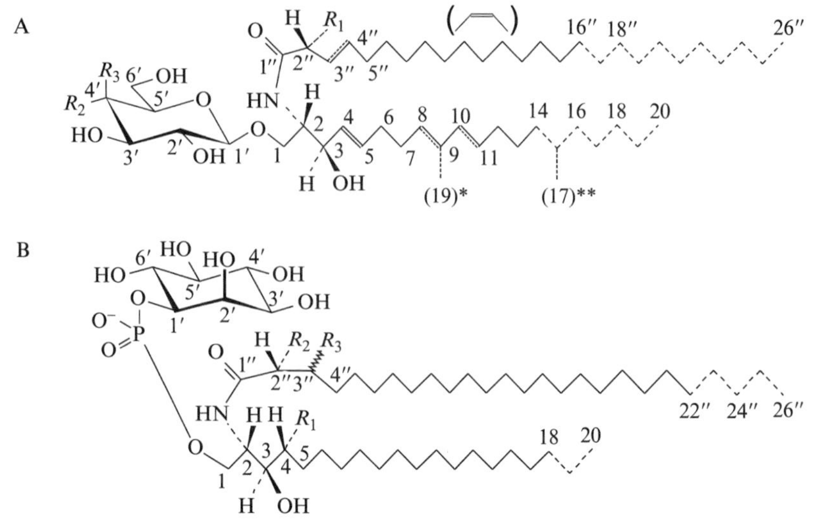

A

R_3 6' OH
R_2 4' 5' O
HO 3' 2' OH 1'
O H R_1 4''
1'' 2'' 3'' 5''
$(\frown)$ 16'' 18'' 26''
HN H
O 2 4 6 8 10 14 16 18 20
3 5 7 9 11
H OH
(19)* (17)**

B

6' HO HO 4'
HO O 5' 3' OH
2' OH
O⁻ 1'
O=P
O H R_2 R_3
1'' 2''3'' 4''
HN H R_3
22'' 24'' 26''
HN H H R_1
O 3 5 18 20
1 2 4
H OH

Scheme 1. Composite structures illustrating a variety of functional modifications incorporated into ceramides of two types of eukaryotic GSLs. (A) Monohexosylceramides with (E)-4-unsaturated sphingoids. Specifics will vary widely with kingdom, phylum, species, tissue or cell type, state of development or differentiation, environmental factors including culture conditions, and pathology; however, overall, the following are true. Predominant fatty-N-acyl group distributions (i) cover chain lengths from C16 to C26 (are usually even numbered); (ii) may be nonhydroxy (R_1 = H) or 2''-hydroxy (R_1 = OH; most common in plants, fungi, nematoda, as well as mammalian nervous tissues and kidney epithelia; (iii) may incorporate mid-chain (Z)-unsaturation (parenthesis), especially in vertebrates; and (iv) may incorporate (3''E)-unsaturation (a feature so far detected almost exclusively in species of euascomycete fungi ("molds"). In addition, predominant hexoses are glucose (R_2 = OH, R_3 = H) and galactose (R_2 = H, R_3 = OH). Last, a sphingoid chain may be, for example, a C18/C20 backbone with (4E)-unsaturation (typical for mammals, often referred to specifically as *sphingosines*); a C18 backbone with (4E,8E/Z)-unsaturations (typical for plants) (Jung *et al.*, 1996; Shibuya *et al.*, 1990; Sperling and Heinz, 2003; Sullards *et al.*, 2000); a C18 backbone with (4E,8E)-unsaturations and 9-Me (19)* branch (typical for fungi) (Barreto-Bergter *et al.*, 2004; Warnecke and Heinz, 2003) but also found in glucosylceramide pools of sea anemones (Karlsson *et al.*, 1979), annelids (Tanaka *et al.*, 1997), marine sponges and starfish (Costantino *et al.*, 1995a,b; Jin *et al.*, 1994; Natori *et al.*, 1994), animal species that may also express additional [10E]-unsaturation (Jin *et al.*, 1994); a C14/C16 with (4E)-unsaturation (typical for dipterans (Dennis and Wiegandt, 1993); and, finally, a C16 backbone with 15 Me (17)** branch (typical for nematode) (Chitwood *et al.*, 1995). (B) Inositol phosphorylceramides with phytoceramides (the obligate intermediates for biosynthesis of glycosylinositol phosphorylceramides, or GIPCs, found in fungi, plants, and some other organisms but not mammals (Lester and Dickson, 1993)). These are made up of C18/C20 sphinganines (R_1 = H) or 4-hydroxysphinganines (R_1 = OH, often referred to as *phytosphingosines*). Phytoceramides can consist of fatty-N-acylated chains that may (i) have typical lengths from C22 to C26 (and are usually even numbered) or (ii) can have nonhydroxy (R_1 = R_1 = H), 2''-hydroxy (R_1 = OH, R_2 = H; most common in fungi and plants), or 2'',3''-dihydroxy (R_1 = OH, R_2 = OH). Phytoceramides found in some plant GIPCs and cerebrosides may also incorporate phosphates, have been observed to be very low in normal tissues and cells. The bulk of cellular sphingolipids, as their derivatives are collectively known, can be found in cell membranes in the form of amphiphilic conjugates, each composed of a polar head group attached to an N-acylated sphingoid, or ceramide. Glycosphingolipids (GSLs), which are the glycosides of ceramide (Cer) or myo-inositol-(1-O)-phosphoryl-(O-1)-ceramide (InsPCer or IPC), are a structurally and functionally diverse sphingolipid subclass ubiquitously distributed among all eukaryotic species (Kolter and Sandhoff, 1999) and also found in some bacteria (Olsen and Jantzen, 2001). Some typical defining features of eukaryotic GSLs are illustrated in Scheme 1A (for monohexosylceramides, commonly referred to as CMHs or cerebrosides) (Barreto-Bergter et al., 2004; Warnecke and Heinz, 2003) and Scheme 1B (for IPC, the obligate intermediate in the biosynthesis of glycosylinositol phosphorylceramides, or GIPCs) (Dickson and Lester, 2002; Funato et al., 2002). As with other carbohydrates and glycoconjugates, even under the constraints of what is biosynthetically possible, the potential for variation of GSL glycan structure is enormous, especially when one includes noncarbohydrate modifications such as sulfation, acetylation, fatty acetylation, cyclic fatty acetylation, and methyl or other etherifications; interpolation of phosphate or myo-inositol-1-phosphate; as well as addition of phosphorylethanolamine, phosphorylcholine, or other such substances. The additional diversity observed in nature with respect to the sphingoid and N-acyl components multiplies again the number of distinct molecular species possible. Thus, even a relatively small GSL can represent a considerable amount of three-dimensional (chemical information. That at least some of this molecular language is essential for proper development and functioning of an organism is supported by the results of transgenic experiments directed at specific steps in GSL biosynthesis (e.g., murine glucosylceramide [GlcCer], galactosylceramide [GalCer], and G_{M2}/G_{D2} ganglioside synthases) (Furukawa et al., 2001).

8- or 10-unsaturation (not shown) (Falsone et al., 1987; Hsieh et al., 1978; Kang et al., 1999). These composites are neither exhaustive nor exclusive—many exceptions and cross-correlations are known (e.g., glycosylceramides of animals and fungi have been observed with phytoceramide features; 8-unsaturation has been observed in minor pools of mammalian GSLs, and minor pools of fungal GlcCer or GalCer may lack one or more structural modifications, such as 9-Me group, or one or more unsaturations of sphingoid). Odd-numbered sphingoids and fatty-N-acylations have also been observed. Ceramide distributions are also frequently observed partitioned according to attached glycan structure—that is, further processing pathways taken by GlcCer, GalCer, or InsPCer may be strongly influenced by ceramide structure where multiple pools are available. An important class of ceramides, found in mammalian epidermal tissues, characterized by long-chain ω-hydroxy-fatty-N-acylation, have been omitted.

Certain features of GSL structure expression are predetermined in a general way by type of organism, varying between the animal, plant, and fungal kingdoms; on the other hand, as has been observed particularly within the animal kingdom (Animalia), some features may be shared between all major phyla. Within these groups, structural and quantitative expression patterns can be highly dependent on phylogeny, all the way down to the species level, and incorporate as well an exquisite degree of programming with respect to organ and tissue, state of development and differentiation, and a variety of other inheritable factors, such as carbohydrate-based blood groups (Kanfer and Hakomori, 1983). Expression is also affected by certain pathological conditions, such as cancer (Hakomori, 1998, 1999; Hakomori and Zhang, 1997). Despite the high level of variability, which may even be manifested in humans at the individual level, there also exists a great deal of commonality (in many cases over the entire range of eukaryotes) based on certain highly conserved enzymes, such as the ceramide and GlcCer synthases (Guillas et al., 2001; Leipelt et al., 2001; Schorling et al., 2001; Venkataraman et al., 2002; Warnecke and Heinz, 2003). Other perhaps unexpected identities exist, such as observed for a sphingoid backbone structure associated with monohexosylceramides of fungi (Barreto-Bergter et al., 2004; Warnecke and Heinz, 2003) but also found in those of sea anemones, for example (Karlsson et al., 1979).

Since GSLs are secondary metabolites, a direct and comprehensive analysis (metabolomics) must be an essential complement to genomic and proteomic approaches for establishing the structural repertoire within an organism and deducing possible functional roles of GSLs. The glycosphingolipidome clearly comprises an important and extensive subset of both the glycome and the lipidome. The complexities of GSL structure, biosynthesis, and function form the outlines of a considerable analytical problem, especially since their structural diversity confers by extension an enormous variability with respect to physicochemical properties. Alongside other methodologies, mass spectrometry (MS) has traditionally played a crucial role in GSL structural analysis and glycosphingolipidomics and will continue to do so. This chapter will cover selected improvements and applications of MS techniques, with a particular emphasis on electrospray ionization, that have contributed to GSL analysis since the previous publication of volume 193 of *Methods in Enzymology* devoted to MS (McCloskey, 1990). Recent characterizations of GSLs from selected non-mammalian eukaryotes will also be discussed. For the most part, MS of oligosaccharides and nonglycosylated sphingolipids, even though related, will not be covered. Due to limitations of time, space, and energy, it is impossible for this review to be comprehensive. The author apologizes for any work of interest that should have been cited herein but was not.

Common Glycosphingolipid Core Sequences and Nomenclature

Mammalian Core Structures. The lacto- (Lc), neolacto- (nLc), ganglio- (Gg), globo- (Gb), and iso-globo- (iGb) series inner core structures are specified by the sequences Galβ1 $\rightarrow$ 3GlcNAcβ1 $\rightarrow$ 3, Galβ1 $\rightarrow$ 4GlcNAcβ1 $\rightarrow$ 3, Galβ1 $\rightarrow$ 3GalNAcβ1 $\rightarrow$ 4, GalNAcβ1 $\rightarrow$ 3Galα1 $\rightarrow$ 4, and GalNAcβ1 $\rightarrow$ 3Galα1 $\rightarrow$ 3, respectively, linked to the terminal galactose of lactosylceramide (LacCer, Galβ1 $\rightarrow$ 4Glcβ1 $\rightarrow$ 1Cer). In all cases, except for Lc/nLc, the series is already determined by addition of the first monosaccharide residue to LacCer. Particularly neolacto-series GSLs are often found repetitively elongated and/or branched by Galβ1 $\rightarrow$ 4GlcNAcβ1 $\rightarrow$ 3/6 units. Hybrid cores may be formed by tandem addition or by branching of defining saccharide units from different series. Subscripted numbers are used to specify the total number of monosaccharide residues; an alternative shorthand sequence code is sometimes used, omitting arrows and anomeric designations, where these are understood (e.g., Gg$_4$Cer = Galβ1 $\rightarrow$ 3GalNAcβ1 $\rightarrow$ 4Galβ1 $\rightarrow$ 4Glcβ1 $\rightarrow$ 1Cer = Galβ3GalNAcβ4Galβ4Glcβ1Cer). In another commonly used nomenclature, branching or terminal noncore substituents are located by Roman numeral for the core residue that is substituted (counting from the nonreducing end), followed by an Arabic superscript designating the hydroxyl group that is substituted (e.g., the sequence Neu5Acα2 $\rightarrow$ 3Galβ1 $\rightarrow$ 4GlcNAcβ1 $\rightarrow$ 3Galβ1 $\rightarrow$ 4GlcNAcβ1 $\rightarrow$ 6Galβ1 $\rightarrow$ 4Glcβ1 $\rightarrow$ 1Cer = Neu5Acα3Galβ4GlcNAcβ3Galβ4GlcNAcβ6Galβ4Glcβ1Cer can be further condensed to VI^{3}Neu5Ac-nLc$_6$Cer). This nomenclature follows (IUPAC) recommendations of 1976 and 1997 (IUPAC, 1978; Chester, 1998). In addition, the branched neolacto-series core, Galβ4GlcNAcβ3 (Galβ4GlcNAcβ6)Galβ4GlcNAcβ3Galβ4Glcβ1Cer, is sometimes referred to as iso-nLc$_8$Cer; the extended globo-series core, Galβ3 GalNAcβ3Galα4Galβ4Glcβ1Cer, is sometimes referred to as Gb$_5$Cer.

Nonmammalian Core Structures. Two other common core structures referred to are (i) the arthro-(Ap) series, wherein the tetraglycosylceramide Ap$_4$Cer is defined by the sequence GalNAcβ4GlcNAcβ6Manβ4Glcβ1Cer, and (ii) the neogala-series (also referred to as the nGa-series), wherein the tetraglycosylceramide nGa$_4$Cer is defined by the sequence Galβ6Galβ6Galβ6Galβ1Cer (Dennis and Wiegandt, 1993). Note that the "At-" designation has been recommended by IUPAC for the arthro-series; no official recommendation has been given for the neogala-series (Chester, 1998).

Lipoforms. The term *lipoforms* is proposed to refer to the microheterogeneity of ceramides on a GSL that has been purified to homogeneity with respect to its glycan moiety. It was suggested (Levery, 1993) as an analogy to Raymond Dwek's coinage of the term *glycoforms*, which refers to the microheterogeneity of glycan structures attached to a homogeneous polypeptide sequence. A single lipoform, homogeneous with respect to sphingosine, fatty acid, and glycan, would be equivalent to what is sometimes referred to as a "molecular species" in GSL literature. It is suggested that such a term might be useful in discussions of profiling of GSLs and the subsequent selection of molecular ion species for MS–CID–MS experiments.

Fragmentation Nomenclature

The glycan fragmentation nomenclature introduced by Domon, Costello, and Vath (Costello and Vath, 1990; Domon and Costello, 1988) (illustrated in Scheme 2) will be used except where noted. For GIPCs, the nomenclature used by Singh et al. (Singh et al., 1991) (illustrated in Scheme 3) will be followed. In positive-ion mode spectra of underivatized GIPCs, salt-forming cations not contributing to charge will be designated by parentheses, while adducted cations contributing to charge will follow a + sign. For ceramide fragments, the revised nomenclature of Adams and Ann (1993) will be used, with some additional ions designated by Hsu and Turk (2001). Nominal, monoisotopic mass-to-charge ratio (m/z) will be used for all illustrative spectra presented herein.

Electrospray Ionization Quadrupole MS of a Permethylated Neolacto-Series Ganglioside: Illustration of Fragment Types and Nomenclature

In-Source Fragmentation. $^{+}$ESI–CID–MS. In-source fragmentation electrospray ionization collision-induced dissociation mass spectrometry ($^{+}$ESI–CID–MS) of permethylated GSLs can be performed on single and triple quadrupole instruments with the orifice-to-skimmer (OR) potential raised to 120–200 V (Levery et al., 1995; Stroud et al., 1998). This technique has been applied as well on electrospray ionization orthogonal acceleration

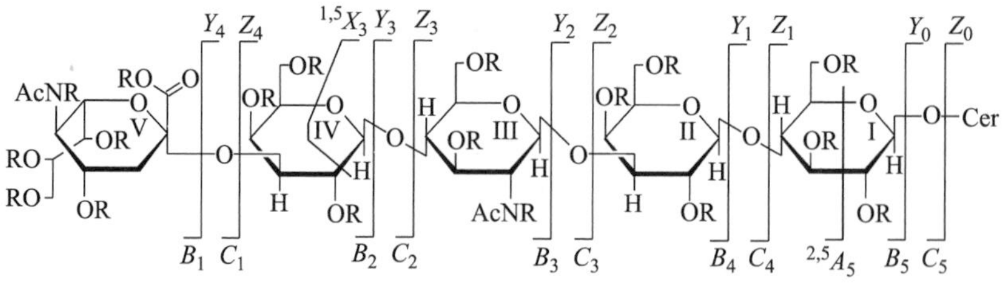

SCHEME 2. Nomenclature for fragmentation of glycosphingolipids, illustrated with per-N,O-methylated IV^{3}NeuAc-nLc$_4$Cer (**1**, R = Me). Standard designations according to Domon and Costello (1988). Hydrogen transfers are omitted.

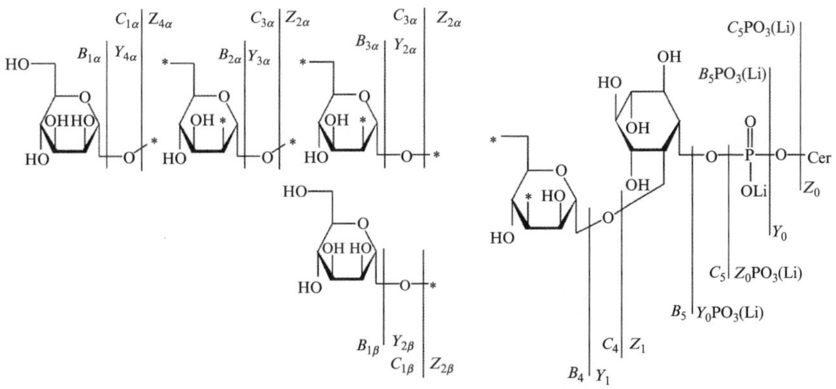

SCHEME 3. Nomenclature for glycosyl, inositol, and phosphate fragmentation in glycosyli-nositol phosphorylceramides, illustrated with lithium cationized **An-5**, possessing the partially indeterminate structure Manα?Manα?Manα3(Manα6)Manα2Ins-P-Cer from *Aspergillus nidulans* (Bennion *et al.*, 2003). Designations adapted from Singh *et al.* (1991). Hydrogen transfers, the adduct designation "+Li", and the charge form have been omitted from the labels for clarity.

quadrupole time-of-flight (ESI–oa-QTOF) instruments (see later sections). As illustrated in Fig. 1, for the simple neolacto-series ganglioside isolated from human placenta, $IV^3Neu5AcnLc_4Cer$ (**1**) (Taki *et al.*, 1988), increasing OR can have several effects: (i) increasing the ratio of lower-charge relative to higher-charge adducts (in this case of monosodiated versus disodiated); (ii) increasing the production of fragment ions, as expected; and (iii) increasing the overall signal-to-noise level, especially at the low mass end of the spectrum (less than m/z 800). Similar results are obtained using one analyzer of a triple quadrupole instrument (Fig. 2).

In the spectra of **1** obtained on either instrument, a cluster of mono-sodiated molecular adduct ions is consistent with a glycan formula Neu5Ac•Hex₃•HexNAc in combination with ceramide lipoforms composed of d18:1 sphingenine and even carbon number saturated fatty acids 16:0–26:0 (nominal, monoisotopic m/z 1820, 1848, 1876, 1904, 1932, 1960), with 16:0, 22:0, and 24:0 predominating (fatty acid analysis was in agreement with this lipoform distribution). In the spectra obtained on the single quadrupole instrument, the corresponding set of doubly charged disodiated ions was observed abundantly from m/z 921.5 to 991.5. The ion observed at m/z 1371 in these spectra is attributed to $[M + Na]^+$ of a small amount of G_{M3} (ceramide with d18:1 sphingosine + 16:0 fatty acid), the major ganglioside of human placenta (Taki *et al.*, 1988), incompletely separated from compound **1** in this sample.

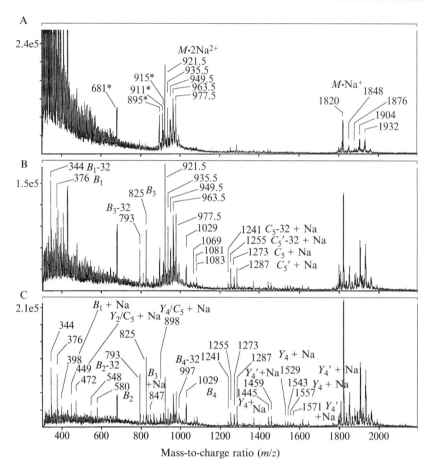

Fig. 1. Single quadrupole (Sciex API-100) +ESI mass spectra of per-*N*,*O*-methylated IV³NeuAc-nLc₄Cer (**1**) acquired at increasing values of orifice-to-skimmer potential (OR). (A) Value at 30 V. (B) Value at 120 V. (C) Value at 200 V. All other other experimental conditions and instrument parameters were identical.

Typical glycosidic cleavage fragments observed in this mode (Scheme 3) can be grouped into three different classes.

Nonsodiated B-Type Fragments. Similar to patterns observed in positive fast-atom bombardment (+FAB) and liquid secondary ion mass spectrometry (+LSI) mass spectra of permethylated oligosaccharides, GSLs, and other glycoconjugates (Dell, 1987; Dell *et al.*, 1993; Domon and Costello, 1988; Egge and Peter-Katalinić, 1987; Peter-Katalinić, 1994; Peter-Katalinić and Egge, 1990), "classical" oxonium *B*-type fragments,

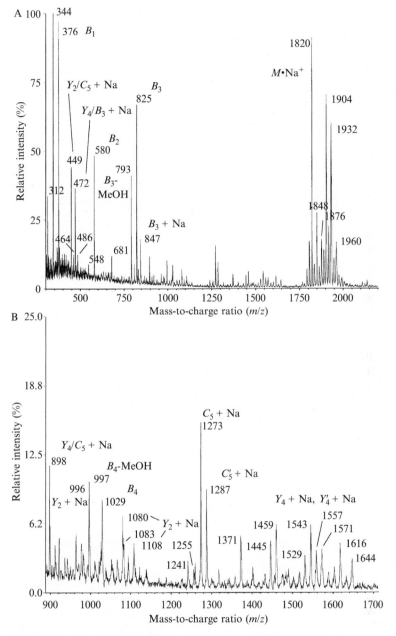

FIG. 2. Single quadrupole (Q1 of Sciex API-III) $^+$ESI–CID–MS of per-N,O-methylated IV^{3}NeuAc-nLc$_4$Cer (**1**) acquired at OR = 200 V. (A) Profile mode mass spectrum m/z 300–2200. (B) Expanded region of spectrum m/z 900–1700.

often accompanied by losses of MeOH, are produced by permethylated GSLs under certain conditions of $^+$ESI, particularly if no salts are added, and can be detected abundantly in single analyzer $^+$ESI–CID–MS at higher orifice potentials (Levery *et al.*, 1995; Stroud *et al.*, 1998). Similar behavior was decribed by Viseux *et al.* (1997, 1998) for permethylated oligosaccharides structurally related to those of GSLs. As with FAB, relative stabilization of B and $[B - \text{MeOH}]$ ions formed by cleavage at HexNAc or Neu5Ac residues is observed (Viseux *et al.*, 1998) (Fig. 2: m/z 376 $[B_1]$, 344 $[B_1 - \text{MeOH}]$, 312 $[B_1 - 2\text{MeOH}]$, 825 $[B_3]$, and 793 $[B_3 - \text{MeOH}]$). However, fragments from cleavage at non-HexNAc residues are also often represented (Fig. 2: m/z 580 $[B_2]$, 548 $[B_2 - \text{MeOH}]$, and 1029 $[B_4]$). An internal ion arising from two B-type cleavages (m/z 450, Scheme 6A) is often observed in these spectra although not in the examples shown. Regardless of the source, interpretation of oxonium B fragments is straightforward (Dell, 1987; Domon and Costello, 1988; Egge and Peter-Katalinić, 1987; Peter-Katalinić and Egge, 1990).

Scheme 4A illustrates a mechanism that formally describes the production of B-type fragments in $^+$FAB–MS and $^+$LSI–MS processes (Dell, 1987; Domon and Costello, 1988; Peter-Katalinić and Egge, 1990; Poulter and Burlingame, 1990; Dell *et al.*, 1993; Peter-Katalinić, 1994). As noted by Dell (1987), this ion is formally similar to the A_1 fragment originally observed in EI–MS (Kochetkov and Chizhov, 1966), in which case its formation is initiated by loss from the gas phase molecule of a single electron (this could be from either the ring or the glycosidic oxygen), followed by loss of the C-1 substituent as a neutral radical. Because of the common structure, the designation A_1- or A-type is still sometimes used for this fragment in

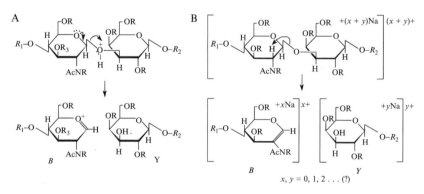

Scheme 4. Comparison of general mechanisms for $B1|2Y$ cleavages. (A) $B|Y$ cleavages in $^+$FAB–MS and $^+$ESI–CID–MS (initiated by protonation). (B) $B|Y$ cleavages in $^+$ESI–CID–MS and $^+$ESI–MS/CID–MS (initiated by sodiation).

discussions of desorption mass spectra (Dell, 1987, 1990; Dell *et al.*, 1993, 1994; Easton *et al.*, 2000).

Interestingly, protonated molecular precursor ions in appreciable abundance have not been observed in $^+$ESI–CID–MS spectra of permethylated gangliosides even though their presumptive products are present (Levery *et al.*, 1995; Stroud *et al.*, 1998). Precursor scanning of *B*-type product ions in tandem $^+$ESI–(CID–MS)^2 mode showed that they originated from protonated molecular precursors, regardless of whether these were appreciably observable, while no sodiated precursors were detectable in their lineage despite the overwhelming predominance of sodiated molecular ions in the profile spectra (Levery *et al.*, unpublished observations). These results suggest that the sodium ion does not catalyze classical *B*-type fragmentation effectively and that *B*-type cleavage may be so rapid a process under these conditions that the lifetime of protonated molecular species formed is short compared to the corresponding sodiated species. This is consistent with the results of Viseux *et al.* (1997), who reported that effective CID fragmentation of sodium adducts of permethylated oligosaccharides required higher collision energies than the protonated molecular ions; the latter underwent extensive fragmentation even under low collision energies (15 eV). This finding is also consistent with the fact that the abundance of *B* ions is increased by addition of a proton source, such as ammonium acetate to the analyte solution. For further comments on the different fragmentation patterns associated with $[M + Na]^+$ and $[M + H]^+$, see Perreault *et al.* (1997).

Sodiated fragments. The second class of ions are sodiated fragments; these have been frequently reported in tandem $^+$ESI–MS/CID–MS of permethylated gangliosides and other glycoconjugates (Metelmann *et al.*, 2001a; Reinhold *et al.*, 1994, 1995; Stroud *et al.*, 1995, 1996a,b; Yohe *et al.*, 1997), but the fragments are also observed in abundance in single-analyzer $^+$ESI–CID–MS at higher orifice potentials (Levery *et al.*, 1995; Stroud *et al.*, 1998; Viseux *et al.*, 1997, 1998). The majority of fragments occurring in $^+$ESI–MS/CID–MS of sodium adducts of permethylated GSLs could be explained by the general mechanism depicted in Scheme 4B (Stroud *et al.*, 1995, 1996a,b; Yohe *et al.*, 1997). The mechanism corresponds formally to that of the β cleavage proposed to occur in FAB–MS of permethylated glycans (Dell, 1987, 1990; Dell *et al.*, 1993) but with one or more sodium ions distributed over the precursor ion and its products. Although the left-hand product has generally been referred to as a *B*, or sodiated *B*, fragment, it corresponds formally to a classical *B* fragment plus sodium minus a proton and mechanistically to the fragment that would normally be lost in the β-elimination process (e.g., pathway *B* of Dell, 1987). This fragment is of course not observed in FAB mass spectra because it has no charge when formed by FAB-initiated processes. Here, attachment of sodium produces the observable ion. The right-hand fragment corresponds to the Y-type ion of Domon and Costello (1988), whether protonated or sodiated. Scheme 4B illustrates a general formal equation for $B|Y$ fragmentation with precursor and product ions carrying any number of sodium ions and corresponding charges. In the $^+$ESI–CID–MS spectrum of permethylated **1** (Fig. 2), numerous fragments arising from $B|Y$-type cleavages are observed (m/z 398 [B_1 + Na], m/z 847 [B_3 + Na], m/z 1255 [B_5 + Na], m/z 1445, 1529, 1557 [Y_4 + Na]; m/z 996, 1080, 1108 [Y_2 + Na]). Multiple $B|Y$ fragmentations will produce internal ions (e.g., m/z 472 [Y_4/B_3 + Na]) (Scheme 6B).

Scheme 5A illustrates a general mechanism for another process observed abundantly in $^+$ESI–MS/CID–MS of permethylated GSLs, which

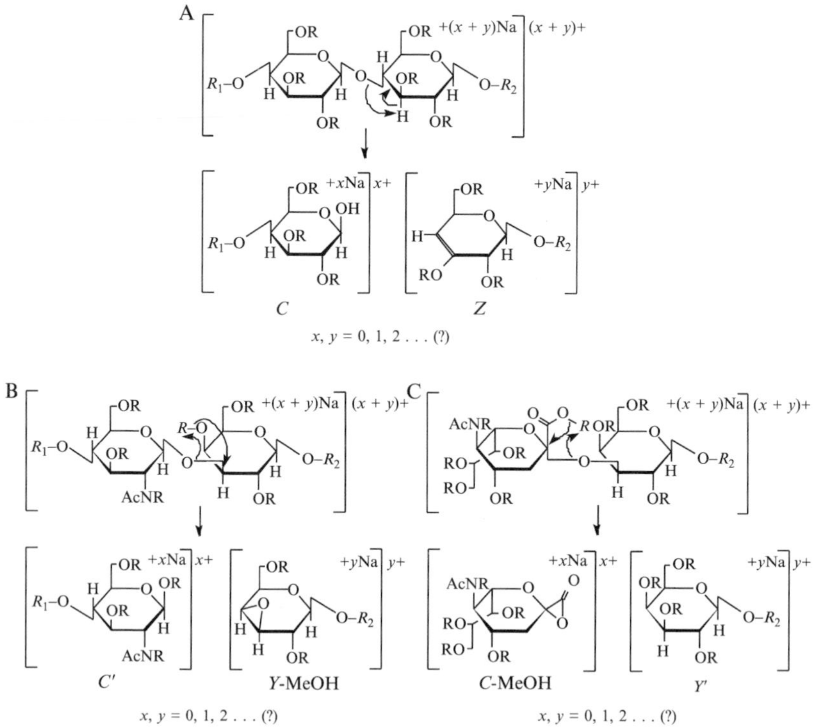

SCHEME 5. Comparison of various general mechanisms. (A) General mechanism for sodium initiated C/Z cleavages in $^+$ESI–CID–MS and $^+$ESI–MS/CID–MS. (B) Possible mechanism for generation of C' (methyl transfer) fragment observed in $^+$ESI–CID–MS. (C) Possible mechanism for generation of Y' (methyl transfer) fragment at nonreducing end of ganglioside observed in $^+$ESI–CID–MS.

also occurs in $^+$ESI–CID–MS: cleavage on the other side of the glycosidic oxygen to produce sodiated C and/or Z ions (Metelmann *et al.*, 2001a; Reinhold *et al.*, 1994, 1995; Stroud *et al.*, 1995, 1996a,b; Yohe *et al.*, 1997). The ion observed at m/z 1273 in the $^+$ESI–CID–MS spectrum of permethylated **1**, which corresponds to the entire glycan plus the glycosidic oxygen, is the $[C_5 + \text{Na}]$ ion resulting from cleavage at the glycan-ceramide linkage, as observed in tandem $^+$ESI–MS/CID–MS (Metelmann *et al.*, 2001a; Reinhold *et al.*, 1994, 1995; Stroud *et al.*, 1995, 1996a,b; Yohe *et al.*, 1997). Other fragments arising from this cleavage, in combination with $B|Y$-type cleavages, are also observed in the spectrum (m/z 898 $[Y_4/C_5 + \text{Na}]$; m/z 449 $[Y_2/C_5 + \text{Na}]$) (Scheme 6C). The ion at m/z 1241 corresponds to $[C_5 + \text{Na} - \text{MeOH}]$.

Methyl Transfer Reactions. In $^+$ESI–CID–MS of permethylated gangliosides, ions of m/z corresponding to $[C_n + \text{Na} + 14]$ and $[Y_n + \text{Na} + 14]$ may sometimes appear; formally, the $[C_n + \text{Na} + 14]$ ion could correspond to a $[B_n + \text{Na}]$ ion plus MeOH or to a $[C_n + \text{Na}]$ ion in which the glycosyl OH has been replaced by OMe; similarly, the $[Y_n + \text{Na} + 14]$ could correspond to a $[Y_n + \text{Na}]$ ion in which the glycosyl OH has been replaced by OMe. Such ions were observed abundantly in $^+$ESI–CID–MS spectra of permethylated sialosylgalactosylgloboside (V^3Neu5AcGb$_5$Cer) and proposed at that time to correspond formally to $[C_n + \text{Na} + \text{CH}_2]$ (Stroud

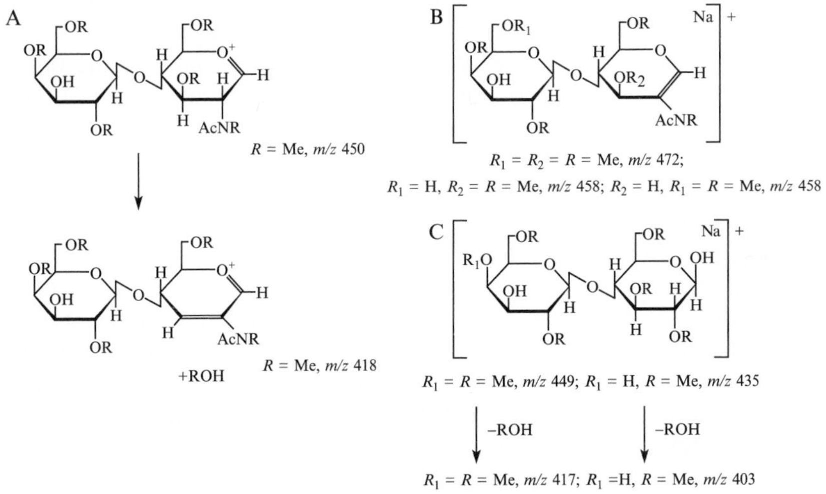

SCHEME 6. Formation of fragments in various modes. (A) Fragments at m/z 450 and 418 in $^+$ESI–CID–MS mode. (B) Fragments at m/z 472 or 458 in $^+$ESI–CID–MS and $^+$ESI–MS/CID–MS modes. (C) Fragments at m/z 449 or 435 and m/z 417 or 403 in $^+$ESI–CID–MS and $^+$ESI–MS/CID–MS modes.

et al., 1998). They have been observed for a variety of permethylated neolacto-series gangliosides under similar conditions but generally not in the $^{+}$ESI–MS/CID–MS of the sodiated molecular precursors (Levery *et al.*, 1995; Levery *et al.*, unpublished observations). Interestingly, however, $[Y_n + Na + 14]$ ions were observed by Metelmann *et al.* (2001a) in $^{+}$ESI–MS/ CID–MS spectra of permethylated ganglio-series gangliosides acquired on an $^{+}$ESI–QTOF instrument. Subsequently, Levery *et al.* (unpublished observations), the author of this chapter and coworkers, have obtained experimental evidence consistent with the explanation that these unusual fragments arise from internal methyl group transfers accompanying glycosidic cleavage, as shown in Schemes 5B and 5C.

Tandem. $^{+}$*ESI–MS/CID–MS.* Tandem $^{+}$ESI–MS/CID–MS product ion experiments on permethylated **1** are shown, for comparison to the $^{+}$ESI–CID–MS spectra, in Fig. 3A–C (Levery *et al.*, 1995). These were acquired by selection of an appropriate sodiated molecular adduct ion with Q1 CID of that ion in Q2, and analysis of the products via Q3. The utility of this procedure for sequence analysis of permethylated GSLs has been demonstrated previously with a variety of analyzer configurations (Costello, 1999; Levery *et al.*, 1995; Metelmann *et al.*, 2001a; Reinhold and Sheeley, 1998; Reinhold *et al.*, 1994, 1995; Solouki *et al.*, 1998; Stroud *et al.*, 1995, 1996a,b, 1998; Yohe *et al.*, 1997).

As illustrated in Fig. 3, one advantage of $^{+}$ESI–MS/CID–MS (or other tandem MS or MS^{n} techniques) is that by acquiring spectra from more than one lipoform, one can easily recognize ions that include the ceramide aglycone because their masses shift by an increment equal to the differences in precursor mass. Scheme 7 shows the assignment of most of these fragments in relation to the known glycan sequence of **1**. It is apparent that the majority of these fragments, with the exception of those that involve loss of the ceramide, can be formally accounted for by *B*|*Y*-type cleavages (Scheme 4B). In the case of a monosodiated precursor, the sodium ion may have a similar chance of remaining with either the *B* or the *Y* fragment. For example, the ions at *m/z* 847 and 996 represent the products of a single [*B*|*Y*] cleavage, with the abundance ratio of the two ions most likely approximating the relative probability for the sodium ion partitioning into either fragment. As with classical oxonium *B*-type cleavage, the preferred [*B*|*Y*] cleavage sites are at HexNAc-|-O-Hex and Neu5Ac-|-O-Hex, whenever these are present.

Another interesting feature of these spectra is the abundance of internal fragments (Reinhold *et al.*, 1995). These can all be formed by multiple [*B*|*Y*] fragmentations or by any combination of [*B*|*Y*] fragmentations with the [*C*$_5$|*Z*$_0$] loss of ceramide. With few exceptions, the products are all either [*Y*/*B* + Na] or [*Y*/*C*$_5$ + Na] ions, accompanied in most cases

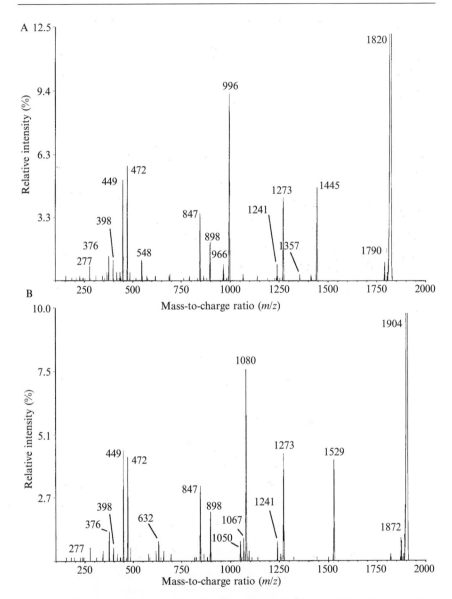

FIG. 3. Tandem quadrupole (Sciex API-III) product ion $^+$ESI–MS/CID–MS of per-*N,O*-methylated IV^{3}NeuAc-nLc$_4$Cer (**1**), with selection in Q1 of monosodiated molecular adducts. (A) Spectrum of products for *m/z* 1820. (B) Spectrum of products for *m/z* 1904.

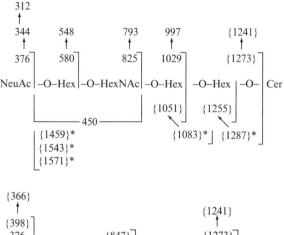

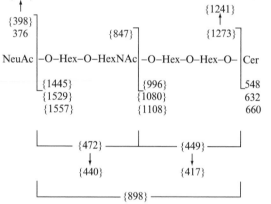

fa	Cer	MW	MH⁺	{M·Na}⁺	{M·2Na}²⁺
16:0	548	1797	1798	{1820}	{921.5}
18:0	576	1825	1826	{1848}	{935.5}
20:0	604	1853	1854	{1876}	{949.5}
22:0	632	1881	1882	{1904}	{963.5}
24:0	660	1909	1910	{1932}	{977.5}
26:0	688	1937	1938	{1960}	{991.5}

SCHEME 7. Fragmentation scheme for per-N,O-methylated IV³NeuAc-nLc₄Cer (**1**). Fragmentation in ⁺ESI–CID–MS mode includes both upper and lower schemes; predominant fragmentation in ⁺ESI–MS/CID–MS mode depicted mainly by lower scheme. In this scheme, brackets "{}" are used to denote Na⁺ adducts; fragments marked in addition with an asterisk are those corresponding in m/z to [C_n + Na + 14] or [Y_n + Na + 14], proposed to result from methyl transfer reactions (see chapter text).

by loss of MeOH. Among the exceptions are nonsodiated Z_0 and B_1 ions. In the $^+$ESI–MS/CID-MS of **1** (Fig. 3), the sodiated internal (double cleavage) fragments m/z 898 $[Y_4/C_5 + \text{Na}]$, m/z 472 $[Y_4/B_3 + \text{Na}]$, and m/z 449 $[Y_2/C_5 + \text{Na}]$, are all abundant. The latter fragment, which includes the two Hex residues of the reducing end lactose unit common to most GSLs, is a useful one for ruling out ganglio-a-series sequences (e.g., derivatives of G_{M1a})—since those gangliosides are distinguished by a characteristic Neu5Ac substituent on Gal II, the mass of the $[Y_2/C_5 + \text{Na}]$ ion is incremented by 361 Th for each Neu5Ac residue added in the series, appearing at m/z 810 for G_{M1a} (Reinhold *et al.*, 1994; Metelmann *et al.*, 2001a). This fragment would be accompanied by a triple cleavage product ion for loss of the Neu5Ac residue $[Y_{2\alpha}/Y_{2\beta}/C_4 + \text{Na}]$, as well as in the case of G_{M1a}; $[Y_{2\alpha}/Y_{2\beta}/C_5 + \text{Na}]$ for G_{D1a}), at m/z 435 (= 449 − 14) (Reinhold *et al.*, 1994; Metelmann *et al.*, 2001a) (Scheme 6C).

Lithium Adduction of Underivatized Glycosphingolipids

The use of Li^+ cationization with high energy CID was investigated by Leary and colleagues for oligosaccharide analysis (Hofmeister *et al.*, 1991; Zhou *et al.*, 1990) and established for analysis of ceramides and simple underivatized GSLs by Ann and Adams (1992, 1993). The method was subsequently applied to characterization of monohexosylceramides (cerebrosides) from sources such as fungi, plants, and corals (Duarte *et al.*, 1998; Pocsfalvi *et al.*, 2000; Sullards *et al.*, 2000). The efficacy of the method was demonstrated at lower collision energies (400 eV) by Olling *et al.* (1998) in a magnetic sector-TOF hybrid fitted with an ESI source and applied to more complex GSLs. In that work, the effects of cationization by K^+, Na^+, and Li^+ were compared, and the latter appeared to yield the best results with respect to overall sensitivity of detection and yield of useful fragments from both the glycan and ceramide moieties. Alkali metal cationization of underivatized oligosaccharides (principally Li^+ versus Na^+) was also revisited by Asam and Glish (1997) with respect to fragmentation behavior in $^+$ESI–QIT–MS. Useful linkage information appears to be contained in cross-ring cleavages of Li^+ cationized oligosaccharides, but this does not seem to have been much exploited in practice. Li^+ cationization, in conjunction with low-energy tandem quadrupole $^+$ESI–MS/CID–MS, was later applied to cerebrosides from a number of different sources, including bovine brain, soybean, and a variety of fungi (Levery *et al.*, 2000); these were applied with a view toward correlating the appearance of characteristic fragmentation modes, under these conditions, with a systematic, cumulative addition of key ceramide functional groups, such as 2-hydroxylation of the fatty-N-acyl group and additional unsaturations of both the sphingoid and fatty acid. Even under low-energy conditions, a substantial increase in both sensitivity and fragmentation was observed on CID of $[M + Li]^+$ versus $[M + Na]^+$ of the same monohexosylceramide (CMH) components analyzed under similar conditions. This made it possible to characterize minor components present at less than 5% relative abundance in the molecular profile, even where these low-abundance molecular adduct peaks represented multiple isomeric lipoforms.

These findings just described are illustrated in Figs. 4 and 5, based on a study of GlcCer variants isolated from the dimorphic fungus *Histoplasma capsulatum* (Toledo *et al.*, 2001a). $^+$ESI–MS profiles of lithium-adducted GlcCer from *H. capsulatum* show abundant monolithiated molecular ion adducts, observed virtually exclusively at *m/z* 760 and *m/z* 762 (Fig. 4A and B, respectively), corresponding to nominal molecular masses of 753 and 755 Da for the mycelium and yeast form GlcCer, respectively. These molecular masses are consistent with lipoforms containing (d19:2) (4*E*,8*E*)-9-methyl-4,8-sphingadienine attached to either *N*-2′-hydroxy-(*E*)-3′-octadecenoate or *N*-2′-hydroxyoctadecanoate, respectively. Confirmation that the observed difference of *m/z* 2 is due to variation in the fatty acid moiety was easily obtained by tandem $^+$ESI–MS/CID–MS experiments (Toledo *et al.*, 2001a). Illustrated in Fig. 5 are tandem $^+$ESI–MS/ CID–MS spectra of *minor* components appearing in the molecular ion profiles, which could correspond either to low-abundance intermediates

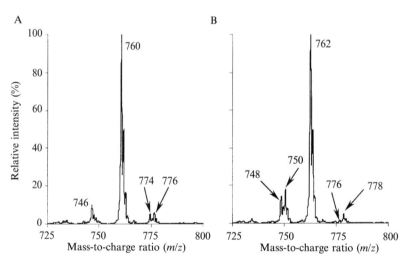

FIG. 4. Molecular ion regions of $^+$ESI–MS spectra for Li$^+$ adducts of cerebrosides from *H. capsulatum*. (A). Yeast-form component. (B) Mycelium-form component. From Toledo *et al.* (2001a); Oxford University Press (2001).

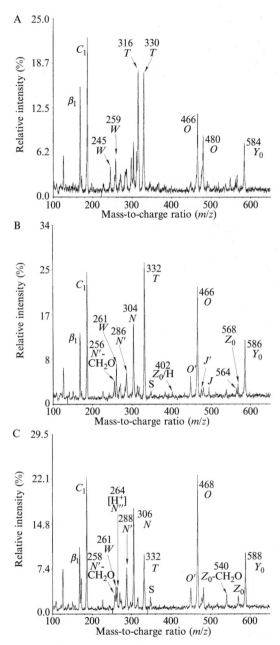

FIG. 5. Tandem $^+$ESI–MS/CID–MS product ion spectra of selected $[M + Li]^+$ from minor *H. capsulatum* cerebroside components (see Fig. 4). (A) Product ion spectrum from m/z 746, mycelium-form cerebroside profile. (B) Product ion spectrum from m/z 748, yeast-form cerebroside profile. (C) Product ion spectrum from m/z 750, yeast-form cerebroside profile. From Toledo *et al.* (2001a); Oxford University Press (2001).

in the biosynthetic pathway or to structural variants differing, for example, in fatty acid chain length. Both of these possibilities are apparent in the product ion spectrum of m/z 746 (Fig. 5A) selected from the molecular ion profile of the mycelium form GlcCer. The spectrum exhibits a single $[M + \text{Li} - \text{hexose}]^+$ (Y_0) ion at m/z 584; as expected, this is 14 Th less than that observed for the major molecular ion (Y_0 at m/z 598). However, at lower m/z, product ions arising from two isobaric components are apparent, characterized by two sets of O and T ions (Scheme 8), one at m/z 480 and 316 and the other at m/z 466 and 330. Confirmatory W ions are present at m/z 245 and 259, respectively. These findings are consistent with two components, present in approximately equal amounts, in which the 14 Th decrement is carried by the fatty N-acyl group in one lipoform and the sphingoid moiety in the other. The former can be clearly attributed to a variant carrying an h17:1 fatty N-acyl group, while the latter most likely corresponds to an h18:1/d18:2 component in which the sphingoid 9-methyl group is missing, which may represent a biosynthetic intermediate.

Unlike the result with the mycelium-form m/z 746 ion, products of the corresponding minor m/z 748 ion in the yeast-form profile (Fig. 5B) are consistent with essentially a single component in which the 14 Th decrement is carried by the sphingoid moiety, as characterized by a single set of O and T ions at m/z 466 and 332. With the exception of a very low-abundance set of O and T ions at m/z 480 and 318, corresponding to the h17:0/d19:2 fatty N-acyl/sphingoid combination, the masses and relative

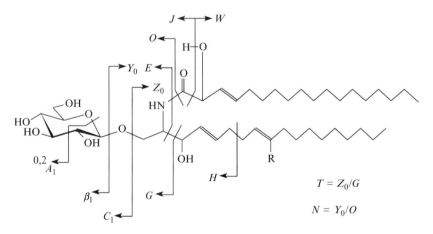

SCHEME 8. Fragmentation of lithium-adducted cerebrosides with nomenclature of Costello and Vath (1990) as modified by Adams and Ann (1993).

abundances of almost all other fragments in the spectrum are consistent with an h18:0/d18:2 component lacking the sphingoid 9-methyl group.

Products of the m/z 750 component (Fig. 5C) were also consistent with essentially one component lacking both the sphingoid 9-methyl group and the Δ^8-unsaturation. In addition to the O and T fragments observed at m/z 468 and 332, respectively, which show the additional 2 Th increment to be carried by the sphingoid moiety, further evidence for the mono-unsaturated (d18:1) sphing-4-enine structure are the obvious changes in the abundance of the T and N (m/z 306) fragments relative to each other and the increase in the abundance of the Y_0 (m/z 588) relative to the O fragment. A number of other characteristic changes in the spectrum are diagnostic for the 2-hydroxy fatty N-alkanoyl/sphing-4-enine combination under these conditions; in fact, the product spectrum is essentially the same as that obtained previously from a bovine brain galactocerebroside with h18:0/d18:1 ceramide (Levery et al., 2000). Interestingly, an analogous component is not observed in significant abundance in the mycelium-form profile (at m/z 748) (Fig. 3A). On the other hand, a molecular ion at m/z 774 was sufficiently abundant in the mycelium-form profile for a product ion spectrum to be acquired (not shown). In this case, the predominant O and T ions were observed at m/z 480 and 344, respectively, consistent with an h19:1/d19:2 fatty N-acyl/sphingoid combination. Other product ions observed in the spectrum were consistent with this structure.

Subsequently, Hsu and Turk (2001) published an extensive investigation of the fragmentation of Li$^+$-adducted GSLs under these conditions, extending their earlier studies of lithiated sphingomyelins (Hsu and Turk, 2000). Use of hydrogen-denterium (H-D) exchange experiments facilitated dissection of decomposition pathways and fragment structures in more detail; the authors also applied tandem MS/CID–MS with prior in-source CID (i.e., product ion scans on primary fragments generated in the orifice region, essentially equivalent to $^+$ESI– [CID–MS]^2), parent ion scanning, and constant neutral loss scanning. This study was followed by a very detailed investigation of the fragmentation of Li$^+$-adducted ceramides (Hsu et al., 2002).

During this time, Li$^+$ adduction was also extended to the characterization of fungal GIPCs (Levery et al., 2001), with observation of a similar increase in sensitivity and fragmentation in product ion $^+$ESI–MS/CID–MS and $^+$ESI–(CID–MS)^2 mode. This paper described extensive applications of $^+$ESI–MS, –CID–MS, –MS/CID–MS, and –(CID–MS)^2 with Li$^+$ adduction to characterization of GIPC fractions of increasing structural complexity from a variety of fungi, including a nonpathogenic basidiomycete (mushroom), Agaricus blazei, and pathogenic Euascomycete species such as Aspergillus fumigatus, H. capsulatum, and Sporothrix schenckii. Similar to the procedure described by Hsu and Turk (2001), $^+$ESI– (CID–MS)^2 was used with selection of the ceramide Z_0 ion by Q1 for collision and analysis in Q3. This was particularly useful for detailed characterization of GIPC ceramides because secondary fragments of phytoceramides were not very abundant in $^+$ESI–MS/CID–MS of the molecular adducts. In addition to differentiating between isobaric lipoforms where variable numbers of CH_2 units could be carried by either the fatty-N-acyl or sphingoid (or both), this technique facilitated detection of dihydroxy fatty-N-acylated species in some GIPC fractions. The $^+$ESI– (CID–MS)^2 technique could also be used with selection of, for example, a phosphoglycan primary fragment by Q1; this would yield clean secondary product spectra, as described previously with Na$^+$-adducted GIPCs (Levery et al., 1998), but usually no more fragments would be apparent than already observable in the $^+$ESI–MS/CID–MS of the molecular adducts. In one subsequent case, however, $^+$ESI– (CID–MS)^2 with selection of the phosphoglycan ion gave results far superior to the results with $^+$ESI–MS/CID–MS of the molecular adducts, facilitating sequencing of the glycan (Toledo et al., 2001b) (see later discussion). Precursor ion scanning from appropriate glycosylinositol phosphate product ions yielded clean molecular ion profiles in the presence of obscuring impurity peaks.

Due to the different composition of the GIPC ceramides that generally contains phytosphingosine (4-hydroxy-sphinganine) in place of the (4E,8E)-sphing-4,8-dienine (found as the major sphingoid of fungal cerebrosides) as well as the presence of the nonglycosidic and anionic inositol phosphate moiety interpolated between glycan and ceramide, a number of differences are observed in the fragmentation patterns of these compounds. Useful fragments could be categorized into five significant types, the first four of which are observed in pairs separated by 18 Th, depending on which side the glycosidic or phosphoryl oxygen is cleaved: (i, ii) phytoceramide with and without phosphate; (iii) glycosylinositol phosphate, accompanied by sequential losses of monosaccharide residues from the nonreducing end; (iv) glycosylinositol without phosphate, accompanied by further glycosidic cleavages from either the reducing or nonreducing end (since underivatized inositol has the same residue m/z as a hexose, it is impossible to distinguish these possibilites without labeling); and (v) glycosidic cleavages from the nonreducing end of entire molecular adduct leading to sequentially smaller ceramide-containing products. These are all illustrated by Scheme 3. The abundant series (iii) is the most useful, since the phosphate group effectively labels the reducing end of the glycosylinositol; all smaller phosphorylated fragments must be derived from one or more additional cleavages along the glycan chains, eliminating increasing numbers glycosidic residues from the nonreducing end.

The lithiation protocol was readily adapted to $^+$ESI–MS analysis of GIPCs on a oa-QTOF instrument, yielding similar spectra but with expected improvements in sensitivity, resolution, and m/z accuracy (Bennion *et al.*, 2003). This is illustrated by $^+$ESI–MS analysis of the largest component characterized, a pentamannosyl IPC with predominant $[M(\text{Li}) + \text{Li}]^+$ at nominal, monoisotopic m/z 1748 and 1776 (Fig. 6). With this component, another sometimes useful feature of the lithiation technique was the consolidation of a molecular profile consisting of mixed $[M(\text{Na}) + \text{Na}]^+$, $[M(\text{K}) + \text{Na}]^+$, and $[M(\text{K}) + \text{K}]^+$ adducts into a single $[M(\text{Li}) + \text{Li}]^+$ adduct for each species (compare Figs. 6A and 6B). A more general advantage is an increase in the abundance of ceramide containing primary fragments in $^+$ESI–MS/CID–MS spectra and in secondary fragments in high orifice-to-skimmer (OR) product ion $^+$ESI–(CID–MS)^2 spectra (Bennion *et al.*, 2003; Levery *et al.*, 1998). The $^+$ESI–MS/CID–MS of this component (Fig. 6C), compared with that of the simpler trimannosyl IPC and other known compounds, provided convincing evidence that it shared the same Manα1 → 3(Manα1 → 6)Manα1 → 2Ins core motif, modified by addition of two Man residues to one branch (Scheme 3). As has been observed previously with some other techniques (Gillece-Castro and Burlingame, 1990; Poulter and Burlingame, 1990; Webb *et al.*, 1988), the branching residue was characterized by the relatively low abundance of fragments requiring at least two glycosidic cleavages for their appearance (Bennion *et al.*, 2003). High OR product ion $^+$ESI–(CID–MS)^2 spectra with selection of the lithiated phytoceramide precursors yielded abundant secondary products diagnostic for both sphingoid and fatty-N-acyl chain lengths (Bennion *et al.*, 2003; Levery *et al.*, 1998). This is particularly efficacious for the saturated and highly hydroxylated phytoceramides found in GIPCs, which, compared to unsaturated analogs, do not appear to produce a wealth of fragment ions in some techniques.

The use of MALDI–MS for analysis of fungal cerebrosides and GIPCs, using a hybrid quadrupole ion trap-TOF analyzer (Axima-QIT, Shimadzu Biotech, Columbia, MD), has also been investigated. An example of MALDI–QIT/TOF–MS of a lithiated fungal cerebroside (from wild-type *N. crassa*) (Park *et al.*, 2005) is shown in Fig. 7. Lithiation was accomplished by doping 2,5-dihydroxybenzoate (DHB) with LiI. The technique of lithiation has been used previously in MALDI–TOF–MS and Fourier transform ion cyclotrone resonance (FTICR)–MS (Botek *et al.*, 2001; Chen *et al.*, 2003; Hoteling *et al.*, 2003; Huffer *et al.*, 2001; Keki *et al.*, 2001; Laine *et al.*, 2003; North *et al.*, 1997; Rashidzadeh *et al.*, 2000; Sakaguchi *et al.*, 2001; Tomlinson *et al.*, 1999; Wang *et al.*, 2000), and it is worth noting that lithium 2,5-DHB has been recommended as a matrix for analysis of lipids and high molecular weight hydrocarbons (Cvacka and Svatos, 2003). The MS^1

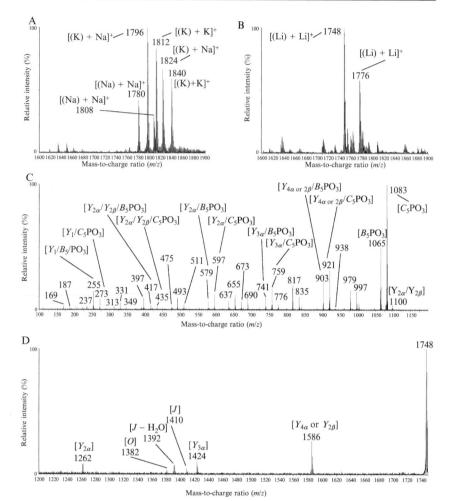

FIG. 6. $^+$ESI–oa-QTOF (Micromass Global) spectra of **An-5**. (A) Profile of molecular ions as mixed $[M(Na) + Na]^+$, $[M(K) + K]^+$, and $[M(Na) + K]^+$ (or $[M(K) + Na]^+$) adducts. (B) Simplified profile of molecular ions as $[M(Li) + Li]^+$ adducts. (C) MS/CID–MS of $[M(Li) + Li]^+$ of m/z 1748, low m/z region. (D) MS/CID–MS of $[M(Li) + Li]^+$ of m/z 1748, high m/z region. The y-axis expansion in 6D relative to 6C is 9x. Ion designations correspond to Schemes 3 and 9. The designations "+Li" and "(Li) + Li" and the charge form have been omitted from the fragment labels for clarity. From Bennion *et al.* (2003); American Society for Biochemistry and Molecular Biology (2003).

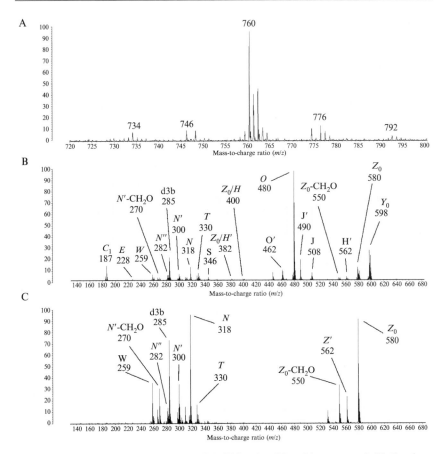

FIG. 7. Positive-ion MALDI–QIT–TOF (Shimadzu Biotech) spectra of GlcCer from *Neurospora crassa*. (A) Profile of molecular ions as Li[+] adducts. (B) CID–MS[2] of $[M + \text{Li}]^+$ at m/z 760. (C) CID–MS[3] of $[M + \text{Li}]^+$ at m/z (760 →) 598 (Y_0 fragment). Matrix is DHB + LiI; collison gas is argon. Labels correspond to Scheme 8.

spectrum (Fig. 7A) shows a typical profile dominated by a lithium adduct at m/z 760, consistent with lipoform consisting of d19:2 (4E,8E)-9-methyl-4,8-sphingadienine attached to N-2′-hydroxy-(E)-3′-octadecenoate. A minor component at m/z 762 is consistent with some saturated acyl component, N-2′-octadecanoate. The MS[2] spectrum of the major m/z 760 ion (Fig. 7B) displays a variety of fragment ions from loss of the hexose residue and further decomposition of the ceramide, as described previously (see Scheme 8). The O ion characteristic for d19:2 (4E,8E)-9-methyl-4,8-sphingadienine is the most abundant; however, ions specific to the fatty-N-acyl moiety, such as T and W, are rather weak. In the MS^3 spectrum of the Y_0 ion (m/z 760 → 598) the m/z W 259 ion diagnostic for N-2′-hydroxy-(E)-3′-octadecenoate is prominent (Levery *et al.*, 2000).

Application to a fungal GIPC, Manα1 → 3(Manα1 → 6) Manα1 → 2InsPCer (**An-3**), from *A. nidulans* (Bennion *et al.*, 2003), is illustrated in Fig. 8. In this case, lithiation was not complete (compare

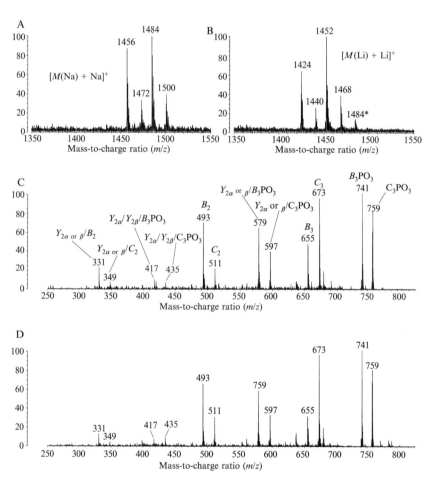

FIG. 8. Positive-ion MALDI–QIT–TOF (Shimadzu Biotech) spectra of GIPC **An-3**. (A) Profile of molecular ions as Na+ adducts (matrix, 2,5-DHB). (B) Profile of molecular ions as Li+ adducts (matrix, 2,5-DHB + LiI). (C) CID–MS₂ of [M(Li) +Li]+ of m/z 1424. (D) CID–MS₂ of [M(Li) +Li]+ at m/z 1452. Collison gas is argon. Labels correspond to Schemes 8 and 9.

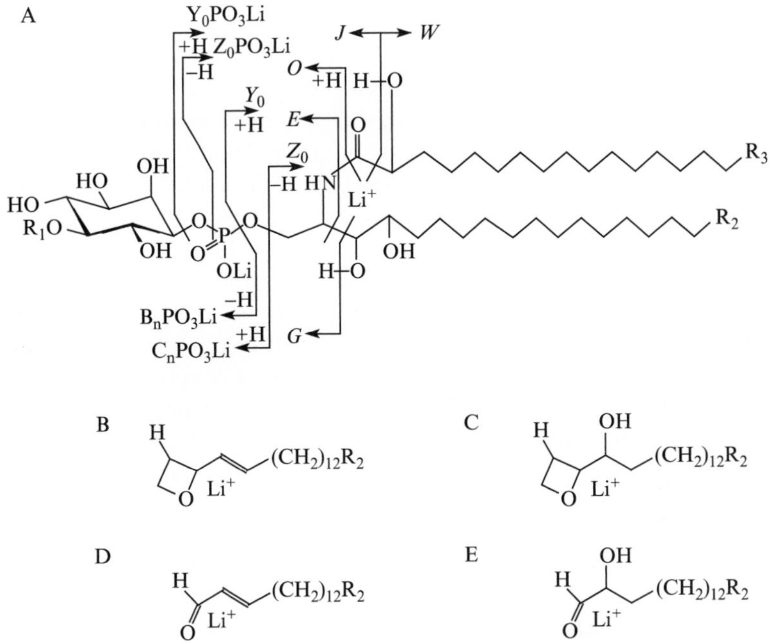

SCHEME 9. Characteristic ceramide fragmentations of lithium-adducted GIPCs. (A) Nomenclature of Adams and Ann (1993) for fragmentation of ceramide moiety and of Singh *et al.* (1991) for *myo*-inositol phosphoryl group. (B) Sphingoid d_{3b} ion, proposed by Hsu and Turk (2001). (C) Hydrated analogue of d_{3b} ion, proposed in this study as a product of t18:0 or t20:0 phytosphingosine-containing ceramides. (D) Sphingoid c_{1b} ion, proposed by Hsu and Turk (2001). (E) Hydrated analogue of c_{1b} ion, proposed in this study as a product of t18:0 or t20:0 phytosphingosine-containing ceramides.

Fig. 8A and B), and some sodiated components were present. It should be pointed out that the increment of 16 Th could also correspond to an extra hydroxylation of the fatty acid (Scheme 1B). In that case, however, MS^2 of the hydroxyl incremented components would yield a single spectrum of glycosylinositol and glycosylinositol phosphate products, while mixed Na^+/Li^+ adducts would produce a mixed product spectrum (the latter situation was verified in this case for the precursors at m/z 1440, 1468, and 1484 in Fig. 8B). Product ion spectra from precursors at m/z 1424 and 1452 are reproduced in Fig. 8C and D, respectively, showing mainly identical series of glycosylinositol+Li^+ and glycosylinositol phosphate(Li^+)+Li^+ product ions.

Emerging Functional, Structural, and Analytical Issues

Three areas of particular biological significance over the last decade include (i) the interaction of microbial adhesins with mammalian cell-surface glycoconjugate receptors, many of which are GSLs (Karlsson, 1989; Karlsson *et al.*, 1992); (ii) defining in more detail tissue, developmental, and pathological patterns of mammalian GSL expression, particularly as a background for correlating knockout or disease phenotypes with deletions of specific structural features, which in turn will help elucidate their contributions to biological function; and (iii) defining the repertoire of GSL expression in other phyla, especially those of insects (including the genomic model organism *Drosophila melanogaster*), nematodes (including parasitic species as well as the model organism *Caenorhabditis elegans*), parasitic cestodes and trematodes, parasitic trypanosomatids (not covered herein), plants (not covered herein), and fungi (including mycopathogens). With respect to parasite species and pathogenic fungi, detailed knowledge of GSL structure and biosynthesis is expected to yield valuable insights into both their functions within the organism and their interactions with the host immune system as well as to provide potential targets for diagnostic and therapeutic agents. MS analysis has made considerable contributions in all of these areas; some of these are highlighted in the following sections.

Selected Applications of Electrospray MALDI Techniques

ESI–MS. Positive-mode electrospray ionization ($^+$ESI–MS, $^+$ESI–CID–MS, $^+$ESI–MS/CID–MS, $^+$ESI–(CID–MS)^2, and $^+$ESI–MS^n) of permethylated neutral GSLs and gangliosides has been used extensively since the early 1990s. Triple quadrupole $^+$ESI–MS and ESI–MS/CID–MS were used along with $^+$FAB–MS in the characterization of a series of GSL antigens with tandem Lewis *a*/Lewis *a*, Lewis *b*/Lewis *a*, and Lewis *a*/Lewis *X* structures (Stroud *et al.*, 1991, 1992, 1994) as well as with an extensive series of polylactosaminoglycan GSLs ("myeloglycans") from myelocytic HL-60 cells and normal human leukocytes, carrying various numbers of Lewis *X* (Galβ1 → 4[Fucα1 → 3]GlcNAcβ1 →) determinants (Stroud *et al.*, 1995, 1996a,b). Subsequently, high orifice-to-skimmer potential $^+$ESI–CID–MS and conventional $^+$ESI–MS/CID–MS, also performed on a triple quadrupole instrument, were used in the analysis of a preferred hexaglycosylceramide receptor for uropathogenic *E. coli* that was isolated from human kidney (Stroud *et al.*, 1998). The receptor was characterized as sialosylgalactosylgloboside (SGG); the former technique yielded a so-diated molecular ion profile consistent with a complex distribution of lipoforms, having both non-hydroxy and 2-hydroxy fatty-*N*-acylation of sphing-4-enine. An extensive suite of fragment ions produced in both modes was fully consistent with the proposed linear glycan sequence, supported also by nuclear magnetic resonance (NMR) spectroscopy and linkage analysis. Tandem $^+$ESI–MS/CID–MS spectra acquired with selection of doubly charged disodiated molecular adducts yielded mainly sodiated B, C, Y, Y/B, and Y/C fragments, as expected. Interestingly, in the $^+$ESI–CID–MS spectrum, classical oxonium B-type fragments were observed along with the sodiated B, C, and internal Y/C fragments; in addition, a series of fragments were also observed at m/z values corresponding to $[C_n + Na + 14]^+$.

Positive-ion electrospray ($^+$ESI–CID–MS) of underivatized GSLs was used in the characterization of two unusual disialogangliosides from renal cell carcinoma (Ito *et al.*, 2001). The more complex ganglioside was shown to have the novel structure of GalNAcβ4[Neu5Acα3]Galβ3[Neu5Acα6]GlcNAcβ3Galβ4Glcβ1Cer. The second ganglioside was missing the terminal GalNAcβ4 residue, yielding a structure previously, although rarely, observed (Fukushi *et al.*, 1986).

A series of articles by Peter-Katalinić and colleagues (Meisen *et al.*, 2003, 2004; Metelmann *et al.*, 2000, 2001b; Müthing *et al.*, 2002, 2004; Vukelic *et al.*, 2001; Zamfir *et al.*, 2002a, 2004a) have examined in great detail the use of nanoESI–MS and MS/CID–MS on a oa-QTOF instrument for analysis of gangliosides from human brain tissue, granulocytes, and other cell types. Brain gangliosides are complex mixtures of G_{M3} (Neu5Acα2 $\rightarrow$ 3Galβ1 $\rightarrow$ 4Glcβ1 $\rightarrow$ 1Cer) and more complex sialylated ganglio-series GSLs. Gangliosides from human granulocytes are mainly G_{M3} and neolacto-series core structures [Galβ1 $\rightarrow$ 4GlcNAcβ1 $\rightarrow$ 3]$_n$Galβ1 $\rightarrow$ 4Glcβ1 $\rightarrow$ 1Cer, terminated by a Neu5Acα2 $\rightarrow$ 3 or Neu5Acα2 $\rightarrow$ 6 residue, and incorporating various numbers of Lewis X (Galβ1 $\rightarrow$ 4[Fucα1 $\rightarrow$ 3]GlcNAcβ1 $\rightarrow$) determinants from 0–n (detected on Neu5Acα2 $\rightarrow$ 3 terminated species). Many of these compounds were previously characterized with the extensive aid of MS and MS/MS techniques, employing especially tandem magnetic sector FAB/LSI–MS and, later, tandem quadrupole ESI–MS. Thus, these compounds make up a good system for testing the use of new capabilities as well as furthering the development and validation of new methodologies, and, as significant improvements in sensitivity and resolution make it possible, mining of these materials for new structural variants will proceed. Some key analytical challenges, encountered routinely over many years with neolacto-series GSLs, have been (i) to determine the linkage of the terminal sialic acid; (ii) to discover which β-GlcNAc residues have α-Fuc attached; and (iii) to verify that the β-Gal residues are linked 1 $\rightarrow$ 4 and not 1 $\rightarrow$ 3 and that the α-Fuc residues are linked 1 $\rightarrow$ 3 and not 1 $\rightarrow$ 4, to GlcNAc, since the latter linkage arrangements are also encountered in many normal and pathological tissues (sometimes on the same GSLs). Ganglio-series gangliosides have formally the same alternating core arrangement of (Hex-O-HexNAc-O-)$_n$, (where usually $n = 1$) attached to Galβ1 $\rightarrow$ 4Glcβ1 $\rightarrow$ 1Cer, but the HexNAc is GalNAcβ1 $\rightarrow$ 4, the terminal Hex is Galβ1 $\rightarrow$ 3, and Neu5Ac may be attached by α2 $\rightarrow$ 3 to either terminal or internal β-Gal, by α2 $\rightarrow$ 6 to internal β-GalNAc, or by α2 $\rightarrow$ 8 to other sialic acid residues. NMR spectroscopy is very good at elucidating most of these structural features, but cannot compete with MS where sample limitations and high through-put are major issues. Metelmann *et al.* (2000, 2001b) observed detection limits in the sub-femtomole range for single species in a mixture of brain gangliosides and sample consumption in the low femtomole range for analysis of multiple components in a mixture of granulocyte gangliosides. The question, then, rests on the ease and reliability of the glycan sequence analysis, especially with respect to the key structural elements outlined above.

To address these questions, systematic studies were carried out using the nanoESI–oa-QTOF instrument in both positive and negative-ion de-tection modes (positive for permethylated compounds, both modes for underivatized species) (Meisen *et al.*, 2003, 2004; Metelmann *et al.*, 2000, 2001a,b; Vukelic *et al.*, 2001). Metelmann *et al.* (2001b) investigated low-abundance gangliosides in mixed fractions from human granulocytes; gang-liosides containing α-Fuc residues were identified and characterized with respect to the attachment of α-Fuc to specific β-GlcNAc residues in the sequence. In this case, *de novo* elucidation of the entire sequence was not an issue because foreknowledge of the neolacto-series core structures could be assumed, and characterization of the Neu5Acα2 $\rightarrow$ linkage was not claimed.

Subsequently, with respect to differentiating between terminal Neu5Acα2 $\rightarrow$ 3 or Neu5Acα2 $\rightarrow$ 6 on neolacto-series core structures, Meisen *et al.* (2003) reported that low-energy negative-ion mode MS/CID–MS of underivatized gangliosides on the nanoESI–oa-QTOF instru-ment enabled a clear distinction between the two linkages attached to nLc$_4$Cer and nLc$_6$Cer cores. Diagnostic decarboxylated ^{0,4}A$_2$ ions at m/z 306.2, ^{2,4}A$_3$ ions at m/z 468.3, and ^{0,2}X$_4$/^{0,2}X$_6$ ions $[M - H - 221.1]$ obtained by sialic acid ring cleavage, accompanied by clearly decreased relative intensity of Y$_4$/Y$_6$ ions $[M - H - 291.1]$, could be reliably assigned to gangliosides with the Neu5Acα2 $\rightarrow$ 6Galβ1 $\rightarrow$ 4GlcNAcβ1 $\rightarrow$ terminal structure. These criteria were subsequently applied to neolacto-series gang-liosides extracted from single spots on immunostained high-performance thin-layer chromatography (HPTLC) plates, where sample amounts on the order of ~1 μg were characterized (Meisen *et al.*, 2004).

In human brain gangliosides, the overwhelming majority of which are variously sialylated species based on the Gg_4Cer core, the existence of isobaric variants differing in the arrangement of Neu5Ac residues is known as a long-standing problem. In a study by Metelmann *et al.* (2001a), underivatized ganglio-series gangliosides could be recognized and partially characterized by negative-ion nanoESI–oa-QTOF MS/CID–MS spectra, but precise linkage patterns could not be deduced. It was found that positive-ion MS/CID–MS spectra yielded considerably more information; however, it remains questionable whether this methodology is sufficient in itself for analysis of isobaric variants should they occur in mixtures. Vukelic *et al.* (2001) described the application of this strategy in combination with HPTLC and immunochemical methods to analysis of gangliosides from histopathologically defined human cerebrum-resembling remnants and cerebellar tissue from 30- and 37-week-old gestational anencephaluses and compared them to normal newborn/fetal and adult brain regions. The method identified neolacto- as well as ganglio-series gangliosides in these tissues. Overall, significant quantitative differences were observed in ganglioside profiles, suggesting the potential for diagnostic applications.

Recently, Zamfir *et al.* (2004a) described application of an automated, chip-based interface to sensitive, high-throughput screening of a complex mixture of human cerebellar gangliosides by negative-ion nanoESI–oa-QTOF MS and MS/CID–MS. In this case, in-source fragmentation was minimized, a condition essential for reliable identifications because the structural diversity of the true molecular profile, especially of low-abundance species, would be compromised by uncontrolled losses of sialic acid residues. In all, 46 ganglioside glycoforms were identified in a sample by this method, and automated data-dependent precursor ion selection and MS/CID–MS fragmentation was demonstrated with one molecular species, G_{T1}.

The MS^n capability of quadrupole ion traps has been exploited for GSL analysis particularly by Reinhold and colleagues (Duk *et al.*, 2001; Reinhold and Sheeley, 1998). Reinhold and Sheeley (1998) demonstrated generation of fragmentation of permethylated GSLs through MS^4, utilizing cross-ring cleavages for dissection of structural features. Some limitations with respect to the dissipation of cross-ring cleavage modes with increasing size, and by Neu5Ac and HexNAc residues, were noted, as has previously been observed on other instruments at low-collision energies. Nevertheless, indications are that the information content available from MS^n of permethylated glycans is considerable; the increase in linkage-specifying fragmentation modes appears to outweigh the extra time and expense of the derivatization step, which can be reliably applied to micro-scale samples. MS^n methods were applied to characterization of a novel pentaglycosylceramide isolated from erythrocytes of the rare polyagglutin-able NOR phenotype (Duk *et al.*, 2001). Gas phase disassembly yielded the glycan sequence, except for anomeric configurations that were supplied by other methods.

MALDI–MS. Applications of MALDI–MS to analysis of carbohy-drates and glycoconjugates, including GSLs, were extensively reviewed by Harvey (1999). Besides many general issues concerning MALDI–MS and aspects of glycan sequencing, the review covers issues important for GSL analysis, such as matrix selection as well as the problem of decarbox-ylation and losses of sialic acid from gangliosides. The interested reader is advised to consult that review as a good starting point for such issues as detection mode, laser power, and matrix selection, which can especially influence success with ganglioside analysis; aside from reiterating a few salient points, the focus herein will be on some subsequent developments.

Native neutral GSLs have generally yielded satisfactory results with 2,5-DHB, 4-hydroxycinnamic acid (HCCA), and 6,7-dihydroxycoumarin (esculetin) (Harvey, 1999). The 2,5-DHB has been employed for positive-ion MALDI–TOF–MS of zwitterionic PCho- and PEtn-containing GSLs from nematoda, but metastable losses of choline and ethanolamine were noted in reflectron mode (Gerdt *et al.*, 1999; Lochnit *et al.*, 1998a; Wuhrer *et al.*, 2000a). Matrices employed in negative-ion MALDI–TOF–MS of acidic fungal GIPCs have been *nor*-harmane (Loureiro y Penha *et al.*, 2001; Heise *et al.*, 2002) and 7-amino-4-methylcoumarin (Aoki *et al.*, 2004). In the latter case, as noted previously, prompt losses (probably of trimethylamine) were observed in linear mode spectra of PCho-containing GIPCs (Aoki *et al.*, 2004).

For ganglioside analysis, a wide variety of matrices have been used (Harvey, 1999), including 2,5-DHB, 1,5-diaminonaphthalene, 4-hydrazino-benzoic acid, 6-aza-2-thiothymine (ATT), 2-(4-hydroxyphenylazo)benzoic acid (HABA), 5-chloro-2-mercaptobenzothiazole, 2-thiohydantoin (escu-letin), and α-cyano-4-HCCA; particularly the latter is not recommended because it has promoted excessive loss of sialic acid and CO_2 (Harvey, 1999). Methyl esterification, as well as permethylation in general, mini-mizes loss of sialic acids (Harvey, 1999); these derivatizations may be compatible with some schemes for further analysis of gangliosides but may preclude their use for other purposes. Metastable decompositions of gangliosides are particularly problematic for MALDI–FTICR–MS analysis due to the extended times between ionization and detection (Penn *et al.*, 1997; O'Connor *et al.*, 2002), obviating the advantages of using this tech-nique. The use of Cs adduction, brought about by doping of the matrix with CsCl, has been suggested to stabilize gangliosides (Penn *et al.*, 1997), but fragmentation is not altogether eliminated (O'Connor *et al.*, 2002). An ionic liquid matrix, 2,5-DHB butylamine (DHBB), has been recommended for enhancement of MALDI–MS analysis of biomolecules, including gangliosides (Mank *et al.*, 2004). In one illustration with the MALDI–TOF in linear mode, DHBB suppressed loss of sialic acid from G_{M1} ganglioside compared to DHB, but the spectrum exhibited considerable loss of resolution. It is unclear whether this is a problem with a straightforward solution.

In another approach, a prototype high-pressure MALDI source (O'Connor and Costello, 2001) developed to improve molecular ion yields of labile molecules in FTICR–MS by collisional cooling, was applied to the analysis of gangliosides (O'Connor and Costello, 2001; O'Connor *et al.*, 2002). It was observed that transient elevation of pressure in the source into the 1–10 mbar range during ionization decreases the metastable fragmentation of gangliosides. This allowed detection of molecular ion species without loss of the highly labile sialic acid residues. Gangliosides with up to three sialic acids were detected in positive-ion mode MALDI–FTICR–MS with little fragmentation (matrix, ATT). Gangliosides with four to five sialic acids produced weak spectra in positive-ion mode but were analyzed successfully in negative-ion mode, with considerable suppression of sialic acid loss. A tendency to form multiple matrix adducts was noted in this mode, but this was deemed less of a practical problem than the metastable losses in FTICR–MS (O'Connor *et al.*, 2002). A newer version of this source has been designed with a movable stage accommodating large targets, such as TLC plates or transfer membranes, enabling coupling of TLC separation methods to MALDI–FTICR–MS (O'Connor *et al.*, 2004). This was demonstrated by desorption of gangliosides with up to three sialic acids directly from a silica gel plate (matrix, sinapinic acid), with minimal sialic acid loss.

Positive-ion reflectron mode MALDI–TOF–MS was used in the characterization of complex branched neolacto-series gangliosides from human A erythrocytes as their permethylated derivatives (Kushi *et al.*, 2001), including a decaglycosylceramide with a complex blood group determinant on one branch (GalNAcα1 → 3[Fucα1 → 2]Galβ1 → 4[Fucα1 → 3] GlcNAcβ1 → 6). MALDI–TOF–MS has been used for characterization of very large dendritic polyglycosylceramides (PGCs) (Miller-Podraza, 2000) from human erythrocytes in conjunction with development of screening methods for possible GSL targets for *Helicobacter pylori* (Karlsson *et al.*, 1999, 2000). Terminal sequence analysis by FAB–MS and GC–MS methods were also employed (Karlsson *et al.*, 2000). These studies employed positive-ion MALDI–TOF–MS (2,5-DHB matrix) for neutral GSLs and oligosaccharides released by endoglycoceramidase (from *Rhodococcus*) or endo-β-galactosidase (from *Bacterioides fragilis*), and negative-ion mode (2,5-DHB matrix) for acidic sialylated species comprising up to 41 monosaccharide units (calculated MW of intact PGC 8058.96 D; observed m/z 8057.4) (Karlsson *et al.*, 1999).

Hunnam *et al.* (2001) applied MALDI–oa-QTOF–MS to analysis of neutral GSLs and gangliosides. Spectral profiling results with a standard mixture of neutral GSLs (tetraglycosyceramides to tetraglycosyceramides detected as Na^+ adducts) with the MALDI–oa-QTOF–MS were compared to those acquired by nanoESI–oa-QTOF–MS on the same instrument and on a MALDI–Reflectron–TOF. In this case, metastable fragmentation was not a problem with any of the instruments, and the results were somewhat comparable, but some loss of sensitivity with higher m/z GSLs was observed in the ESI profile. Results with G_{M1} ganglioside appeared to be less satisfactory, with significant losses of sialic acid in both MALDI profiles—both prompt (in-source) and metastable losses with the MALDI–Reflectron–TOF, and in-source loss exclusively with the MALDI–oa-QTOF–MS. With the neutral GSLs, MS/CID–MS spectra acquired from selected molecular adduct precursors on the oa-QTOF–MS, with nanoESI and MALDI sources, respectively, were compared with PSD spectra on the MALDI–Reflectron–TOF. Both MALDI and ESI–MS/CID–MS data with the oa-QTOF–MS appeared comparable and superior to the results with MALDI–Reflectron–PSD–TOF.

Remove the Ceramide or Leave it Intact?

The availability of ceramide glycanases (Li *et al.*, 1994; Zhou *et al.*, 1989) or endoglycoceramidases (Ito and Yamagata, 1986, 1989a,b) provides a convenient method for cleaving GSL glycans from their ceramides in high yield (except in the case of monohexosylceramides, which are poor substrates for these enzymes). Previously, it was demonstrated that such glycans, containing up to 11 sugars, released from complex GSL mixtures, could be analyzed by GC–MS of their per-N,O-methyl derivatives (Hansson *et al.*, 1989). Monohexosylceramides, along with the released ceramides, from the same sample could be separated by solid phase extraction on a C18-silica cartridge, permethylated, and analyzed in a second GC–MS run. Release of glycans also opens up the possibility of employing LC applications developed for oligosaccharide separation, with the option of interfacing to suitable MS detection; in addition, the released glycans could potentially be derivatized with any of the available reducing end tags by reductive amination; the end tags have been developed in conjunction with a wide variety of separation and MS protocols. An additional application is the release of GSL glycans from intact cells (Ito *et al.*, 1991, 1993a,b; Rasilo *et al.*, 1989). Glycosylinositol phosphorylceramides from plants and fungi can also be released from their ceramides by treatment with aqueous ammonia at high temperature (Barr and Lester, 1984; Levery *et al.*, 1996). In this case, the use of reducing end tags is not applicable, but MS or LC–MS of intact glycosylinositols is possible as well as MS following permethylation (Loureiro y Penha *et al.*, 2001). Intact ceramides are not recoverable following the ammonia release protocol.

A report by Friedl *et al.* (2000) described structure elucidation of zwitterionic glycans released from GSLs of the nematode *Ascaris suum*. Release of glycans with recombinant *Rhodococcus* species endoglycocer-amidase was not hindered by the substitution of the nematode/trematode/insect core sequence GlcNAcβ1 → 3Manβ1 → 4Glcβ1 → 1Cer in place of the mammalian GlcNAcβ1 → 3Galβ1 → 4Glcβ1 → 1Cer structural motif. Positive- and negative-ion ESI–MS analyses performed with an ion trap mass spectrometer were used for determination of glycan monosaccharide (as Hex and HexNAc) sequence and linkage as well as the location and linkage of phosphocholine and phosphoethanolamine substituents. Both deprotonated and chloride attachment ions were analyzed by negative-ion mode ESI–MS^n experiments. However, it was not made completely clear why performing ion trap ESI–MS^n experiments on released glycans was a superior strategy to direct analysis of the intact glycophingolipids.

Subsequently, Wing *et al.* (2001) reported an HPLC assay based on release of glycans from gangliosides and neutral GSLs, followed by labeling of the reducing end with 2-aminopyridine for fluorescence detection. The labeling and separation scheme was demonstrated to be compatible with subsequent MALDI–MS analysis (2,5-DHB matrix) to confirm the identity of the glycans and, in principle, modifications thereto. Sensitivity of detection was observed to be favorable compared to that of a previously developed scheme employing *p*-aminobenzoic acid ethyl ester as a UV-absorbing chromophore. Gangliosides were generally cleaved at greater than or equal to 90%. A limitation of the method is that neutral GSLs appear to be cleaved somewhat less well than gangliosides under some conditions, and, as observed previously, GlcCer is very refractory (the highest activity observed yielded 34% hydrolysis of GlcCer).

A possible disadvantage of glycan release methods applied to GSL mixtures is that it becomes impossible to associate glycan structures with their specific ceramide profiles, which may differ between components. Complete profiles of all intact molecular species in a GSL mixture can provide important information on their biosynthesis, as cases are known where both the initial and subsequent glycosylation steps can be influenced by ceramide structure. For cases in which this information is deemed important, it is better to first purify GSL components with respect to glycan structure and determine their ceramide compositions before cleaving them off. Profiling of intact GSL mixtures may also reveal ceramide profile differences for components differing in the number of sugars, but the presence of different components having the same monosaccharide composition may be missed in the absence of other criteria. In some applications, the analyst may find information about ceramide profiles of less importance, outweighed by the gains in sensitivity from eliminating a source of molecular ion partitioning and from attaching an optically sensitive tag. Other negative considerations include the possibility of preferential cleavage of certain GSL structures over others and the additional expense of using these enzymes. The analyst should consider whether the effort and expense of removing what is already a fairly good reducing end tag, with charge-nucleating and lipophilic characteristics favorable for a variety of techniques, is worth the work even if the purpose is to replace it with a better one.

TLC Blotting and TLC–MS

The first attempts to couple TLC with soft-ionization methods for GSL analysis were made by Handa, Kushi, and coworkers (Kushi and Handa, 1985; Kushi et al., 1988, 1990). LSI–MS was performed after application of liquid matrix to a small section cut from the original HPTLC plate; both chromatographic ion profiles (mass chromatograms) as well as R_f-resolved mass spectra could be acquired by moving the plate manually through the secondary ion beam. Karlsson et al. (1991) later described the use of a motorized TLC–FAB probe. Major advantages of the method derived from the ease and high chromatographic resolution possible with TLC; in addition, TLC–MS fits well with TLC-overlay strategies for characterizing the specificity of interactions of proteins or microbes with cell surface GSL receptors (Karlsson et al., 1991).

MS of GSLs transferred by blotting from a developed TLC plate to a polyvinylidene fluoride (PVDF) membrane was introduced by Taki and Ishikawa (Taki and Ishikawa, 1997; Taki et al., 1994, 1995). The membrane is more convenient to cut; in particular, the ability to punch out a relatively small piece of membrane obviates the need for a special secondary ion mass spectrometry (SIMS) or FAB–MS probe and also means the technique can be extended to the analysis of GSLs resolved by TLC in two dimensions. Johansson and Miller-Podraza (1998) incorporated the technique of Taki and Ishikawa (1997) into a protocol to differentiate between terminal Neu5Acα2 → 3 or Neu5Acα2 → 6 on neolacto-series gangliosides by blotting of HPTLC-separated mixtures to PVDF membranes, followed by direct negative-ion FAB–MS analysis (triethanolamine matrix) and membrane overlay binding assay with Neu5Acα2 → 3 or Neu5Acα2 → 6 specific lectins (*Maackia amurensis* and *Sambuca nigra*, respectively). Additional structural techniques used were FAB–MS and a newly developed trifluoroacetolysis/GC–MS analysis (Johansson and Karlsson, 1998) of the gangliosides washed from small sections of the PVDF membrane. These results were correlated with those of a third assay, HPTLC overlay with radiolabeled *Helicobacter pylori*. In this way, the preferred ganglioside binding affinities of various *H. pylori* strains could be screened, and, conceivably, previously unknown binding targets could be identified.

Guittard *et al.* (1999) further adapted the technique to use with MALDI–TOF–MS analysis, achieving very high sensitivity but with some inherent limits on mass resolution and accuracy deriving from irregularities in the membrane surface. Use of delayed extraction did not appear to compensate fully for this effect. The research group compared results with both nitrogen and Er-YAG lasers, as well as with a number of different membrane types and desorption matrices, and observed optimal sample deposition and analysis with the combination of Immobilon P (PVDF) membrane, the nitrogen laser, and 2,5-DHP/ANP (1:1) matrix.

Flow-Injection Analysis, Liquid Chromatography, and
    Capillary Electrophoresis

Several groups have undertaken development of LC–MS methods for quantitative GSL analysis and extension of sphingolipidomic LC–MS protocols to GSLs. The natural interface between LC and ESI–MS techniques was exploited by Gu *et al.* (1997), who used ESI–MS/CID–MS in product ion scanning mode to detect characteristic fragments for sphingosine and dihydrosphingosine and in precursor scanning mode to detect ceramides incorporating those sphingoids in LC eluent. An unnatural internal standard was employed for quantitation. The method was applied to extracts of three related lymphocyte cell lines. A product ion spectrum of a fungal cerebroside was included, showing that (in principle at least) this method could be extended to glycosphingolipids, even those of other phyla. It should be noted that methods relying on detection of one or two products must be viewed with caution because they will certainly miss any variant lipoforms incorporating unexpected sphingoids—the history of this field has provided ample reason to expect the unexpected. An example is in a study by Colsch *et al.* (2004), wherein GSLs from mouse brain were characterized in detail using ESI–MS/CID–MS; the research group found significant incorporation of d18:2 sphingoid in galactosylceramide and sulfatide fractions but not in ganglioside fractions.

Improved protocols have employed multiple reaction monitoring (MRM) to detect a wider variety of sphingoid/ceramide-containing precursors. Sullards and coworkers (Sullards, 2000; Sullards and Merrill, 2001;

Sullards *et al.*, 2003) have focused on development of a comprehensive sphingolipidomic LC–MS protocol extendable in principle to glycosphingolipidomics although only simple GSLs have been detected so far. Internal standards of unnatural sphingolipids are included for quantitation. Boscaro *et al.* (2002) have developed a protocol based on flow-injection analysis (FIA) for high-throughput detection and quantitation of Gb$_3$Cer, which is characteristically elevated in urine and serum of patients with Anderson-Fabry disease. A practical application is for monitoring the effectiveness of therapeutic treatments for this disease. Internal standards for this protocol are composed of stable-isotope labeled ceramides (Mills *et al.*, 2002). Some protocols should allow for monitoring based on characteristic neutral losses because these would not depend on detection of specific ceramide fragments; however, as already shown, some structural features can stimulate or suppress certain fragmentation modes, a possibility that must be anticipated to lead to variable ion yields and resulting differential sensitivity of detection.

Several groups investigating the interfacing of capillary electrophoresis to ESI sources have applied this technique to ganglioside analysis. Zamfir *et al.* (2002b, 2003) initially developed an off-line approach to glycoscreening, including ganglioside analysis (Zamfir *et al.*, 2002a), but more recently have applied a capillary electrophoresis electrospray ionization (CE–ESI) interface (Zamfir and Peter-Katalinić, 2001) to on-line glycoscreening by CE–nanoESI–oa-QTOF–MS (Zamfir *et al.*, 2004b), an approach that could potentially be applied to gangliosides as well, provided additional technical problems can be overcome, primarily the development of a suitable buffer system compatible with both acceptable component resolution and MS analysis. For example, an alternative system developed by Ju *et al.* (1997) has shown good sensitivity of detection and electrophoretic resolution of a variety of gangliosides, but GD1a and GD1b coeluted in this system (Zamfir and Peter-Katalinić, 2004). A robust interface for linking micellar electrokinetic chromatography (MEKC) to ESI–MS while employing nonvolatile buffer systems has also been developed (Tseng *et al.*, 2004). The system appeared to accomplish at least partial resolution of GD1a and GD1b.

GSLs of Insects and Worms

GSL studies have defined among insect and worm species a number of divergent core glycan motifs as well as some commonly distributed peripheral modifications that provide additional challenges not encountered in the structural analysis of their mammalian counterparts. MS has proved especially useful for defining the number and location of these modifications. Studies of insect GSLs have so far been concentrated on dipteran (fly) species, which are now particularly relevant with the completion of the *D. melanogaster* genome sequence. These studies have defined an "arthro-" (Ap-) series core structure based on extensions of the neolacto-series analogue GalNAcβ4GlcNAcβ3Manβ4Glcβ1Cer (Dennis and Wiegandt, 1993). This core sequence is apparently shared with GSLs of nematodes, including a variety of mammalian parasites (Lochnit *et al.*, 1997, 1998a; Wuhrer *et al.*, 2000a), and *C. elegans* (Gerdt *et al.*, 1997, 1999), another model organism for which the genome sequence has been completed and that appears to have a particularly rich glycobiology. In GSLs of both diptera and nematoda, extended, unbranched glycan structures are encountered, along with modification by zwitterionic substituents, phosphorylethanolamine (PEtn), and phosphorylcholine (PCho).

With respect to complex GSLs of insects, only fly species seem to have been examined so far, and a large time gap is apparent between studies of the early 1990s and investigation of GSL expression in the genetic model organism, *D. melanogaster*. Post-1990 characterizations of zwitterionic PEtn-containing GSLs from *Lucilia caesar* (Itonori *et al.*, 1991) and *Calliphora vicina* (Helling *et al.*, 1991) employed positive- and negative-ion mode FAB–MS of native, peracetylated, and permethylated fractions along with a variety of other structural techniques, including NMR spectroscopy. In both studies, GSLs with up to seven monosaccharide residues were analyzed; almost all structures characterized fell into an identical series of linear glycans based on extension of the arthro GalNAcβ4Glc-NAcβ3Manβ4Glcβ1Cer core. In all zwitterionic structures, PEtn was found linked to the sixth position of the (internal) GlcNAcβ3 residue; the heptaglycosylceramide possessed an additional nonreducing terminal GlcNAcβ3 residue that was not modified. Similar to previous observations, the predominant ceramides were composed of characteristically short sphingoids; d14:1 tetradecasphing-4-enine (mainly) and d16:1 hexadecasphing-4-enine; and fatty-*N*-acylated with saturated, nonhydroxylated 20:0 (mainly) and 22:0 acids.

In a study of complex GSLs from *D. melanogaster* (Seppo *et al.*, 2000), negative-ion LSI–MS and positive-ion ESI–MS were used in conjunction mainly with linkage analysis and sequential exoglycosidase digestions. In addition to a similar series of zwitterionic "dipteran" linear glycans, based on the arthro core structure modified with PEtn on the internal GlcNAcβ3 residue, an octaglycosylceramide was characterized with an additional PEtn linked to the now penultimate GlcNAcβ3 residue. As also found in previous studies, complex acidic zwitterionic GSLs were also isolated; these have been observed to have terminal L2/HNK1 glycan, composed of the GlcAβ3Galβ3 disaccharide sequence although the mammalian HNK1 epitope is sulfated (Chou *et al.*, 1986). This is an important adhesion recognition ligand (Jungalwala, 1994), and the nonsulfated epitope was recognized by some antibodies on GSLs and glycoproteins of *Calliphora vicina* (Dennis *et al.*, 1988).

In addition to a previously isolated hexaglycosylceramide with one PEtn modification, Seppo *et al.* (2000) isolated a pair of nonaglycosylceramides having one and two PEtn modifications, respectively. In these characterizations, the linkage positions and configurations of the PEtn and GlcA residues were not rigorously determined, and the structural proposals relied somewhat on structures more rigorously elucidated in prior studies (Dennis and Wiegandt, 1993). Molecular ions observed were generally consistent with incorporation of characteristic "dipteran" ceramides, predominantly 20:0/d14:1, as noted earlier. Altogether, these characterizations of the dipteran glycosphingolipidome, alongside the *Drosophila* genome, now form the basis for further investigations of GSL biosynthesis and functions in the fly (Muller *et al.*, 2002; Seppo and Tiemeyer, 2000; Schwientek *et al.*, 2002; Wandall *et al.*, 2003, 2005).

The first studies of *C. elegans* GSLs were performed by Chitwood *et al.* (1995), who isolated and characterized GlcCer species containing an unusual branched chain sphingoid, d17:1 15-methylhexadecasphing-4-enine (15-methyl-2-aminohexadec-4-en-1,3-diol), that was fatty-*N*-acylated predominantly with 2-hydroxylated C20-C26 acids. Initial profiling of predominant molecular species was carried out by CI–MS using methane, NH_3, $[^{15}N]H_3$, and $N[^2H_3]$ as reagent gases. This process was followed by further structural characterization of neutral GSLs containing up to three monosaccharide residues (Gerdt *et al.*, 1997) and larger GSLs elaborated with antigenic PCho substituents linked to the sixth position of GlcNAcβ3, containing up to six monosaccharide residues (Gerdt *et al.*, 1999). Similar structures have been isolated and characterized from the parasitic nematodes *A. suum* and *Onchocerca volvulus* (Lochnit *et al.*, 1997, 1998a; Wuhrer *et al.*, 2000a); however, in the case of *A. suum*, a complex zwitterionic GSL was found to contain PEtn attached to Man as well as the more usual PCho attached to GlcNAc. Ceramides similar to those characterized initially by Chitwood *et al.* (1995), incorporating the branched d17:1 sphingoid and mainly h22:0-h24:0 fatty-N-acylation, have been observed in these GSLs although *A. suum* components also incorporated a significant amount of d17:0 sphingoid (15-methylhexadecasphinganine) as well (Lochnit *et al.*, 1997, 1998a). In characterizing these nematode GSLs, extensive use was made of MALDI–TOF–MS analysis of intact, underivatized GSLs as well as of GSL glycans released from their ceramides by an endoglycoceramidase and derivatized with 2-aminopyridine by reductive amination (Gerdt *et al.*, 1999; Lochnit *et al.*, 1998a). PA-oligosaccharides were released from a GSL mixture and then fractionated by HPLC prior to MS and permethylation linkage analysis. Structure elucidations were aided by specific exoglycosidase treatments of either intact GSLs or PA-oligosaccharides, with progress of the reactions monitored by $^+$MALDI–TOF–MS; additional exoglycosidase reactions were performed on target. Additional data were obtained from monosaccharide, fatty acid, and sphingoid component analysis by GC–MS, and, in some studies, NMR spectroscopy. HF treatment was also used for safe removal of PCho and PEtn residues from the GSL cores.

Interestingly, an acidic fraction of glycolipids from *A. suum* was found to contain an unusual GIPC, Galα1 $\rightarrow$ 2Ins-P-Cer (Sugita *et al.*, 1996); negative-ion FAB–MS was used to support this characterization. Later, this finding was confirmed by Lochnit *et al.* (1998b), who also found small amounts of a sulfatide, 3-sulfo-GalCer, in the *A. suum* acidic fraction. These components were found with the major ceramides having typical nematode structures.

Phosphorylcholine substituents were again encountered in GSLs of cestodes and annelids, but, in these cases, they were on "neoGala-" (nGa-) series core structures based on repetitive extensions of Galβ1Cer by one or more Galβ6 residues (Dennis and Wiegandt, 1993; Dennis *et al.*, 1992). Additional modifications include substitution of the Galβ6 units by Fucα3 or Manα4 as either terminal or branching residues (Dennis and Wiegandt, 1993; Sugita *et al.*, 1995).

Investigations on the trematode parasite *Fasciola hepatica* showed that it expresses mammalian GSLs; that is, this parasite expresses Gb$_3$Cer and iGb$_3$Cer, Forssmann pentaglycosylceramide antigens based on those cores, and highly antigenic Galβ6- and Galβ6Galβ6-terminated GSL antigens also based on the same mammalian cores (Wuhrer *et al.*, 2004). Phosphocholine substituents have not been reported so far on GSLs of trematodes.

The trematode parasite *Schistosoma mansoni*, unlike any other species examined so far, presents a characteristic "schisto-" series core structure (Makaaru *et al.*, 1992) based on extension of the sequence Gal-NAcβ4Glcβ1Cer by repeating units consisting of one or two Fuc residues attached to GlcNAc. The characterization of these highly fucosylated, highly immunogenic structures presents an interesting demonstration of both the power of MS and some of its limitations. Initial attempts to define the structure of GSL antigens from eggs of *S. mansoni* were made by $^+$FAB–MS following permethylation of a high molecular weight fraction, combined with monosaccharide and fatty acid consituent analysis, methylation linkage analysis, and $^-$FAB–MS of several underivatized fractions (Levery *et al.*, 1992). The highest MH$^+$ of the permethylated fraction were observed at nominal, monoisotopic *m/z* 3706/3736, corresponding to deoxy-Hex$_8$•HexNAc$_6$•Hex•Cer; MH$^+$ for deletions of up to four deoxyHex residues were also observed, with the most abundant, at *m/z* 3532/3562, corresponding to deoxyHex$_7$•HexNAc$_6$•Hex•Cer. Prominent B-series ions at m/z 608, 1201, and 1794 (deoxyHex$_{2n}$•HexNAc$_n$; n = 1, 2, 3), along with the linkage analysis and other data, led to consideration of two alternate basic repeating units: (i) → 2Fuc1 → 4(Fuc1 → 3)GlcNAc1 → or (ii) → 4 (Fuc1 → 2Fuc1 → 3)GlcNAc1 → attached to a → 3GalNAc1 → 3GalNAc1 → 4 Glc1 → 1Cer core structure. Both repeating unit structures allowed for deletions of one or more Fuc residues along the glycan, consistent with a variety of other B- and Y-series ions in the $^+$FAB mass spectrum; neither could be excluded by any of the data at hand. The first structure was thought to be most likely (Levery *et al.*, 1992), but a subsequent reinvestigation by Khoo *et al.* (1997) showed in fact that the second structure is correct; in addition, the second → 3GalNAc1 → residue was relocated to the periphery as a terminal Fuc1 → 2Fuc1 → 3GalNAc1 → 4 unit, yielding GlcNAcβ1 → 3GalNAcβ1 → 4Glcβ1 → 1Cer as the likely core structure. In the study by Khoo *et al.* (1997), $^+$FAB–MS of perdeuteromethylated native and periodate oxidized antigens, along with analysis of *S. mansoni* egg glycoprotein glycans, provided the necessary desiderata. The initial proposal (Levery *et al.*, 1992) of a unique core disaccharide consisting of GalNAc1 → 4 linked to Glc1 → 1Cer was subsequently confirmed by Makaaru *et al.* (1992), Khoo *et al.* (1997), and Wuhrer *et al.* (2000b,c); overall, the revised structure is consistent with what is known about glycoconjugates and glycosyltransferase activities from *S. mansoni* and other parasite species (Makaaru *et al.*, 1992) as well as from the snail *Lymnea stagnalis* (snails are the intermediate hosts for schistosomes).

Investigations of GSLs from *S. mansoni* cercariae have indicated a stage-specific shift in glycan biosynthesis, with predominant expression of a mammalian Lewis X (Galβ4[Fucα3]GlcNAcβ3) and a pseudo-Lewis Y (Fucα3Galβ4[Fucα3]GlcNAcβ3) antigen based on Fucα3 transfers to a neolacto-series core analogue Galβ4GlcNAcβ3GalNAcβ4Glcβ1Cer as well as to a novel pentaglycosylceramide core structure containing a GlcNAcβ3GlcNAcβ3 repeat (Wuhrer *et al.*, 2000b). Possibly a shift away from the predominant egg GSL structures with repeating (GlcNAc)$_n$ backbone toward these truncated antigens could for the most part be achieved through activation of a single glycosyltransferase, a UDP-Gal:β-GlcNAc β4Gal-T, for which a large family of homologues is already known; however, this would not explain the presence of the GSL with the single GlcNAcβ3GlcNAcβ3 repeat. The structural characterization employed methods similar to those summarized previously for nematode GSLs. Additional detailed studies have been carried out on monoglycosyl- and diglycosyl-ceramide fractions of *S. mansoni* eggs, cercariae, and adults, chiefly to examine stage-associated alterations in ceramide structure (Wuhrer *et al.*, 2000c). These studies were carried out using monosaccharide,

fatty acid, and sphingoid component analysis in combination with
$^{+}$MALDI–TOF–MS of HPLC fractionated GSLs, both underivatized and
following peracetylation.

Glycosphingolipids of Mycopathogens and Other Fungi

Increasing occurences of life-threatening systemic mycosis have paral-
leled the growth in populations of immunosuppressed or immunocompromised individuals, including those with acquired immunodeficiency
syndrome (AIDS), recipients of organ and tissue transplants, and patients
with leukemias and other cancers (Durden and Elewski, 1997; Lortholary
et al., 1999; Wade, 1997; Walsh et al., 1996). The emergence of fungal
strains resistant to existing therapeutic treatments is an additional cause
for concern (Latgé and Calderone, 2002; Lortholary et al., 1999; Sanglard,
2002; van Burik and Magee, 2001). To facilitate development of new
diagnostic and therapeutic agents, continuing studies directed toward dis-
secting the relationship between fungal life cycles, processes of mycotic
infection, and factors contributing to virulence and antibiotic resistance
are urgently needed.

Several promising lines of research have focused on the function of
fungal GSLs, particularly of GIPCs. The synthesis of GIPCs appears to
be required for the survival of fungi, but this class of GSLs has not been
found in mammalian cells (Dickson and Lester, 1999, 2002; Funato et al.,
2002). Inhibitors of inositol phosphorylceramide (IPC) synthase are highly
toxic to many fungi but exhibit low toxicity in mammals (Mandala and
Harris, 2000; Mandala et al., 1997, 1998; Nagiec et al., 1997; Takesako et al.,
1993). A number of studies have also pointed to potential interactions of
GIPCs of fungi with the mammalian immune system (Jennemann et al.,
1999, 2001a; Straus et al., 1995; Suzuki et al., 1997; Toledo et al., 1995,
2001c). Elucidation of these interactions as well as the role of GIPCs in the
fungal life cycle and in virulence calls for detailed knowledge of their
structures. Compared to the mammalian GSL structural database, the
fungal database has been confined to a handful of entries with none that
has represented a complete structure elucidated without unambiguities
(the best characterized fungal GIPCs were those of H. capsulatum, for
which the precise core linkage Manα1 → 2/6Ins remained formally unde-
termined [Barr et al., 1984a,b]). In the last few years, however, careful
studies of fungal GIPCs, aided by both NMR and MS analysis, have greatly
expanded knowledge of their structures, to the extent that certain patterns
have begun to emerge (Singh et al., 2003). Most of these studies have
made extensive use of product ion mode ESI–MS and ES–MS/CID–MS,
but ESI– (CID–MS)^{2} and precursor ion scanning have also been applied.

For example, Jennemann *et al.* (2001b), in analyzing GIPCs of a wide variety of basidiomycete (mushroom) species, profiled molecular ion species of native compounds in negative-ion mode ESI–MS as $[M - H]^-$; the research group also applied for this purpose parent ion scanning from m/z 79 (phosphate anion). In the same work, $^-$ESI–MS/CID–MS analysis was used to obtain glycosylphosphorylinositol ions and glycosidic fragments thereof as well as ceramide phosphate ions. Levery *et al.* (1998), in a detailed study of GIPCs from the dimorphic mycopathogen *Paracoccidioides brasiliensis*, employed molecular species profiling of sodium adducts $[M(Na) + Na]^+$ by $^+$ESI–MS on a triple-quadrupole (API–III) mass spectrometer, followed by $^+$ESI–-MS/CID–MS on these molecular precursors, to obtain phosphorylceramide ions as well as both glycosylphosphorylinositol and glycosylinositol ions and fragments thereof. High orifice-to-skimmer (OR) potential $^+$ESI– $(CID–MS)^2$ was also employed to enhance formation of glycosylphosphorylinositol and glycosylinositol precursor ions for the subsequent CID–MS step (Levery *et al.*, 1998). In a later study of novel GIPCs from another dimorphic mycopathogen *S. schenckii* (Toledo *et al.*, 2001d), lithium adduction was employed with considerable advantage in the detection of molecular adducts in $^+$ESI–MS profiles as well as an increased abundance of significant fragments in high OR $^+$ESI–CID–MS mode. Considerably improved molecular ion profiles were obtained by precursor ion scanning in $^+$ESI–MS/CID–MS mode with selection of a suitable glycosylphosphorylinositol product in Q3. An unexpected result was the apparent detection of additional molecular species at increments of m/z 134 from each of the major adducts. After initial speculation that some previously uncharacterized modification had been detected, it was realized that these peaks represented attachment of an additional LiI ($M = 133.85$ u). At about the same time, a more extensive systematic study of the fragmentation behavior of lithiated GIPCs under these conditions was undertaken (Levery *et al.*, 2001). In addition to the results described earlier in the section on lithium adduction, an interesting observation was that GIPCs based on isomeric Manα1 → 2Ins-P-Cer and Manα1 → 6Ins-P-Cer cores (both of which are present in *S. schenckii* but differentially expressed between the mycelial and yeast forms) could be distinguished from each other by major differences in the relative abundance of glycosylphosphorylinositol product ion pairs, $[C_nPO_3(Li) + Li]^+$ and $[B_nPO_3(Li) + Li]^+$, in $^+$ESI–MS/CID–MS mode. These criteria were applied to confirm that the core linkage in a complex GIPC from the mycelial form of *H. capsulatum*, originally proposed by Barr *et al.* (1984b) to have the structure Manα1 → 3(Manβ1 → 4)Manα1 → 2/6Ins-P-Cer, was indeed Manα1 → 2 and not Manα1 → 6Ins (Levery *et al.*, 2001). An identical structure was proposed for one of the GIPC components found in the mycelial form of *S. schenckii* (Toledo *et al.* 2001d), and consistent with this finding was the observed similarity of their $^+$ESI–MS/ CID–MS mode fragmentations (Levery *et al.*, 2001).

The subsequent elucidation of another *S. schenckii* GIPC component, a triglycosylinositol phosphorylceramide containing a nonacetylated glu- cosamine (GlcN) residue, also showed the utility of these techniques. In this case, the presence of a free amino group was first confirmed by specific *N*-acetylation of the native GIPC, incrementing the molecular adducts, as well as the primary glycosylphosphorylinositol fragment produced from them in high orifice-to-skimmer potential (OR) $^+$ESI–CID–MS, uniformly by *m/z* 42. Molecular ion profiles were obtained by both normal mode $^+$ESI–MS and, more selectively, by precursor ion scanning in $^+$ESI–MS/ CID–MS mode with selection of the glycosylphosphorylinositol product *m/z* 800 in Q3. High OR product ion $^+$ESI– (CID–MS)^2 of the primary glycosylphosphorylinositol fragment then yielded a virtually complete array of glycosidic fragments from both ends of the chain, including both phosphorylated and nonphosphorylated series (Toledo *et al.*, 2001b). It was observed that prior *N*-acetylation improved the sensitivity of detec- tion of these fragments considerably, compared with the non-*N*-acetylated native GIPC, as well as improved the chances of defining the location of the HexN residue by the more unmistakable HexNAc decrement (useful particularly under rather low resolution conditions employed for this work).

These data provided unambiguous confirmation that the GlcN residue is linked directly to the *myo*-inositol moiety in the GIPC, which was found only in the yeast form of *S. schenckii*. Taken together with data from NMR spectroscopy, a highly novel structure of Manα1 $\rightarrow$ 3Manα1 $\rightarrow$ 6GlcNα1 $\rightarrow$ 2 Ins-P-Cer was demonstrated for this GIPC. It is noteworthy that the GlcNα1 $\rightarrow$ 2Ins core motif is isomeric to the common core linkage, GlcNα1 $\rightarrow$ 6Ins, found in glycosylinositol phospholipids (GIPLs) of para- sites and in the GPI protein membrane anchors widely distributed among eukaryotes. The occurrence of non-*N*-acetylated glucosamine in another linkage to *myo*-inositol is particularly interesting because it is known from studies with mammalian and trypanosomatid cells that two steps are required for formation of GlcNα1 $\rightarrow$ 6Ins found in GPI anchors, transfer of GlcNAc from UDP-GlcNAc to Ins of phosphatidylinositol, and subsequent de-*N*-acetylation. That both reactions could occur together in forming an alternate glycosyl linkage is an intriguing possibility.

The lithiation protocol was readily adapted to $^+$ESI–MS analysis of GIPCs on a oa-QTOF instrument (Bennion *et al.*, 2003). Using this technique, three major GIPC components from *Aspergillus nidulans*, a nonpathogenic model species, were analyzed using $^+$ESI–MS, $^+$ESI–MS/CID–MS, and high-OR product ion $^+$ESI– (CID–MS)^2, together with NMR spectroscopy and other techniques. The apparently nonantigenic *A. nidulans* GIPCs were characterized as a related series having oligo-α-mannosyl-phosphorylinositol headgroups with increasing numbers of Man residues. The major component was a novel trimannosyl IPC with a branched structure similar to that found in the trimannose core of eukaryotic glycoprotein *N*-glycans, Manα1 → 3(Manα1 → 6)Manα1 → 2Ins-P-Cer, predominant $[M(Li) + Li]^+$ at nominal, monoisotopic *m/z* 1424 and 1452. The largest component characterized was a pentamannosyl IPC with predominant $[M(Li) + Li]^+$ at nominal, monoisotopic *m/z* 1748 and 1776. The $^+$ESI–MS/CID–MS of this component, compared with that of the simpler trimannosyl IPC and other known compounds, provided convincing evidence that it shared the same Manα1 → 3(Manα1 → 6)Manα1 → 2Ins core motif modified by addition of two Man residues to one branch.

Aoki *et al.* (2004) isolated and characterized from an *Acremonium* species a series of GIPCs incorporating the GlcNα1 → 2Ins core linkage, indicating that the modification is not confined to a single species. Remarkably, significant fractions of the *Acremonium* GIPCs were also characterized by further modification with phosphorylcholine. Aoki *et al.* (2004) used negative-ion linear mode MALDI–TOF–MS in their characterization; detection of $[M - H]^-$ confirmed ceramide as well as glycan size, while the phosphorylcholine modification on some fractions was indicated by GIPC pseudomolecular masses incremented *m/z* 165 over the expected values. The $[M - H]^-$ for modified GIPCs were accompanied by abundant ions that had lost *m/z* 59, consistent with prompt losses of $(CH_3)_3N$ from the PC group (the research groups' explanation of these accompanying peaks corresponding to a loss of $CH_2 = N(CH_3)_2$, *m/z* 58, seems dubious and appears not to be congruent with their data as presented).

Negative-ion MALDI–TOF–MS was used for initial profiling of an extensive series of oligo-α-mannosyl-IPCs from the yeast form of *S. schenckii*, characterized by increasing numbers of α-Man residues and incorporating the less commonly observed Manα1 → 6Ins-P-Cer core motif (Loureiro y Penha *et al.*, 2001). This novel series included two GIPC components, containing one and two α-Man residues, also found in the mycelial form of *S. schenckii* by Toledo *et al.* (2001d), as well as additional tetramannosyl-IPCs and pentamannosyl-IPCs not previously characterized. An interesting heterogeneity in ceramide structure was observed in particular for the yeast form GIPCs, regardless of structure; mass spectral evidence led both groups to conclude that a significant portion incorporated dihydroxy-fatty-*N*-acylation although the groups' results differed with respect to predominant chain length (Levery *et al.*, 2001; Loureiro y Penha *et al.*, 2001). Loureiro y Penha *et al.* (2001) demonstrated more rigorously a 2,3-dihydroxy fatty acid structure. Both groups detected the presence of 2-hydroxy as well as a small amount of nonhydroxy, fatty-N-acylation as well (Loureiro y Penha *et al.*, 2001; Toledo *et al.*, 2001d). In contrast to the marked heterogeneity in fatty-N-acylation, only t18:0 4-hydroxysphinganine (phytosphingosine) was detected as the sphingoid base in *S. schenckii* GIPCs.

Negative-ion MALDI–TOF–MS was also used for the profiling and preliminary characterization of novel complex GIPCs from wild-type *Cryptococcus neoformans*, an opportunistic basidiomycetous mycopathogen, and an acapsular mutant of *Cr. neoformans* with attenuated virulence (Heise *et al.*, 2002). By far, the major GIPC of the wild type had the pentaglycosylinositol structure Manα1 → 3(Xylβ1 → 2)Manα1 → 4Galβ1 → 6Manα1 → 2Ins attached to a phosphorylceramide with t18:0 4-hydroxysphinganine and h24:0 fatty-N-acylation (nominal, monoisotopic m/z for intact GIPC, $[M - H]^- = 1704$). The major GIPCs of the acapsular mutant were more complex, having the same pentaglycosylinositol as their core structure but with additional α-Man residues (highest nominal, monoisotopic mass-to-charge ratios for the intact GIPC component with the same ceramide structure, $[M - H]^- = 2352$). Very little of the mass-to-charge ratio 1704 component was apparent in the mutant GIPC profile. Following permethylation of GIPC component by standard procedures (Ciukanu and Kerek, 1984; Paz Parente *et al.*, 1985), it was observed that the ceramide portions had been eliminated, leaving per-O-methylated glycosylinositols retaining the phosphate group. Fortunately, these were amenable to $^+$ESI–MS/CID–MS analysis, which was performed on a oa-QTOF instrument, yielding $[Y_nPO_3 + Na]^+$ and $[B_n + Na]^+$ series fragments from the cationized molecular ions, consistent with the proposed structures. The potential loss of ceramide during permethylation of GIPCs should be noted, particularly as it may result in unexpected product behavior during workup procedures.

In addition to GIPCs, fungi also express monohexosylceramides (especially GlcCer) (Barreto-Bergter *et al.*, 2004; Warnecke and Heinz, 2003) that have distinctive structural modifications of the ceramide moiety, some of which are also found in GSLs of plants and certain marine invertebrates but not in those of mammals (Ballio *et al.*, 1979; Costantino *et al.*, 1995a,b; Fogedal *et al.*, 1986; Fujino and Ohnishi, 1976; Natori *et al.*, 1994; Sawabe *et al.*, 1994; Shibuya *et al.*, 1990; Sitrin *et al.*, 1988). These modifications include addition of a characteristic Δ^8-unsaturation and a branching 9-methyl group to the sphingoid base (Scheme 1A). Despite an accumulation of evidence suggesting that such additional variations may have functional importance in growth, life cycle, morphogenesis, and host/pathogen interactions (Barreto-Bergter *et al.*, 2004; Kawai, 1989; Kawai *et al.*, 1985;

Mizushina *et al.*, 1998; Tanaka *et al.*, 1997; Toledo *et al.*, 1999, 2001a,c; Umemura *et al.*, 2000, 2002; Warnecke and Heinz, 2003), very little is known about the true functions, biosynthesis, or metabolic fate of fungal cerebrosides *in vivo*. In any case, aside from determining the identity of the single hexose residue (usually β-Glc, in some species also β-Gal), the analytical problem lies mainly in determining which ceramide species are present and in what proportions.

While NMR spectroscopy is very useful for identifying many of the characteristic structural features of the ceramides described and others, its use is particularly limited with respect to defining relative or absolute chain lengths of sphingoid and fatty-*N*-acyl moieties, especially where these occur in heterogeneous molecular species distributions and even more so when it comes to detecting and defining minor lipoforms. Although fatty-*N*-acyl and sphingoid component analysis by GC–MS can be helpful in this regard, this comes with a loss of information about intact lipoform distributions. On the other hand, MS and MS/MS analyses of intact cerebrosides yield detailed information on precisely those features that are refractory to NMR analysis as well as provide confirmatory evidence for the others at high sensitivity.

Since 1990, MS has been used in the characterization of cerebrosides from the following organisms: mushroom *Hypsizigus marmoreus* (Sawabe *et al.*, 1994); *Aspergillus fumigatus, A. versicolor, A. niger,* and *A. nidulans* (Levery *et al.*, 2000, 2002; Toledo *et al.*, 1999; Villas Boas *et al.*, 1994); *Magnaporthe grisea* (Koga *et al.*, 1998); *Fusarium* spp. (Duarte *et al.*, 1998); *Pseudallescheria boydii* (Pinto *et al.*, 2002); *Candida albicans* (Levery *et al.*, 2000); *Cryptococcus* spp. (Levery *et al.*, 2002; Rodrigues *et al.*, 2000); *Termitomyces albuminosus* (Qi *et al.*, 2000, 2001); *Absydia corymbifera* (Batrakov *et al.*, 2003); *Ganoderma lucidum* (Mizushina *et al.*, 1998); *Paracoccidioides brasiliensis* (Levery *et al.*, 2002; Toledo *et al.*, 1999); *Sporothrix schenckii* (Toledo *et al.*, 2000); *Histoplasma capsulatum* (Toledo *et al.*, 2001a); *Colletotrichum gloeosporioides* (da Silva *et al.*, 2004); *Pichia pastoris* (Sakaki *et al.*, 2001); *Rhynchosporium secalis* (Sakaki *et al.*, 2001); *Fonsecaea pedrosoi* (Nimrichter *et al.*, 2004); and *Neurospora crassa* (Park *et al.*, 2005). Many of these are endemic or opportunistic human mycopathogens, and several are model species whose genome sequences have been characterized. In a number of studies, cerebrosides had demonstrated physiological or immunological activities. Analysis of fungal cerebrosides has been justified to aid dissection of their biosynthesis, their apparent elicitor or other physiological activities; their putative roles in host/pathogen interactions, fungal life cycle, and infectivity (Barreto-Bergter *et al.*, 2004; Warnecke and Heinz, 2003); or their potential as targets for antifungal drugs (Levery *et al.*, 2002).

Techniques most commonly employed for fungal cerebroside characterization have been FAB–MS, along with tandem MS/CID–MS and ESI–MS, also with tandem MS/CID–MS. Particularly in several studies using FAB–MS, prior per-O-acetylation has been employed, but this is far from essential. In a study by Sawabe *et al.* (1994), positive- and negative-ion FAB–MS were used for the analysis of underivatized cerebroside on a double-focusing instrument; CID with constant B/E linked scanning was presented as a method particularly useful for assigning the positions of double bonds in the sphingoid base. Duarte *et al.* (1998) employed these methods as well but made effective use of lithium cationization of the cerebroside, as pioneered for sphingolipid applications by Ann and Adams (1992, 1993). Positive-ion mode ESI–MS on quadrupole instruments has been used extensively for fungal cerebroside analysis, with low-energy tandem MS/CID–MS applied first to sodium adducted molecular ions (Toledo *et al.*, 1999, 2000) and later to lithium adducts (Levery *et al.*, 2000; Park *et al.*, 2005; Toledo *et al.*, 2001a). Some results from these studies have been summarized in a previous section.

Moreover, ESI–MS/CID–MS of lithium adducts has also been applied to the analysis of cerebrosides from the model filamentous fungus *N. crassa*; these studies were carried out on an ESI–oa-QTOF–MS instrument, yielding spectra with similar characteristics but with substantial gains in sensitivity and resolution compared with the previous triple and quadrupole studies (Park *et al.*, 2005). Although *N. crassa* is not a phytopathogen, it has been used as model for plant/phytopathogen interactions, and its GlcCer has been proposed as a membrane binding target of antifungal plant peptides called defensins (Thevissen *et al.*, 2003; Thomma *et al.*, 2002). In these studies, mutant *N. crassa* strains selected for resistance to a GlcCer-recognizing plant defensin were found to express GlcCer with altered ceramide structures (Ferket *et al.*, 2003). ESI–MS/CID–MS was a crucial technique used to show that the alterations included not only ablation of fatty-N-acyl (E)-Δ^3-unsaturation but also shortening of the predominant fatty acid from 18 to 16 carbons (Park *et al.*, 2005). Analysis of *N. crassa* cerebrosides as lithium adducts has also been carried out on a MALDI–QIT–TOF instrument (unpublished, Figure 7).

Sulfatides

Sulfatides, which contain one or more strongly acidic sulfate groups, represent a somewhat special case with some interesting challenges for the analyst. Sulfated galactosylceramides are important components of the myelin in brain, spinal cord, and peripheral nerve; along with more complex sulfated GSLs, they are are also found in significant concentrations in mammalian kidney tissue and, to a lesser extent, in intestinal and other tissues. They have also been implicated as targets of self-recognizing antibodies in autoimmune demyelinating neuropathies and in insulin-dependent diabetes mellitus. Very early studies with LSI–MS examined molecular ion yield and fragmentation of sulfated glyceroglycolipids in both positive- and negative-ion modes, and although molecular ions could be detected in the positive mode (i.e., $[M - H + Na_2]^+$ species), the S/N was clearly much higher for detection of negatively charged species (i.e., $[M - H]^-$) (Kushi et al., 1985). For this reason, only the negative mode was used for subsequent analysis of sulfated GSLs, with detection of molecular species as $[M - H]^-$ for monosulfated and $[M - 2H + Na]^-$ for bis-sulfated compounds. Aside from molecular species, the spectra were dominated by sulfate-containing fragments, both from the nonreducing end (C series more abundant than B series), and from the reducing end (only sulfated Y series were detected).

A later study of monosulfated galactosylceramides by FAB–MS included more extensive characterization of both positive- and negative-ion spectra as well as CID–MS experiments using constant-B/E-linked scanning in a double focusing sector instrument of EB geometry (Ohashi and Nagai, 1991). In this case, molecular $[M - H + Na_2]^+$ species were clearly detected in the positive mode; moreover, they were also accompanied by ions 102 lower in m/z. These were clearly proven to be products of the molecular ions, not arising simply from contamination by already desulfated galactosylceramide. However, these fragments were not observed in CID–MS of molecular ions, leading to the conclusion that they are not formed by unimolecular decomposition (via O-S bond cleavage) in the gas phase but only formed where a source of protons is available (i.e., from matrix, moisture, or other sulfatide molecules at the point of FAB ionization [selvedge]). An abundant fragment at m/z 143, on the other hand, appearing both in the primary and in the CID spectra, corresponds to $[NaHSO_4 + Na]^+$, which would have to arise via C-O bond cleavage, in the latter case with concurrent intramolecular transfer of a proton from the sugar moiety. Simple modeling showed that a six-membered transition state is possible for this unimolecular elimination reaction. Interestingly, neither ceramide or galactose sulfate fragments, products of glycosidic bond cleavage, were very abundant in the positive-ion FAB spectra, being particularly lacking in the CID spectra; in contrast, a dehydrogenated galactose sulfate anion (m/z 257) was prominent in both the primary FAB–MS and the CID product ion spectra from $[M - H]^-$ in the negative mode. Both negative-ion spectra displayed abundant ions representing HSO_4^- and SO_3^- (m/z 97 and 80, respectively). In addition, a fragment m/z 540 was prominent in spectra of 2-hydroxy-fatty-N-acylated sulfatide components, representing elimination of the fatty acid to produce a "lyso-sulfatide" product ion. In the CID spectrum, a series of ions from charge-remote fragmentation of the ceramide moiety also facilitated location of the double bond in fatty-N-acylated chains of unsaturated components.

From these results, as well as those described in the previous study, showing ample glycosidic fragmentation of complex sulfated GSLs, one could conclude that the negative mode is clearly preferable for this class; virtually all of the subsequent studies of these compounds have been in this mode.

An investigation by Tadano-Aritomi et al. (1995) compared negative-ion LSI–MS/CID–MS of monosulfated and bis-sulfated GSLs with Gg_3Cer and Gg_4Cer core structures under high- and low-energy CID conditions. The high-energy tandem sector CID spectra were characterized by exten-sive fragmentation, including cross-ring cleavage of the β-Glc residue attached to ceramide; the low-energy CID spectra acquired on a triple-quadrupole instrument were by contrast very simple, with few glycosidic cleavages represented although these were useful. It was concluded that negative charge-remote fragmentation mechanisms might take place under the high energy, but not low energy, conditions with sulfatides.

Negative-ion LSI–MS was used in characterization of novel sulfated GSLs throughout the 1990s, including sulfated tetraglycosylceramides and pentaglycosylceramides with iGg4Cer core structures (Tadano-Aritomi et al., 1992, 1994), a sulfated Gg_3Cer (Tadano-Aritomi et al., 1996), and a sulfated G_{M1a} ganglioside (Tadano-Aritomi et al., 1998), all from rat kidney. This type of analysis was also used in a study of the expression of sulfatides in erythrocytes and platelets of bovine origin (Kushi et al., 1996).

Other techniques have been explored. Extensive investigations of sulfatide fragmentation with negative-ion ESI–MS/CID–MS at low energy in both triple quadrupole and quadrupole ion trap instruments have been carried out by Hsu and Turk (Hsu and Turk, 2004; Hsu et al., 1998). The technique was applied to characterization of sulfatide molecular species in bovine brain and rat pancreas (Hsu et al., 1998). A signature fragmentation was observed with sulfatide ceramides having 2-hydroxy-fatty-N-acylation, allowing these lipoforms to be easily identi-fied. Scanning for precursors of selected characteristic product ions (e.g., m/z 97) allowed for reliable profiling of all sulfatide molecular species in crude mixtures in the presence of nonsulfated impurities; this pro-cedure could be refined to profile sulfatides with specific ceramide features (e.g., m/z 540) and used to identify all precursor lipoforms with 2-hydroxy-fatty-N-acylation of d18:1 sphing-4-enine. In their following systematic ion trap study, Hsu and Turk (2004) reported an apparent loss of an internal galactose residue that could only be accounted for by rearrangement; negative charge-remote mechanisms were also invoked to explain extensive fragmentation of ceramide.

Han *et al.* (2002) used ESI–MS for a comparative analysis of sulfatides in the brain's white matter for healthy individuals and patients with Alzheimer's disease and concluded that a marked decrease in sulfatide content correlated with Alzheimer's disease pathology. Marbois *et al.* (2000) developed an ESI–MS/CID–MS triple quadrupole strategy (product and precursor ion scanning, multiple reaction monitoring) to analyze sulfatides from rat cerebellum and to compare sulfatide content in white matter from a control group and from multiple sclerosis patients. For patients with multiple sclerosis, significant decreases were observed in the proportion of lipoforms having 2-hydroxy-fatty-*N*-acylation. Sandhoff *et al.* (2002) used nanoESI tandem MS/CID–MS to study sulfatides in mouse models of inherited GSL processing disorders. Tandem ESI–MS/CID–MS was used to characterize sulfatides extracted from the urine of patients with metachromatic leukodystrophy (deficiency of lysosomal arylsulfatase A) and the samples were compared with urine from age-matched controls (Whitfield *et al.*, 2001); quantitation was based on an internal sulfatide standard. MALDI–TOF–MS has also been used for sulfatide analysis although fewer examples have appeared. Sugiyama *et al.* (1999) performed quantitative analysis of serum sulfatides using hydrogenated *N*-acetyl lysosulfatide as an internal standard (essentially sulfatide having an unnatural ceramide composed of d18:0 sphinganine with 2:0 *N*-acylation). The method should be suitable for high-throughput applications.

Conclusion: Toward Comprehensive (Glyco)sphingolipidomics

Three trends hold great promise for improvements in GSL structure analysis and the goal of comprehensive, quantitative, high-throughput glycosphingolipidomics: (i) continued development of automated on-line interfacing of LC and CE with ESI–MS as well as coupling of solid phase media (HPTLC and transfer membranes) with MALDI–MS and overlay methods; (ii) ion trap applications that include rapid automated, data-dependent MS^n capabilities; and (iii) use of computational bioinformatics in conjunction with extensive fragment databases and pattern search algorithms (Ashline *et al.*, 2005; Lapadula *et al.*, 2005; Zhang *et al.*, 2005). To a great extent, most of these capabilities, including automation, are already in place, and some will already have been highlighted by new, significant publications by the time this volume becomes available. On the other hand, due to the wide variance in physico-chemical properties, current methods of GSL extraction and fractionation are still labor intensive, and high-throughput potential, where it exists, is confined to subsets of the total

GSL expression profile. Application, as expected, is most efficient where the biological system in question has already been characterized in great detail or where only limited answers are required. Sample processing and preparation will continue to present the most challenging problems.

Acknowledgments

The author gratefully acknowledges the financial support of the National Institutes of Health/National Center for Research Resources (NIH R21 RR020355). Also appreciated are the financial support and instrumental resources provided by the NIH/NCRR-funded Resource Center for Biomedical Complex Carbohydrates (NIH P41 RR05351) and the New Hampshire Biological Research Infrastructure Network—Center for Structural Biology (NIH P20 RR16459).

References

Adams, J., and Ann, Q. (1993). Structure determination of sphingolipids by mass spectrometry. *Mass Spectrom. Rev.* **12**, 51–85.

Ann, Q., and Adams, J. (1992). Structure determination of ceramides and neutral glycosphingolipids by collisional activation of $[M + Li]^+$ ions. *J. Am. Soc. Mass Spectrom* **3**, 260–263.

Ann, Q., and Adams, J. (1993). Structure-specific collision-induced fragmentations of ceramides cationized with alkali-metal ions. *Anal. Chem.* **65**, 7–13.

Aoki, K., Uchiyama, R., Itonori, S., Sugita, S., Che, F.-S., Isogai, A., Hada, N., Takeda, T., Kumagai, H., and Yamamoto, K. (2004). Structural elucidation of novel phosphocholine-containing glycosylinositol-phosphoceramide in filamentous fungi and their induction of cell death of cultured rice cells. *Biochem. J.* **378**, 461–472.

Asam, M. R., and Glish, G. L. (1997). Tandem mass spectrometry of alkali cationized polysaccharides in a quadrupole ion trap. *J. Am. Soc. Mass Spectrom* **8**, 987–995.

Ashline, D., Singh, S., Hanneman, A., and Reinhold, V. (2005). Congruent strategies for carbohydrate sequencing. 1. Mining structural details by MS^n. *Anal. Chem.* **77**, 6250–6262.

Ballio, A., Casinovi, C. G., Framondino, M., Marino, G., Nota, G., and Santurbano, B. (1979). A new cerebroside from *Fusicoccum amygdali* Del. *Biochim. Biophys. Acta* **573**, 51–60.

Barr, K., and Lester, R. L. (1984a). Occurrence of novel antigenic phosphoinositol-containing sphingolipids in the pathogenic yeast *Histoplasma capsulatum*. *Biochemistry* **23**, 5581–5588.

Barr, K., Laine, R. A., and Lester, R. L. (1984b). Carbohydrate structures of three novel phosphoinositol-containing sphingolipids from the yeast *Histoplasma capsulatum*. *Biochemistry* **23**, 5589–5596.

Barreto-Bergter, E., Pinto, M. R., and Rodrigues, M. L. (2004). Structure and biological functions of fungal cerebrosides. *An. Acad. Bras. Cienc.* **76**, 67–84.

Batrakov, S. G., Konova, I. V., Sheichenko, V. I., and Galanina, L. A. (2003). Glycolipids of the filamentous fungus *Absidia corymbifera* F-295. *Chem. Phys. Lipids* **123**, 157–164.

Bennion, B., Park, C., Fuller, M., Lindsey, R., Momany, M., Jennemann, R., and Levery, S. B. (2003). Glycosphingolipids of the model fungus *Aspergillus nidulans*: Characterization of GIPCs with oligo-α-mannose-type glycans. *J. Lipid Res.* **44**, 2073–2088.

Boscaro, F., Pieraccini, G., la Marca, G., Bartolucci, G., Luceri, C., Luceri, F., and Moneti, G. (2002). Rapid quantitation of globotriaosylceramide in human plasma and urine: A potential application for monitoring enzyme replacement therapy in Anderson-Fabry disease. *Rapid Commun. Mass Spectrom* **16**, 1507–1514.

Botek, E., Debrun, J. L., Hakim, B., and Morin-Allory, L. (2001). Attachment of alkali cations on beta-D-glucopyranose: Matrix-assisted laser desorption/ionization time-of-flight studies and *ab initio* calculations. *Rapid Commun. Mass Spectrom* **15**, 273–276.

Chen, H., He, M., Wan, X., Yang, L., and He, H. (2003). Matrix-assisted laser desorption/ionization study of cationization of PEO-PPP rod-coil diblock polymers. *Rapid Commun. Mass Spectrom* **17**, 177–182.

Chester, M. A. (1998). IUPAC-IUB Joint Commission on Biochemical Nomenclature (JCBN). Nomenclature of glycolipids—Recommendations 1997. *Eur. J. Biochem.* **257**, 293–298.

Chitwood, D. J., Lusby, W. R., Thompson, M. J., Kochansky, J. P., and Howarth, O. W. (1995). The glycosylceramides of the nematode *Caenorhabditis elegans* contain an unusual, branched-chain sphingoid base. *Lipids* **30**, 567–573.

Chou, D. K., Ilyas, A. A., Evans, J. E., Costello, C., Quarles, R. H., and Jungalwala, F. B. (1986). Structure of sulfated glucuronyl glycolipids in the nervous system reacting with HNK-1 antibody and some IgM paraproteins in neuropathy. *J. Biol. Chem.* **261**, 11717–11725.

Ciukanu, I., and Kerek, F. (1984). A simple and rapid method for the permethylation of carbohydrates. *Carbohydr. Res.* **131**, 209–217.

Colsch, B., Afonso, C., Popa, I., Portoukalian, J., Fournier, F., Tabet, J. C., and Baumann, N. (2004). Characterization of the ceramide moieties of sphingoglycolipids from mouse brain by ESI–MS/MS: Identification of ceramides containing sphingadienine. *J. Lipid Res.* **45**, 281–286.

Costantino, V., Fattorusso, E., and Mangoni, A. (1995a). Glycolipids from sponges. I. Glycosyl ceramide composition of the marine sponge *Agelas clathrodes*. *Liebigs Ann.* **1995**, 1471–1475.

Costantino, V., Fattorusso, E., and Mangoni, A. (1995b). Glycolipids from sponges. III. Glycosyl ceramides from the marine sponge *Agelas conifera*. *Liebigs Ann.* **1995**, 2133–2136.

Costello, C. E. (1999). Bioanalytic applications of mass spectrometry. *Curr. Opin. Biotechn.* **10**, 22–28.

Costello, C. E., and Vath, J. E. (1990). Tandem mass spectrometry of glycolipids. *Methods Enzymol.* **193**, 738–768.

Cvacka, J., and Svatos, A. (2003). Matrix-assisted laser desorption/ionization analysis of lipids and high molecular weight hydrocarbons with lithium 2,5-dihydroxybenzoate matrix. *Rapid Commun. Mass Spectrom* **17**, 2203–2207.

da Silva, A. F., Rodrigues, M. L., Farias, S. E., Almeida, I. C., Pinto, M. R., and Barreto-Bergter, E. (2004). Glucosylceramides in *Colletotrichum gloeosporioides* are involved in the differentiation of conidia into mycelial cells. *FEBS Lett.* **561**, 137–143.

Dell, A. (1987). FAB–mass spectrometry of carbohydrates. *Adv. Carbohydr. Chem. Biochem.* **45**, 19–72.

Dell, A. (1990). Preparation and desorption mass spectrometry of permethyl and peracetyl derivatives of oligosaccharides. *Methods Enzymol.* **193**, 647–660.

Dell, A., Khoo, K.-H., Panico, M., McDowell, R. A., Etienne, A. T., Reason, A. J., and Morris, H. R. (1993). FAB–MS and ES–MS of glycoproteins. Glycobiology: A practical approach (M. Fukuda and A. Kobata, eds.). IRL Press, New York.

Dell, A., Reason, A. J., Khoo, K.-H., Panico, M., McDowell, R. A., and Morris, H. R. (1994). Mass spectrometry of carbohydrate-containing polymers. *Methods Enzymol.* **230**, 108–132.

Dennis, R. D., and Wiegandt, H. (1993). Glycosphingolipids of the invertebrata as exemplified by a cestode platyhelminth, *Taenia crassiceps*, and a dipteran insect, *Calliphora vicina*. *Adv. Lipid Res.* **26**, 321–351.

Dennis, R. D., Antonicek, H., Wiegandt, H., and Schachner, M. (1988). Detection of the L2/HNK-1 carbohydrate epitope on glycoproteins and acidic glycolipids of the insect *Calliphora vicina*. *J. Neurochem.* **51,** 1490–1496.

Dennis, R. D., Baumeister, S., Geyer, R., Peter-Katalinić, J., Hartmann, R., Egge, H., Geyer, E., and Wiegandt, H. (1992). Glycosphingolipids in cestodes. Chemical structures of ceramide monosaccharide, disaccharide, trisaccharide, and tetrasaccharide from metacestodes of the fox tapeworm, *Taenia crassiceps* (Cestoda: Cyclophyllidea). *Eur. J. Biochem.* **207,** 1053–1062.

Dickson, R. C., and Lester, R. L. (1999). Metabolism and selected functions of sphingolipids in the yeast *Saccharomyces cerevisiae*. *Biochim. Biophys. Acta* **1438,** 305–321.

Dickson, R. C., and Lester, R. L. (2002). Sphingolipid functions in *Saccharomyces cerevisiae*. *Biochim. Biophys. Acta* **1583,** 13–25.

Domon, B., and Costello, C. E. (1988). A systematic nomenclature for carbohydrate fragmentations in FAB–MS/MS spectra of glycoconjugates. *Glycoconj. J.* **5,** 397–409.

Duarte, R. S., Polycarpo, C. R., Wait, R., Hartmann, R., and Bergter, E. B. (1998). Structural characterization of neutral glycosphingolipids from *Fusarium* species. *Biochim. Biophys. Acta* **1390,** 186–196.

Duk, M., Reinhold, B. B., Reinhold, V. N., Kusnierz-Alejska, G., and Lisowska, E. (2001). Structure of a neutral glycosphingolipid recognized by human antibodies in polyagglutinable erythrocytes from the rare NOR phenotype. *J. Biol. Chem.* **276,** 40574–40582.

Durden, F. M., and Elewski, B. (1997). Fungal infections in HIV-infected patients. *Semin. Cutan. Med. Surg.* **16,** 200–212.

Easton, R. L., Patankar, M. S., Lattanzio, F. A., Leaven, T. H., Morris, H. R., Clark, G. F., and Dell, A. (2000). Structural analysis of murine zona pellucida glycans. Evidence for the expression of core 2-type *O*-glycans and the Sd[a] antigen. *J. Biol. Chem.* **275,** 7731–7742.

Egge, H., and Peter-Katalinić, J. (1987). Fast-atom bombardment mass spectrometry for structural elucidation of glycoconjugates. *Mass Spectrom. Rev.* **6,** 331–393.

Falsone, G., Budzikiewicz, H., and Wendisch, D. (1987). Constituents of *Euphorbaceae*. 9. Communication 1. New Cerebrosides from *Euphorbia biglandulosa Desf. Z. Naturforsch.* **42b,** 1476–1480.

Ferket, K. K. A., Levery, S. B., Park, C., Cammue, B. P. A., and Thevissen, K. (2003). Isolation and characterization of *Neurospora crassa* mutants resistant to antifungal plant defensins. *Fungal. Genet. Biol.* **40,** 176–185.

Fogedal, M., Mickos, H., and Norberg, T. (1986). Isolation of *N*-2′-hydroxyhexadecanoyl-1-*O*-β-D-glucopyranosyl-9-methyl-D-erythro-sphingadienine from fruiting bodies of two *Basidiomycetes* fungi. *Glycoconj. J.* **3,** 233–237.

Friedl, C. H., Lochnit, G., Geyer, R., Karas, M., and Bahr, U. (2000). Structural elucidation of zwitterionic sugar cores from glycosphingolipids by nanoelectrospray ionization-ion-trap mass spectrometry. *Anal. Biochem.* **284,** 279–287.

Fujino, Y., and Ohnishi, M. (1976). Structure of cerebroside in *Aspergillus oryzae*. *Biochim. Biophys. Acta* **486,** 161–171.

Funato, K., Vallee, B., and Riezman, H. (2002). Biosynthesis and trafficking of sphingolipids in the yeast *Saccharomyces cerevisiae*. *Biochemistry* **41,** 15105–15114.

Fukushi, Y., Nudelman, E., Levery, S. B., Higuchi, T., and Hakomori, S. (1986). A novel disialoganglioside (IV^{3}NeuAcIII^{6}NeuAcLc$_4$) of human adenocarcinoma and the monoclonal antibody (FH9) defining this disialosyl structure. *Biochemistry* **25,** 2859–2866.

Furukawa, K., Takamiya, K., Okada, M., Inoue, M., Fukumoto, S., and Furukawa, K. (2001). Novel functions of complex carbohydrates elucidated by the mutant mice of glycosyltransferase genes. *Biochim. Biophys. Acta* **1525,** 1–12.

Gerdt, S., Lochnit, G., Dennis, R. D., and Geyer, R. (1997). Isolation and structural analysis of three neutral glycosphingolipids from a mixed population of *Caenorhabditis elegans* (Nematoda: Rhabditida). *Glycobiology* **7,** 265–275.

Gerdt, S., Dennis, R. D., Borgonie, G., Schnabel, R., and Geyer, R. (1999). Isolation, characterization and immunolocalization of phosphorylcholine-substituted glycolipids in developmental stages of *Caenorhabditis elegans*. *Eur. J. Biochem.* **266,** 952–963.

Gillece-Castro, B. L., and Burlingame, A. L. (1990). Oligosaccharide characterization with high-energy collision-induced dissociation mass spectrometry. *Methods Enzymol.* **193,** 689–712.

Gu, M., Kerwin, J. L., Watts, J. D., and Aebersold, R. (1997). Ceramide profiling of complex lipid mixtures by electrospray ionization mass spectrometry. *Anal. Biochem.* **244,** 347–356.

Guillas, I., Kirchman, P. A., Chuard, R., Pfefferli, M., Jiang, J. C., Jazwinski, S. M., and Conzelmann, A. (2001). C26-CoA-dependent ceramide synthesis of *Saccharomyces cerevisiae* is operated by Lag1p and Lac1p. *EMBO J.* **20,** 2655–2665.

Guittard, J., Hronowski, X. L., and Costello, C. E. (1999). Direct matrix-assisted laser desorption/ionization mass spectrometric analysis of glycosphingolipids on thin-layer chromatographic plates and transfer membranes. *Rapid. Commun. Mass Spectrom* **13,** 1838–1849.

Hakomori, S. (1998). Cancer-associated glycosphingolipid antigens: Their structure, organization, and function. *Acta Anat.* **161,** 79–90.

Hakomori, S. (1999). Antigen structure and genetic basis of histo-blood groups A, B, and O: Their changes associated with human cancer. *Biochim. Biophys. Acta* **1473,** 247–266.

Hakomori, S., and Zhang, Y. (1997). Glycosphingolipid antigens and cancer therapy. *Chem. Biol.* **4,** 97–104.

Han, X., Holtzman, M., McKeel, D. W., Jr., Kelley, J., and Morris, J. C. (2002). Substantial sulfatide deficiency and ceramide elevation in very early Alzheimer's disease: Potential role in disease pathogenesis. *J. Neurochem.* **82,** 809–818.

Hansson, G. C., Li, Y. T., and Karlsson, H. (1989). Characterization of glycosphingolipid mixtures with up to ten sugars by gas chromatography and gas chromatography–mass spectrometry as permethylated oligosaccharides and ceramides released by ceramide glycanase. *Biochemistry* **28,** 6672–6678.

Harvey, D. J. (1999). Matrix-assisted laser desorption/ionization mass spectrometry of carbohydrates. *Mass Spectrom. Rev.* **18,** 349–451.

Heise, N., Gutierrez, A. L. S., Mattos, K. A., Jones, C., Wait, R., Previato, J. O., and Mendonca-Previato, L. (2002). Molecular analysis of a novel family of complex glycoinositolphosphoryl ceramides from *Cryptococcus neoformans*: Structural differences between encapsulated and acapsular yeast forms. *Glycobiology* **12,** 409–420.

Helling, F., Dennis, R. D., Weske, B., Nores, G., Peter-Katalinić, J., Dabrowski, U., Egge, H., and Wiegandt, H. (1991). Glycosphingolipids in insects. The amphoteric moiety, *N*-acetylglucosamine-linked phosphoethanolamine, distinguishes a group of ceramide oligosaccharides from the pupae of *Calliphora vicina* (Insecta: Diptera). *Eur. J. Biochem.* **200,** 409–421.

Hofmeister, G. E., Zhou, Z. H., and Leary, J. A. (1991). Linkage position determination in lithium cationized disaccharides: Tandem mass spectrometry and semiempirical calculations. *J. Am. Chem. Soc.* **113,** 5964–5970.

Hoteling, A. J., Kawaoka, K., Goodberlet, M. C., Yu, W. M., and Owens, K. G. (2003). Optimization of matrix-assisted laser desorption/ionization time-of-flight collision-induced dissociation using poly(ethylene glycol). *Rapid Commun. Mass Spectrom* **17,** 1671–1676.

Hsieh, T. C. Y., Kaul, K., Laine, R. A., and Lester, R. L. (1978). Structure of a major glycophosphoceramide from tobacco leaves, PSL-I: 2-Deoxy-2-acetamido-D-glucopyranosyl

(α1 $\rightarrow$ 4)-D-glucuronopyranosyl(α1 $\rightarrow$ 2)myoinositol-1-O-phosphoceramide. *Biochemistry* **17,** 3575–3581.

Hsu, F. F., and Turk, J. (2000). Structural determination of sphingomyelin by tandem mass spectrometry with electrospray ionization. *J. Am. Soc. Mass. Spectrom* **11,** 437–449.

Hsu, F. F., and Turk, J. (2001). Structural determination of glycosphingolipids as lithiated adducts by electrospray ionization mass spectrometry using low-energy collisional-activated dissociation on a triple stage quadrupole instrument. *J. Am. Soc. Mass. Spectrom* **12,** 61–79.

Hsu, F. F., and Turk, J. (2004). Studies on sulfatides by quadrupole ion-trap mass spectrometry with electrospray ionization: Structural characterization and the fragmentation processes that include an unusual internal galactose residue loss and the classical charge-remote fragmentation. *J. Am. Soc. Mass Spectrom* **15,** 536–546.

Hsu, F. F., Bohrer, A., and Turk, J. (1998). Electrospray ionization tandem mass spectrometric analysis of sulfatide. Determination of fragmentation patterns and characterization of molecular species expressed in brain and in pancreatic islets. *Biochim. Biophys. Acta* **1392,** 202–216.

Hsu, F. F., Turk, J., Stewart, M. E., and Downing, D. T. (2002). Structural studies on ceramides as lithiated adducts by low-energy collisional-activated dissociation tandem mass spectrometry with electrospray ionization. *J. Am. Soc. Mass Spectrom* **13,** 680–695.

Huffer, D. M., Chang, H. F., Cho, B. P., Zhang, L. K., and Chiarelli, M. P. (2001). Product ion studies of diastereomeric benzo[ghi]fluoranthene tetraols by matrix-assisted laser desorption/ionization time-of-flight mass spectrometry and post-source decay. *J. Am. Soc. Mass Spectrom* **12,** 376–380.

Hunnam, V., Harvey, D. J., Priestman, D. A., Bateman, R. H., Bordoli, R. S., and Tyldesley, R. (2001). Ionization and fragmentation of neutral and acidic glycosphingolipids with a QTOF mass spectrometer fitted with a MALDI ion source. *J. Am. Soc. Mass Spectrom* **12,** 1220–1225.

Ito, A., Levery, S. B., Saito, S., Satoh, M., and Hakomori, S. (2001). A novel ganglioside isolated from renal cell carcinoma. *J. Biol. Chem.* **276,** 16695–16703.

Ito, M., and Yamagata, T. (1986). A novel glycosphingolipid-degrading enzyme cleaves the linkage between the oligosaccharide and ceramide of neutral and acidic glycosphingolipids. *J. Biol. Chem.* **261,** 14278–14282.

Ito, M., and Yamagata, T. (1989a). Endoglycoceramidase from *Rhodococcus* species G-74-2. *Methods Enzymol.* **179,** 488–496.

Ito, M., and Yamagata, T. (1989b). Purification and characterization of glycosphingolipid-specific endoglycosidases (endoglycoceramidases) from a mutant strain of *Rhodococcus* sp. Evidence for three molecular species of endoglycoceramidase with different specificities. *J. Biol. Chem.* **264,** 9510–9519.

Ito, M., Ikegami, Y., and Yamagata, T. (1991). Activator proteins for glycosphingolipid hydrolysis by endoglycoceramidases. Elucidation of biological functions of cell-surface glycosphingolipids in situ by endoglycoceramidases made possible using these activator proteins. *J. Biol. Chem.* **266,** 7919–7926.

Ito, M., Ikegami, Y., Tai, T., and Yamagata, T. (1993a). Specific hydrolysis of intact erythrocyte cell-surface glycosphingolipids by endoglycoceramidase. Lack of modulation of erythrocyte glucose transporter by endogenous glycosphingolipids. *Eur. J. Biochem.* **218,** 637–643.

Ito, M., Ikegami, Y., and Yamagata, T. (1993b). Kinetics of endoglycoceramidase action toward cell-surface glycosphingolipids of erythrocytes. *Eur. J. Biochem.* **218,** 645–649.

Itonori, S., Nishizawa, M., Suzuki, M., Inagaki, F., Hori, T., and Sugita, M. (1991). Polar glycosphingolipids in insect: Chemical structures of glycosphingolipid series containing 2'-aminoethylphosphoryl-(→ 6)-N-acetylglucosamine as a polar group from larvae of the green-bottle fly, *Lucilia caesar*. *J. Biochem. (Tokyo)* **110**, 479–485.

IUPAC (1978). The nomenclature of lipids (Recommendations 1976) IUPAC-IUB Commission on Biochemical Nomenclature. *Biochem. J.* **171**, 21–35.

Jennemann, R., Bauer, B. L., Bertalanffy, H., Selmer, T., and Wiegandt, H. (1999). Basidiolipids from *Agaricus* are novel immune adjuvants. *Immunobiology* **200**, 277–289.

Jennemann, R., Sandhoff, R., Grone, H.-J., and Wiegandt, H. (2001a). Human heterophile antibodies recognizing distinct carbohydrate epitopes on basidiolipids from different mushrooms. *Immunol. Invest.* **30**, 115–129.

Jennemann, R., Geyer, R., Sandhoff, R., Gschwind, R. M., Levery, S. B., Gröne, H.-J., and Wiegandt, H. (2001b). Glycoinositolphosphosphingolipids (basidiolipids) of higher mushrooms. *Eur. J. Biochem.* **268**, 1190–1205.

Jin, W., Rinehart, K. L., and Jares-Erijman, E. A. (1994). Ophidiacerebrosides: Cytotoxic glycosphingolipids containing a novel sphingosine from a sea star. *J. Org. Chem.* **59**, 144–147.

Johansson, L., and Karlsson, K. A. (1998). Selective binding by *Helicobacter pylori* of leucocyte gangliosides with 3-linked sialic acid, as identified by a new approach of linkage analysis. *Glycoconj. J.* **15**, 713–721.

Johansson, L., and Miller-Podraza, H. (1998). Analysis of 3- and 6-linked sialic acids in mixtures of gangliosides using blotting to polyvinylidene difluoride membranes, binding assays, and various mass spectrometry techniques with application to recognition by *Helicobacter pylori*. *Anal. Biochem.* **265**, 260–268.

Ju, D. D., Lai, C. C., and Her, G. R. (1997). Analysis of gangliosides by capillary zone electrophoresis and capillary zone electrophoresis-electrospray mass spectrometry. *J. Chromatogr. A* **779**, 195–203.

Jung, J. H., Lee, C.-O., Kim, Y. C., and Kang, S. S. (1996). New bioactive cerebrosides from *Arisema amurense*. *J. Nat. Prod.* **59**, 319–322.

Jungalwala, F. B. (1994). Expression and biological functions of sulfoglucuronyl glycolipids (SGGLs) in the nervous system–a review. *Neurochem. Res.* **19**, 945–957.

Kanfer, J. N., and Hakomori, S. (1983). "Handbook of Lipid Research, Volume 3: Sphingolipid Biochemistry." Plenum Press, New York and London.

Kang, S. S., Kim, J. S., Xu, Y. N., and Kim, Y. H. (1999). Isolation of a new cerebroside from the root bark of *Aralia elata*. *J. Nat. Prod.* **62**, 1059–1060.

Karlsson, H., Johansson, L., Miller-Podraza, H., and Karlsson, K. A. (1999). Fingerprinting of large oligosaccharides linked to ceramide by matrix-assisted laser desorption/ ionization time-of-flight mass spectrometry: Highly heterogeneous polyglycosylceramides of human erythrocytes with receptor activity for *Helicobacter pylori*. *Glycobiology* **9**, 765–778.

Karlsson, H., Larsson, T., Karlsson, K.-A., and Miller-Podraza, H. (2000). Polyglycosylceramides recognized by *Helicobacter pylori*: Analysis by matrix-assisted laser desorption/ ionization mass spectrometry after degradation with endo-β-galactosidase and by fast-atom bombardment mass spectrometry of permethylated undegraded material. *Glycobiology* **10**, 1291–1309.

Karlsson, K.-A. (1989). Animal glycosphinglipids as membrane attachment sites for bacteria. *Annu. Rev. Biochem.* **58**, 309–350.

Karlsson, K.-A., Leffler, H., and Samuelsson, B. E. (1979). Characterization of cerebroside (monoglycosylceramide) from the sea anemone, *Metridium senile*. *Biochim. Biophys. Acta* **574**, 79–93.

Karlsson, K.-A., Lanne, B., Pimlott, W., and Teneberg, S. (1991). The resolution into molecular species on desorption of glycolipids from thin-layer chromatograms, using combined thin-layer chromatography and fast-atom-bombardment mass spectrometry. *Carbohydr. Res.* **221: 49–61**, 49–61.

Karlsson, K.-A., Ängstrom, J., Bergstrom, J., and Lanne, B. (1992). Microbial interaction with animal cell surface carbohydrates. *APMIS* **100**, 71–83.

Kawai, G. (1989). Molecular species of cerebrosides in fruiting bodies of *Lentinus edodes* and their biological activity. *Biochim. Biophys. Acta* **1001**, 185–190.

Kawai, G., Ikeda, Y., and Tubaki, K. (1985). Fruiting of *Schizophyllum commune* induced by certain ceramides and cerebrosides from *Penicillium funiculosum. Agric. Biol. Chem.* **49**, 2137–2146.

Keki, S., Deak, G., and Zsuga, M. (2001). Fragmentation study of rutin, a naturally occurring flavone glycoside cationized with different alkali metal ions, using post-source decay matrix-assisted laser desorption/ionization mass spectrometry. *J. Mass Spectrom* **36**, 1312–1316.

Khoo, K.-H., Chatterjee, D., Caulfield, J. P., Morris, H. R., and Dell, A. (1997). Structural characterization of glycophingolipids from the eggs of *Schistosoma mansoni* and *Schistosoma japonicum. Glycobiology* **7**, 653–661.

Kochetkov, N. K., and Chizhov, O. S. (1966). Mass spectrometry of carbohydrate derivatives. *Adv. Carbohydr. Chem.* **21**, 39–93.

Koga, J., Yamauchi, T., Shimura, M., Ogawa, N., Oshima, K., Umemura, K., Kikuchi, M., and Ogasawara, N. (1998). Cerebrosides A and C, Sphingolipid Elicitors of Hypersensitive Cell Death and Phytoalexin Accumulation in Rice Plants. *J. Biol. Chem.* **273**, 31985–31991.

Kolter, T., and Sandhoff, K. (1999). Sphingolipids–their metabolic pathways and the patho-biochemistry of neurodegenerative diseases. *Angew. Chem. Intl. Ed. Engl.* **38**, 1532–1568.

Kushi, Y., and Handa, S. (1985). Direct analysis of lipids on thin layer plates by matrix-assisted secondary ion mass spectrometry. *J. Biochem. (Tokyo)* **98**, 265–268.

Kushi, Y., Handa, S., and Ishizuka, I. (1985). Secondary ion mass spectrometry for sulfoglycolipids: Application of negative ion detection. *J. Biochem. (Tokyo)* **97**, 419–428.

Kushi, Y., Rokukawa, C., and Handa, S. (1988). Direct analysis of glycolipids on thin-layer plates by matrix-assisted secondary ion mass spectrometry: Application for glycolipid storage disorders. *Anal. Biochem.* **175**, 167–176.

Kushi, Y., Ogura, K., Rokukawa, C., and Handa, S. (1990). Blood group A-active glycosphingolipids analysis by the combination of TLC-immunostaining assay and TLC/SIMS mass spectrometry. *J. Biochem. (Tokyo)* **107**, 685–688.

Kushi, Y., Arita, M., Ishizuka, I., Kasama, T., Fredman, P., and Handa, S. (1996). Sulfatide is expressed in both erythrocytes and platelets of bovine origin. *Biochim. Biophys. Acta* **1304**, 254–262.

Kushi, Y., Shimizu, M., Watanabe, K., Kasama, T., Watarai, S., Ariga, T., and Handa, S. (2001). Characterization of blood group ABO(H)-active gangliosides in type AB erythrocytes and structural analysis of type A-active ganglioside variants in type A human erythrocytes. *Biochim. Biophys. Acta* **1525**, 58–69.

Laine, O., Trimpin, S., Rader, H. J., and Mullen, K. (2003). Changes in post-source decay fragmentation behavior of poly(methyl methacrylate) polymers with increasing molecular weight studied by matrix-assisted laser desorption/ionization time-of-flight mass spec-trometry. *Eur. J. Mass Spectrom. (Chichester, Eng)* **9**, 195–201.

Lapadula, A. J., Hatcher, P. J., Hanneman, A. J., Ashline, D. J., Zhang, H., and Reinhold, V. N. (2005). Congruent studies for carbohydrate sequencing. 3. OSCAR: An algorithm for assigning oligosaccharide topology from MS^n data. *Anal. Chem.* **77**, 6271–6279.

Latgé, J. P., and Calderone, R. (2002). Host-microbe interactions: Fungi invasive human fungal opportunistic infections. *Curr. Opin. Microbiol.* **5**, 355–358.

Leipelt, M., Warnecke, D., Zahringer, U., Ott, C., Muller, F., Hube, B., and Heinz, E. (2001). Glucosylceramide synthases, a gene family responsible for the biosynthesis of gluco-sphingolipids in animals, plants, and fungi. *J. Biol. Chem.* **276**, 33621–33629.

Lester, R. L., and Dickson, R. C. (1993). Sphingolipids with inositolphosphate-containing head groups. *Adv. Lipid Res.* **26,** 253–274.

Levery, S. B. (1993). Studies of the primary and secondary structure of glycosphingolipids. Ph.D. Dissertation, University of Washington.

Levery, S. B., Weiss, J. B., Salyan, M. E., Roberts, C. E., Hakomori, S., Magnani, J. L., and Strand, M. (1992). Characterization of a series of novel fucose-containing glyco-sphingolipid immunogens from eggs of Schistosoma mansoni. *J. Biol. Chem.* **267,** 5542–5551.

Levery, S. B., Salyan, M. E. K., and Stroud, M. R. (1995). ESI–MS and ESI–CID–MS/MS of complex neolacto-series glycosphingolipids. *Proceedings of the 43nd ASMS Conference on Mass Spectrometry and Allied Topics, Atlanta, GA,* p. 1178.

Levery, S. B., Toledo, M. S., Suzuki, E., Salyan, M. E., Hakomori, S., Straus, A. H., and Takahashi, H. K. (1996). Structural characterization of a new galactofuranose-containing glycolipid antigen of Paracoccidioides brasiliensis. *Biochem. Biophys. Res. Commun.* **222,** 639–645.

Levery, S. B., Toledo, M. S., Straus, A. H., and Takahashi, H. K. (1998). Structure elucidation of sphingolipids from the mycopathogen *Paracoccidioides brasiliensis*: An immunodomi-nant β-galactofuranose residue is carried by a novel glycosylinositol phosphorylceramide antigen. *Biochemistry* **37,** 8764–8775.

Levery, S. B., Toledo, M. S., Doong, R. L., Straus, A. H., and Takahashi, H. K. (2000). Comparative analysis of ceramide structural modification found in fungal cerebrosides by electrospray tandem mass spectrometry with low-energy collision-induced dissociation of Li$^+$ adduct ions. *Rapid. Commun. Mass Spectrom* **14,** 551–563.

Levery, S. B., Toledo, M. S., Straus, A. H., and Takahashi, H. K. (2001). Comparative analysis of glycosylinositol phosphorylceramides from fungi by electrospray tandem mass spectrometry with low-energy collision-induced dissociation of Li$^+$ adduct ions. *Rapid. Commun. Mass. Spectrom* **15,** 2240–2258.

Levery, S. B., Momany, M., Lindsey, R., Toledo, M. S., Shayman, J. A., Fuller, M., Brooks, K., Doong, R. L., Straus, A. H., and Takahashi, H. K. (2002). Disruption of the glucosylceramide biosynthesis pathway in *Aspergillus nidulans* and *Aspergillus fumigatus* by inhibitors of UDP-Glc: Ceramide glucosyltransferase strongly affects spore germina-tion, cell cycle, and hyphal growth. *FEBS Lett.* **525,** 59–64.

Li, Y. T., and Li, S. C. (1994). Ceramide glycanase from the leech *Macrobdella decora* and oligosaccharide-transferring activity. *Methods Enzymol.* **242,** 146–158.

Lochnit, G., Dennis, R. D., Zahringer, U., and Geyer, R. (1997). Structural analysis of neutral glycosphingolipids from *Ascaris suum* adults (Nematoda: Ascaridida). *Glycoconj. J.* **14,** 389–399.

Lochnit, G., Dennis, R. D., Ulmer, A. J., and Geyer, R. (1998a). Structural elucidation and monokine-inducing activity of two biologically active zwitterionic glycosphingo-lipids derived from the porcine parasitic nematode *Ascaris suum. J. Biol. Chem.* **273,** 466–474.

Lochnit, G., Nispel, S., Dennis, R. D., and Geyer, R. (1998b). Structural analysis and immunohistochemical localization of two acidic glycosphingolipids from the porcine, parasitic nematode, *Ascaris suum. Glycobiology* **8,** 891–899.

Lortholary, O., Denning, D. W., and Dupont, B. (1999). Endemic mycoses: A treatment update. *J. Antimicrob. Chemother.* **43,** 321–331.

Loureiro y Penha, C. V., Todeschini, A. R., Lopes-Bezerra, L. M., Wait, R., Jones, C., Mattos, K. A., Heise, N., Mendonca-Previato, L., and Previato, J. O. (2001). Characterization of novel structures of mannosylinositolphosphorylceramides from the yeast forms of *Sporothrix schenckii. Eur. J. Biochem.* **268,** 4243–4250.

Makaaru, C. K., Damian, R. T., Smith, D. F., and Cummings, R. D. (1992). The human blood fluke *Schistosoma mansoni* synthesizes a novel type of glycosphingolipid. *J. Biol. Chem.* **267**, 2251–2257.

Mandala, S. M., and Harris, G. H. (2000). Isolation and characterization of novel inhibitors of sphingolipid synthesis: Australifungin, viridiofungins, rustmicin, and khafrefungin. *Methods Enzymol.* **311**, 335–348.

Mandala, S. M., Thornton, R. A., Rosenbach, M., Milligan, J., Garcia-Calvo, M., Bull, H. G., and Kurtz, M. B. (1997). Khafrefungin, a novel inhibitor of sphingolipid synthesis. *J. Biol. Chem.* **272**, 32709–32714.

Mandala, S. M., Thornton, R. A., Milligan, J., Rosenbach, M., Garcia-Calvo, M., Bull, H. G., Harris, G., Abruzzo, G. K., Flattery, A. M., Gill, C. J., Bartizal, S., and Kurtz, M. B. (1998). Rustmicin, a potent antifungal agent, inhibits sphingolipid synthesis at the inositol phosphoceramide synthase. *J. Biol. Chem.* **273**, 14942–14949.

Mank, M., Stahl, B., and Boehm, G. (2004). 2,5-dihydroxybenzoic acid butylamine and other ionic liquid matrixes for enhanced AMALDI–MS analysis of biomolecules. *Anal. Chem.* **76**, 2938–2950.

Marbois, B. N., Faull, K. F., Fluharty, A. L., Raval-Fernandes, S., and Rome, L. H. (2000). Analysis of sulfatide from rat cerebellum and multiple sclerosis white matter by negative ion electrospray mass spectrometry. *Biochim. Biophys. Acta* **1484**, 59–70.

McCloskey, J. A. (1990). Mass spectrometry. *Meth. Enzymol.* **193**.

Meisen, I., Peter-Katalinić, J., and Müthing, J. (2003). Discrimination of neolacto-series gangliosides with alpha2-3- and alpha2-6-linked N-acetylneuraminic acid by nanoelectrospray ionization low-energy collision-induced dissociation tandem quadrupole TOF–MS. *Anal. Chem.* **75**, 5719–5725.

Meisen, I., Peter-Katalinić, J., and Müthing, J. (2004). Direct analysis of silica gel extracts from immunostained glycosphingolipids by nanoelectrospray ionization quadrupole time-of-flight mass spectrometry. *Anal. Chem.* **76**, 2248–2255.

Merrill, A. H. J., Schmelz, E. M., Dillehay, D. L., Spiegel, S., Shayman, J. A., Schroeder, J. J., Riley, R. T., Voss, K. A., and Wang, E. (1997). Sphingolipids—The enigmatic lipid class: Biochemistry, physiology, and pathophysiology. *Toxicol. Appl. Pharmacol.* **142**, 208–225.

Metelmann, W., Müthing, J., and Peter-Katalinić, J. (2000). Nanoelectrospray ionization quadrupole time-of-flight tandem mass spectrometric analysis of a ganglioside mixture from human granulocytes. *Rapid. Commun. Mass Spectrom* **14**, 543–550.

Metelmann, W., Vukelic, Z., and Peter-Katalinić, J. (2001a). Nanoelectrospray ionization time-of-flight mass spectrometry of gangliosides from human brain tissue. *J. Mass Spectrom* **36**, 21–29.

Metelmann, W., Peter-Katalinić, J., and Müthing, J. (2001b). Gangliosides from human granulocytes: A nano-ESI QTOF mass spectrometry fucosylation study of low abundance species in complex mixtures. *J. Am. Soc. Mass Spectrom* **12**, 964–973.

Miller-Podraza, H. (2000). Polyglycosylceramides, poly-*N*-acetyllactosamine-containing glycosphingolipids: Methods of analysis, structure, and presumable biological functions. *Chem. Rev.* **100**, 4663–4682.

Mills, K., Johnson, A., and Winchester, B. (2002). Synthesis of novel internal standards for the quantitative determination of plasma ceramide trihexoside in Fabry disease by tandem mass spectrometry. *FEBS Lett.* **515**, 171–176.

Mizushina, Y., Hanashima, L., Yamaguchi, T., Takemura, M., Sugawara, F., Saneyoshi, M., Matsukage, A., Yoshida, S., and Sakaguchi, K. (1998). A mushroom fruiting body-inducing substance inhibits activities of replicative DNA polymerases. *Biochem. Biophys. Res. Commun.* **249**, 17–22.

Muller, R., Altmann, F., Zhou, D., and Hennet, T. (2002). The *Drosophila melanogaster* brainiac protein is a glycolipid-specific beta 1,3N-acetylglucosaminyltransferase. *J. Biol. Chem.* **277**, 32417–32420.

Müthing, J., Burg, M., Mockel, B., Langer, M., Metelmann-Strupat, W., Werner, A., Neumann, U., Peter-Katalinić, J., and Eck, J. (2002). Preferential binding of the anticancer drug rViscumin (recombinant mistletoe lectin) to terminally alpha2–6-sialylated neolacto-series gangliosides. *Glycobiology* **12**, 485–497.

Müthing, J., Meisen, I., Bulau, P., Langer, M., Witthohn, K., Lentzen, H., Neumann, U., and Peter-Katalinić, J. (2004). Mistletoe lectin I is a sialic acid-specific lectin with strict preference to gangliosides and glycoproteins with terminal Neu5Ac alpha 2–6Gal beta 1–4GlcNAc residues. *Biochemistry* **43**, 2996–3007.

Nagiec, M. M., Nagiec, E. E., Baltisberger, J. A., Wells, G. B., Lester, R. L., and Dickson, R. C. (1997). Sphingolipid synthesis as a target for antifungal drugs. Complementation of the inositol phosphorylceramide synthase defect in a mutant strain of *Saccharomyces cerevisiae* by the *AUR1* gene. *J. Biol. Chem.* **272**, 9809–9817.

Natori, T., Morita, M., Akimoto, K., and Koezuka, Y. (1994). Agelasphins, novel antitumor and immunostimulatory cerebrosides from the marine sponge *Agelas mauritianus*. *Tetrahedron* **50**, 2771–2784.

Nimrichter, L., Barreto-Bergter, E., Mendonca-Filho, R. R., Kneipp, L. F., Mazzi, M. T., Salve, P., Farias, S. E., Wait, R., Alviano, C. S., and Rodrigues, M. L. (2004). A monoclonal antibody to glucosylceramide inhibits the growth of *Fonsecaea pedrosoi* and enhances the antifungal action of mouse macrophages. *Microbes. Infect* **6**, 657–665.

North, S., Okafo, G., Birrell, H., Haskins, N., and Camilleri, P. (1997). Minimizing cationization effects in the analysis of complex mixtures of oligosaccharides. *Rapid Commun. Mass Spectrom* **11**, 1635–1642.

O'Connor, P. B., and Costello, C. E. (2001). A high-pressure matrix-assisted laser desorption/ ionization Fourier transform mass spectrometry ion source for thermal stabilization of labile biomolecules. *Rapid Commun. Mass Spectrom* **15**, 1862–1868.

O'Connor, P. B., Mirgorodskaya, E., and Costello, C. E. (2002). High-pressure matrix-assisted laser desorption/ionization Fourier transform mass spectrometry for minimization of ganglioside fragmentation. *J. Am. Soc. Mass Spectrom* **13**, 402–407.

O'Connor, P. B., Budnik, B. A., Ivleva, V. B., Kaur, P., Moyer, S. C., Pittman, J. L., and Costello, C. E. (2004). A high-pressure matrix-assisted laser desorption ion source for Fourier transform mass spectrometry designed to accommodate large targets with diverse surfaces. *J. Am. Soc. Mass Spectrom* **15**, 128–132.

Ohashi, Y., and Nagai, Y. (1991). Fast-atom bombardment chemistry of sulfatide (3'-sulfogalactosylceramide). *Carbohydr. Res.* **221**, 235–243.

Olling, A., Breimer, M. E., Samuelsson, B. E., and Ghardashkhani, S. (1998). Electrospray ionization and collision-induced dissociation time-of-flight mass spectrometry of neutral glycosphingolipids. *Rapid. Commun. Mass Spectrom* **12**, 637–645.

Olsen, I., and Jantzen, E. (2001). Sphingolipids in bacteria and fungi. *Anaerobe* **7**, 103–112.

Park, C., Bennion, B., Francois, I. E., Ferket, K. K., Cammue, B. P., Thevissen, K., and Levery, S. B. (2005). Characterization of neutral glycolipids of the model filamentous fungus, *Neurospora crassa*: Altered expression in plant defensin-resistant mutants. *J. Lipid Res.* **46**, 759–768.

Paz Parente, J., Cardon, P., Montreuil, J., and Fournet, B. (1985). A convenient method for methylation of glycoprotein glycans in small amounts by using lithium methylsulfinyl carbanion. *Carbohydr. Res.* **141**, 41–47.

Penn, S. G., Cancilla, M. T., Green, M. K., and Lebrilla, C. B. (1997). Direct comparison of matrix-assisted laser desorption/ionization and electrospray ionization in the analysis of gangliosides by Fourier transform mass spectrometry. *Eur. Mass Spectrom* **3,** 67–79.

Perreault, H., Hronowski, X. L., Koul, O., Street, J., McCluer, R. H., and Costello, C. E. (1997). High-sensitivity mass spectral characterization of glycosphingolipids from bovine erythrocytes, mouse kidney and fetal calf brain. *Intl. J. Mass Spectrom Ion Process* **169/170,** 351–370.

Peter-Katalinić, J. (1994). Analysis of glycoconjugates by fast-atom bombardment mass spectrometry and related techniques. *Mass Spectrom. Rev.* **13,** 77–98.

Peter-Katalinić, J., and Egge, H. (1990). Desorption mass spectrometry of glycosphingolipids. *Methods Enzymol.* **193,** 713–733.

Pinto, M. R., Rodrigues, M. L., Travassos, L. R., Haido, R. M., Wait, R., and Barreto-Bergter, E. (2002). Characterization of glucosylceramides in *Pseudallescheria boydii* and their involvement in fungal differentiation. *Glycobiology* **12,** 251–260.

Pocsfalvi, G., Malorni, A., Mancini, I., Guella, G., and Pietra, F. (2000). Molecular characterization of a highly heterogeneous mixture of glucosylceramides from a deep-water Mediterranean scleractinian coral *Dendrophyllia cornigera*. *Rapid. Commun. Mass. Spectrom* **14,** 2247–2259.

Poulter, L., and Burlingame, A. L. (1990). Desorption mass spectrometry of oligosaccharides coupled with hydrophobic chromophores. *Methods Enzymol.* **193,** 661–689.

Qi, J., Ojika, M., and Sakagami, Y. (2000). Termitomycesphins A-D, novel neuritogenic cerebrosides from the edible Chinese mushroom *Termitomyces albuminosus*. *Tetrahedron* **56,** 5835–5841.

Qi, J., Ojika, M., and Sakagami, Y. (2001). Neuritogenic cerebrosides from an edible Chinese mushroom. Part 2: Structure of two additional termitomycesphins and activity enhancement of an inactive cerebroside by hydroxylation. *Bioorg. Med. Chem.* **9,** 2171–2177.

Rashidzadeh, H., Wang, Y., and Guo, B. (2000). Matrix effects on selectivities of poly (ethylene glycol)s for alkali metal ion complexation in matrix-assisted laser desorption/ionization. *Rapid Commun. Mass Spectrom* **14,** 439–443.

Rasilo, M. L., Ito, M., and Yamagata, T. (1989). Liberation of oligosaccharides from glycosphingolipids on PC12 cell surface with endoglycoceramidase. *Biochem. Biophys. Res. Commun.* **162,** 1093–1099.

Reinhold, B. B., Chan, S.-Y., Chan, S., and Reinhold, V. N. (1994). Profiling glycosphingolipid structural detail: Periodate oxidation, electrospray, collision-induced dissociation, and tandem mass spectrometry. *Org. Mass Spectrom* **29,** 736–746.

Reinhold, V. N., and Sheeley, D. M. (1998). Detailed characterization of carbohydrate linkage and sequence in an ion trap mass spectrometer: Glycosphingolipids. *Anal. Biochem.* **259,** 28–33.

Reinhold, V. N., Reinhold, B. B., and Costello, C. E. (1995). Carbohydrate molecular weight profiling, sequence, linkage, and branching data: ES–MS and CID. *Anal. Chem.* **67,** 1772–1784.

Rodrigues, M. L., Travassos, L. R., Miranda, K. R., Franzen, A. J., Rozental, S., de Souza, W., Alviano, C. S., and Barreto-Bergter, E. (2000). Human antibodies against a purified glucosylceramide from *Cryptococcus neoformans* inhibit cell budding and fungal growth. *Infect. Immun.* **68,** 7049–7060.

Sakaguchi, H., Watanabe, M., Ueoka, C., Sugiyama, E., Taketomi, T., Yamada, S., and Sugahara, K. (2001). Isolation of reducing oligosaccharide chains from the chondroitin/

dermatan sulfate-protein linkage region and preparation of analytical probes by fluorescent labeling with 2-aminobenzamide. *J. Biochem. (Tokyo)* **129,** 107–118.

Sakaki, T., Zahringer, U., Warnecke, D. C., Fahl, A., Knogge, W., and Heinz, E. (2001). Sterol glycosides and cerebrosides accumulate in *Pichia pastoris, Rhynchosporium secalis,* and other fungi under normal conditions or under heat shock and ethanol stress. *Yeast* **18,** 679–695.

Sandhoff, R., Hepbildikler, S. T., Jennemann, R., Geyer, R., Gieselmann, V., Proia, R. L., Wiegandt, H., and Grone, H. J. (2002). Kidney sulfatides in mouse models of inherited glycosphingolipid disorders: Determination by nanoelectrospray ionization tandem mass spectrometry. *J. Biol. Chem.* **277,** 20386–20398.

Sanglard, D. (2002). Resistance of human fungal pathogens to antifungal drugs. *Curr. Opin. Microbiol.* **5,** 379–385.

Sawabe, A., Morita, M., Okamoto, T., and Ouchi, S. (1994). The location of double bonds in a cerebroside from edible fungi (mushroom) estimated by *B/E* linked scan fast-atom bombardment mass spectrometry. *Biol. Mass Spectrom* **23,** 660–664.

Schorling, S., Vallee, B., Barz, W. P., Riezman, H., and Oesterhelt, D. (2001). Lag1p and Lac1p are essential for the Acyl-CoA-dependent ceramide synthase reaction in *Saccharomyces cerevisae. Mol. Biol. Cell* **12,** 3417–3427.

Schwientek, T., Keck, B., Levery, S. B., Jensen, M. A., Pedersen, J. W., Wandall, H., Stroud, M., Cohen, S. M., Amado, M., and Clausen, H. (2002). The *Drosophila* gene brainiac encodes a glycosyltransferase putatively involved in glycosphingolipid synthesis. *J. Biol. Chem.* **277,** 32421–32429.

Seppo, A., and Tiemeyer, M. (2000). Function and structure of *Drosophila* glycans. *Glycobiology* **10,** 751–760.

Seppo, A., Moreland, M., Schweingruber, H., and Tiemeyer, M. (2000). Zwitterionic and acidic glycosphingolipids of the *Drosophila melanogaster* embryo. *Eur. J. Biochem.* **267,** 3549–3558.

Shibuya, H., Kawashima, K., Sakagami, M., Kawanishi, H., Shimomura, M., Ohashi, K., and Kitagawa, T. (1990). Sphingolipids and glycerolipids. I. Chemical structures and Ionophoretic activities of soya-cerebrosides I and II from soybean. *Chem. Pharm. Bull. (Tokyo)* **38,** 2933–2938.

Singh, B. N., Costello, C. E., and Beach, D. H. (1991). Structures of glycophospho-sphingolipids of *Tritrichomonas foetus*: A novel glycophosphosphingolipid. *Arch. Biochem. Biophys.* **286,** 409–418.

Singh, S., Reinhold, V. N., Bennion, B., and Levery, S. B. (2003). Applications of ion trap MS^n strategies to structure elucidation of diverse glycosylinositols derived from fungal glycosphingolipids. *Glycobiology* **13**(11), 829–830.

Sitrin, R. D., Chan, G., Dingerdissen, J., DeBrosse, C., Mehta, R., Roberts, G., Rottschaefer, S., Staiger, D., Valenta, J., Snader, K. M., Stedman, R. J., and Hoover, J. R. E. (1988). Isolation and structure determination of *Pachybasium* cerebrosides which potentiate the antifungal activity of aculeacin. *J. Antibiot. (Tokyo)* **41,** 469–480.

Solouki, T., Reinhold, B. B., Costello, C. E., O'Malley, M., Guan, S., and Marshall, A. G. (1998). Electrospray ionization and matix-assisted laser desorption/ionization Fourier transform ion cyclotron resonance mass spectrometry of permethylated oligosaccharides. *Anal. Chem.* **70,** 857–864.

Sperling, P., and Heinz, E. (2003). Plant sphingolipids: Structural diversity, biosynthesis, first genes and functions. *Biochim. Biophys. Acta* **1632,** 1–15.

Straus, A. H., Suzuki, E., Toledo, M. S., Takizawa, C. M., and Takahashi, H. K. (1995). Immunochemical characterization of carbohydrate antigens from fungi, protozoa, and mamals by monoclonal antibodies directed to glycan epitopes. *Braz. J. Med. Biol. Res.* **28,** 919–923.

Stroud, M. R., Levery, S. B., Nudelman, E. D., Salyan, M. E., Towell, J. A., Roberts, C. E., Watanabe, M., and Hakomori, S. (1991). Extended type-1 chain glycosphingolipids: Dimeric Le^a ($III^4V^4Fuc_2Lc_6$) as human tumor-associated antigen. *J. Biol. Chem.* **266**, 8439–8446.

Stroud, M. R., Levery, S. B., Salyan, M. E., Roberts, C. E., and Hakomori, S. (1992). Extended type-1 chain glycosphingolipid antigens. Isolation and characterization of trifucosyl-Le^b antigen ($III^4V^4VI^2Fuc_3Lc_6$). *Eur. J. Biochem.* **203**, 577–586.

Stroud, M. R., Levery, S. B., Martensson, S., Salyan, M. E., Clausen, H., and Hakomori, S. (1994). Human tumor-associated Le^a-Le^x hybrid carbohydrate antigen IV^3(Galβ1 → 3 [Fucα1 → 4]GlcNAc)III^3FucnLc$_4$ defined by monoclonal antibody 43-9F: Enzymatic synthesis, structural characterization, and comparative reactivity with various antibodies. *Biochemistry* **33**, 10672–10680.

Stroud, M. R., Handa, K., Ito, K., Salyan, M. E., Fang, H., Levery, S. B., Hakomori, S., Reinhold, B. B., and Reinhold, V. N. (1995). Myeloglycan, a series of E-selectin-binding polylactosaminolipids found in normal human leukocytes and myelocytic leukemia HL60 cells. *Biochem. Biophys. Res. Commun.* **209**, 777–787.

Stroud, M. R., Handa, K., Salyan, M. E., Ito, K., Levery, S. B., Hakomori, S., Reinhold, B. B., and Reinhold, W. N. (1996a). Monosialogangliosides of human myelogenous leukemia HL60 cells and normal human leukocytes. 1. Separation of E-selectin binding from nonbinding gangliosides and absence of sialosyl-Le^x having tetraosyl to octaosyl core. *Biochemistry* **35**, 758–769.

Stroud, M. R., Handa, K., Salyan, M. E., Ito, K., Levery, S. B., Hakomori, S., Reinhold, B. B., and Reinhold, V. N. (1996b). Monosialogangliosides of human myelogenous leukemia HL60 cells and normal human leukocytes. 2. Characterization of E-selectin binding fractions, and structural requirements for physiological binding to E-selectin. *Biochemistry* **35**, 770–778.

Stroud, M. R., Stapleton, A. E., and Levery, S. B. (1998). The P histo-blood group-related glycosphingolipid sialosyl galactosyl globoside as a preferred binding receptor for uropathogenic *Escherichia coli*: Isolation and structural characterization from human kidney. *Biochemistry* **37**, 17420–17428.

Sugita, M., Fujii, H., Dulaney, J. T., Inagaki, F., Suzuki, M., Suzuki, A., and Ohta, S. (1995). Structural elucidation of two novel amphoteric glycosphingolipids from the earthworm, *Pheretima hilgendorfi*. *Biochim. Biophys. Acta* **1259**, 220–226.

Sugita, M., Mizunoma, T., Aoki, K., Dulaney, J. T., Inagaki, F., Suzuki, M., Suzuki, A., Ichikawa, S., Kushida, K., Ohta, S., and Kurimoto, A. (1996). Structural characterization of a novel glycoinositolphospholipid from the parasitic nematode, *Ascaris suum*. *Biochim. Biophys. Acta* **1302**, 185–192.

Sugiyama, E., Hara, A., and Uemura, K. (1999). A quantitative analysis of serum sulfatide by matrix-assisted laser desorption ionization time-of-flight mass spectrometry with delayed ion extraction. *Anal. Biochem.* **274**, 90–97.

Sullards, M. C. (2000). Analysis of sphingomyelin, glucosylceramide, ceramide, sphingosine, and sphingosine 1-phosphate by tandem mass spectrometry. *Methods Enzymol.* **312**, 32–45.

Sullards, M. C., and Merrill, A. H., Jr. (2001). Analysis of sphingosine 1-phosphate, ceramides, and other bioactive sphingolipids by high-performance liquid chromatography-tandem mass spectrometry. *Sci. STKE.* **2001**, L1.

Sullards, M. C., Lynch, D. V., Merrill, A. H., Jr., and Adams, J. (2000). Structure determination of soybean and wheat glucosylceramides by tandem mass spectrometry. *J. Mass Spectrom* **35**, 347–353.

Sullards, M. C., Wang, E., Peng, Q., and Merrill, A. H., Jr. (2003). Metabolomic profiling of sphingolipids in human glioma cell lines by liquid chromatography tandem mass spectrometry. *Cell Mol. Biol. (Noisy -le-grand)* **49**, 789–797.

Suzuki, E., Toledo, M. S., Takahashi, H. K., and Straus, A. H. (1997). A monoclonal antibody directed to terminal residue of β-galactofuranose of a glycolipid antigen isolated from *Paracoccidioides brasiliensis*: Cross-reactivity with *Leishmania major* and *Trypanosoma cruzi*. *Glycobiology* **7**, 463–468.

Tadano-Aritomi, K., Kasama, T., Handa, S., and Ishizuka, I. (1992). Isolation and structural characterization of a monosulfated isoglobotetraosylceramide, the first sulfoglycosphingolipid of the isoglobo-series, from rat kidney. *Eur. J. Biochem.* **209**, 305–313.

Tadano-Aritomi, K., Okuda, M., Ishizuka, I., Kubo, H., and Ireland, P. (1994). A novel monosulfated pentaglycosylceramide with the isoglobo-series core structure in rat kidney. *Carbohydr. Res.* **265**, 49–59.

Tadano-Aritomi, K., Kubo, H., Ireland, P., Okuda, M., Kasama, T., Handa, S., and Ishizuka, I. (1995). Structural analysis of mono- and bis-sulfated glycosphingolipids by negative liquid secondary ion mass spectrometry with high- and low-energy collision-induced dissociation. *Carbohydr. Res.* **273**, 41–52.

Tadano-Aritomi, K., Kubo, H., Ireland, P., Kasama, T., Handa, S., and Ishizuka, I. (1996). Structural characterization of a novel monosulfated gangliotriaosylceramide containing a 3-O-sulfated *N*-acetylgalactosamine from rat kidney. *Glycoconj. J.* **13**, 285–293.

Tadano-Aritomi, K., Kubo, H., Ireland, P., Hikita, T., and Ishizuka, I. (1998). Isolation and characterization of a unique sulfated ganglioside, sulfated GM1a, from rat kidney. *Glycobiology* **8**, 341–350.

Takesako, K., Kuroda, H., Inoue, T., Haruna, F., Yoshikawa, Y., and Kato, I. (1993). Biological properties of Aureobasidin A, a cyclic depsipeptide antifungal antibiotic. *J. Antibiot.* **49**, 1414–1420.

Taki, T., and Ishikawa, D. (1997). TLC blotting: Application to microscale analysis of lipids and as a new approach to lipid-protein interaction. *Anal. Biochem.* **251**, 135–143.

Taki, T., Matsuo, K., Yamamoto, K., Matsubara, T., Hayashi, A., Abe, T., and Matsumoto, M. (1988). Human placenta gangliosides. *Lipids* **23**, 192–198.

Taki, T., Kasama, T., Handa, S., and Ishikawa, D. (1994). A simple and quantitative purification of glycosphingolipids and phospholipids by thin-layer chromatography blotting. *Anal. Biochem.* **223**, 232–238.

Taki, T., Ishikawa, D., Handa, S., and Kasama, T. (1995). Direct mass spectrometric analysis of glycosphingolipid transferred to a polyvinylidene difluoride membrane by thin-layer chromatography blotting. *Anal. Biochem.* **225**, 24–27.

Tanaka, R., Ishizaki, H., Kawano, S., Okuda, H., Miyahara, K., and Noda, N. (1997). Fruiting-inducing activity and antifungal properties of lipid components in members of *Annelida*. *Chem. Pharm. Bull. (Tokyo)* **45**, 1702–1704.

Thevissen, K., Ferket, K. K., Francois, I. E., and Cammue, B. P. (2003). Interactions of antifungal plant defensins with fungal membrane components. *Peptides* **24**, 1705–1712.

Thomma, B. P., Cammue, B. P., and Thevissen, K. (2002). Plant defensins. *Planta* **216**, 193–202.

Toledo, M. S., Suzuki, E., Straus, A. H., and Takahashi, H. K. (1995). Glycolipids from *Paracoccidioides brasiliensis*. Isolation of a galactofuranose-containing glycolipid reactive with sera of patients with paracoccidioidomycosis. *J. Med. Vet. Mycol.* **33**, 247–251.

Toledo, M. S., Levery, S. B., Straus, A. H., Suzuki, E., Momany, M., Glushka, J., Moulton, J. M., and Takahashi, H. K. (1999). Characterization of sphingolipids from mycopathogens: Factors correlating with expression of 2-hydroxy fatty acyl (E)-Δ^3-unsaturation in cerebrosides of *Paracoccidioides brasiliensis* and *Aspergillus fumigatus*. *Biochemistry* **38**, 7294–7306.

Toledo, M. S., Levery, S. B., Straus, A. H., and Takahashi, H. K. (2000). Dimorphic expression of cerebrosides in the mycopathogen *Sporothrix schenckii*. *J. Lipid Res.* **41**, 797–806.

Toledo, M. S., Levery, S. B., Suzuki, E., Straus, A. H., and Takahashi, H. K. (2001a). Characterization of cerebrosides from the thermally dimorphic mycopathogen *Histoplasma capsulatum*: Expression of 2-hydroxy fatty N-acyl (E)-Δ^3-unsaturation correlates with the yeast-mycelium phase transition. *Glycobiology* **11**, 113–124.

Toledo, M. S., Levery, S. B., Straus, A. H., and Takahashi, H. K. (2001b). Sphingolipids of the mycopathogen *Sporothrix schenckii*: Identification of a glycosylinositol phosphorylceramide novel core GlcNH2 α1–>2Ins motif. *FEBS Lett.* **493**, 50–56.

Toledo, M. S., Suzuki, E., Levery, S. B., Straus, A. H., and Takahashi, H. K. (2001c). Characterization of monoclonal antibody MEST-2 specfic to glucosylceramide of fungi and plants. *Glycobiology* **11**, 105–112.

Toledo, M. S., Levery, S. B., Glushka, J., Straus, A. H., and Takahashi, H. K. (2001d). Structure elucidation of sphingolipids from the mycopathogen *Sporothrix schenckii*: Identification of novel glycosylinositol phosphorylceramides with core Manα1–>6Ins linkage. *Biochem. Biophys. Res. Commun.* **280**, 19–24.

Tomlinson, M. J., Scott, J. R., Wilkins, C. L., Wright, J. B., and White, W. E. (1999). Fragmentation of an alkali metal-attached peptide probed by collision-induced dissociation fourier transform mass spectrometry and computational methodology. *J. Mass Spectrom* **34**, 958–968.

Tseng, M.-C., Chen, Y.-R., and Her, G. R. (2004). A low-makeup beveled tip capillary electrophoresis/electrospray ionization mass spectrometry interface for micellar electrokinetic chromatography and nonvolatile buffer capillary electrophoresis. *Anal. Chem.* **76**, 6303–6312.

Umemura, K., Ogawa, N., Yamauchi, T., Iwata, M., Shimura, M., and Koga, J. (2000). Cerebroside elicitors found in diverse phytopathogens activate defense responses in rice plants. *Plant Cell Physiol.* **41**, 676–683.

Umemura, K., Ogawa, N., Koga, J., Iwata, M., and Usumi, H. (2002). Elicitor activity of cerebroside, a sphingolipid elicitor, in cell suspension cultures of rice. *Plant Cell Physiol.* **43**, 778–784.

van Burik, J. A., and Magee, P. T. (2001). Aspects of fungal pathogenesis in humans. *Annu. Rev. Microbiol.* **55**, 743–772.

Venkataraman, K., Riebeling, C., Bodennec, J., Riezman, H., Allegood, J. C., Sullards, M. C., Merrill, A. H., Jr., and Futerman, A. H. (2002). Upstream of growth and differentiation factor 1 (uog1), a mammalian homolog of the yeast longevity assurance gene 1 (LAG1), regulates N-stearoyl-sphinganine (C18-(dihydro)ceramide) synthesis in a fumonisin B1-independent manner in mammalian cells. *J. Biol. Chem.* **277**, 35642–35649.

Villas Boas, M. H., Egge, H., Pohlentz, G., Hartmann, R., and Bergter, E. B. (1994). Structural determination of N-2'-hydroxyoctadecenoyl-1-O-beta-D-glucopyranosyl-9-methyl-4,8-sphingadienine from species of *Aspergillus*. *Chem. Phys. Lipids* **70**, 11–19.

Viseux, N., de Hoffmann, E., and Domon, B. (1997). Structural analysis of permethylated oligosaccharides by electrospray tandem mass spectrometry. *Anal. Chem.* **69**, 3193–3198.

Viseux, N., de Hoffmann, E., and Domon, B. (1998). Structural assignment of permethylated oligosaccharide subunits using sequential tandem mass spectrometry. *Anal. Chem.* **70**, 4951–4959.

Vukelic, Z., Metelmann, W., Müthing, J., Kos, M., and Peter-Katalinić, J. (2001). Anencephaly: Structural characterization of gangliosides in defined brain regions. *Biol. Chem.* **382**, 259–274.

Wade, J. C. (1997). Treatment of fungal and other opportunistic infections in immunocompromised patients. *Leukemia* **11**(Suppl 4), S38–S39.

Walsh, T. J., Hiemenz, J. W., and Anaissie, E. (1996). Recent progress and current problems in treatment of invasive fungal infections in neutropenic patients. *Infect. Dis. Clin. North Am.* **10**, 365–400.

Wandall, H. H., Pedersen, J. W., Park, C., Levery, S. B., Pizette, S., Cohen, S. M., Schwientek, T., and Clausen, H. (2003). *Drosophila* egghead encodes a ß1,4 mannosyltransferase predicted to form the immediate precursor glycosphingolipid substrate for brainiac. *J. Biol. Chem.* **278**, 1411–1414.

Wandall, H. H., Pizette, S., Pedersen, J. W., Eichert, H., Levery, S. B., Mandel, U., Cohen, S. M., and Clausen, H. (2005). Egghead and brainiac are essential for glycosphingolipid biosynthesis *in vivo*. *J. Biol. Chem.* **280**, 4858–4863.

Wang, Y., Rashidzadeh, H., and Guo, B. (2000). Structural effects on polyether cationization by alkali metal ions in matrix-assisted laser desorption/ionization. *J. Am. Soc. Mass Spectrom* **11**, 639–643.

Warnecke, D., and Heinz, E. (2003). Recently discovered functions of glucosylceramides in plants and fungi. *Cell Mol. Life Sci.* **60**, 919–941.

Webb, J. W., Jiang, K., Gillece-Castro, B. L., Tarentino, A. L., Plummer, T. H., Byrd, J. C., Fisher, S. J., and Burlingame, A. L. (1988). Structural characterization of intact, branched oligosaccharides by high-performance liquid chromatography and liquid secondary ion mass spectrometry. *Anal. Biochem.* **169**, 337–349.

Whitfield, P. D., Sharp, P. C., Johnson, D. W., Nelson, P., and Meikle, P. J. (2001). Characterization of urinary sulfatides in metachromatic leukodystrophy using electrospray ionization-tandem mass spectrometry. *Mol. Genet. Metab* **73**, 30–37.

Wing, D. R., Garner, B., Hunnam, V., Reinkensmeier, G., Andersson, U., Harvey, D. J., Dwek, R. A., Platt, F. M., and Butters, T. D. (2001). High-performance liquid chromatography analysis of ganglioside carbohydrates at the picomole level after ceramide glycanase digestion and fluorescent labeling with 2-aminobenzamide. *Anal. Biochem.* **298**, 207–217.

Wuhrer, M., Rickhoff, S., Dennis, R. D., Lochnit, G., Soboslay, P. T., Baumeister, S., and Geyer, R. (2000a). Phosphocholine-containing, zwitterionic glycosphingolipids of adult *Onchocerca volvulus* as highly conserved antigenic structures of parasitic nematodes. *Biochem. J.* **348**(Part 2), 417–423.

Wuhrer, M., Dennis, R. D., Doenhoff, M. J., Lochnit, G., and Geyer, R. (2000b). *Schistosoma mansoni* cercarial glycolipids are dominated by Lewis *X* and pseudo-Lewis *Y* structures. *Glycobiology* **10**, 89–101.

Wuhrer, M., Dennis, R. D., Doenhoff, M. J., and Geyer, R. (2000c). Stage-associated expression of ceramide structures in glycosphingolipids from the human trematode parasite *Schistosoma mansoni*. *Biochim. Biophys. Acta* **1524**, 155–161.

Wuhrer, M., Grimm, C., Dennis, R. D., Idris, M. A., and Geyer, R. (2004). The parasitic trematode *Fasciola hepatica* exhibits mammalian-type glycolipids as well as Gal(beta1–6) Gal-terminating glycolipids that account for cestode serological cross-reactivity. *Glycobiology* **14**, 115–126.

Yohe, H. C., Ye, S., Reinhold, B. B., and Reinhold, V. N. (1997). Structural characterization of the disialogangliosides of murine peritoneal macrophages. *Glycobiology* **7**, 1215–1227.

Zamfir, A., and Peter-Katalinić, J. (2001). Glycoscreening by on-line sheathless capillary electrophoresis/electrospray ionization-quadrupole time of flight-tandem mass spectrometry. *Electrophoresis* **22**, 2448–2457.

Zamfir, A., and Peter-Katalinić, J. (2004). Capillary electrophoresis-mass spectrometry for glycoscreening in biomedical research. *Electrophoresis* **25,** 1949–1963.

Zamfir, A., Vukelic, Z., and Peter-Katalinić, J. (2002a). A capillary electrophoresis and off-line capillary electrophoresis/electrospray ionization-quadrupole time-of-flight-tandem mass spectrometry approach for ganglioside analysis. *Electrophoresis* **23,** 2894–2903.

Zamfir, A., Seidler, D. G., Kresse, H., and Peter-Katalinić, J. (2002b). Structural characterization of chondroitin/dermatan sulfate oligosaccharides from bovine aorta by capillary electrophoresis and electrospray ionization quadrupole time-of-flight tandem mass spectrometry. *Rapid Commun. Mass Spectrom* **16,** 2015–2024.

Zamfir, A., Seidler, D. G., Kresse, H., and Peter-Katalinić, J. (2003). Structural investigation of chondroitin/dermatan sulfate oligosaccharides from human skin fibroblast decorin. *Glycobiology* **13,** 733–742.

Zamfir, A., Vukelic, Z., Bindila, L., Peter-Katalinić, J., Almeida, R., Sterling, A., and Allen, M. (2004a). Fully automated chip-based nanoelectrospray tandem mass spectrometry of gangliosides from human cerebellum. *J. Am. Soc. Mass Spectrom* **15,** 1649–1657.

Zamfir, A., Seidler, D. G., Schonherr, E., Kresse, H., and Peter-Katalinić, J. (2004b). On-line sheathless capillary electrophoresis/nanoelectrospray ionization-tandem mass spectrometry for the analysis of glycosaminoglycan oligosaccharides. *Electrophoresis* **25,** 2010–2016.

Zhang, H., Singh, S., and Reinhold, V. N. (2005). Congruent strategies for carbohydrate sequencing. 2. FragLib: An MS^n spectral library. *Anal. Chem.* **77,** 6263–6270.

Zhou, B., Li, S. C., Laine, R. A., Huang, R. T., and Li, Y. T. (1989). Isolation and characterization of ceramide glycanase from the leech, *Macrobdella decora. J. Biol. Chem.* **264,** 12272–12277.

Zhou, Z., Ogden, S., and Leary, J. A. (1990). Linkage position determination in oligosaccharides: MS/MS study of lithium-cationized carbohydrates. *J. Org. Chem.* **55,** 5444–5446.

[13] Mapping Bacterial Glycolipid Complexity Using Capillary Electrophoresis and Electrospray Mass Spectrometry

By J. Li, A. Martin, A. D. Cox, E. R. Moxon, J. C. Richards, and P. Thibault

Abstract

This chapter presents the application of capillary electrophoresis coupled to electrospray mass spectrometry (CE–ES–MS) for the analysis of complex bacterial lipopolysaccharides (LPS) from pathogenic strains of *Haemophilus influenzae* and *Neisseria meningitidis*. A discussion is included of the development of electrophoretic conditions conducive to trace-level enrichment and separation of closely related glycoforms and

METHODS IN ENZYMOLOGY, VOL. 405
Copyright 2005, Elsevier Inc. and Her Majesty the Queen in Right of Canada

0076-6879/05 $35.00
DOI: 10.1016/S0076-6879(05)05013-5

isoforms, which provided sensitive detection of glycolipids from as little as five bacterial colonies. The chapter also describes the use of mixed MS scanning functions to aid the identification of specific functionalities and immunodeterminants of LPS, such as pyrophosphoethanolamine, phospho-choline, and N-acetyl neuraminic acid (Neu5Ac), which represent less than 2% of the overall LPS population. The combination of high-resolution capillary electrophoresis with sensitive tandem mass spectrometry (MS/MS) provides a unique analytical tool to probe the subtle structural changes resulting from oligosaccharide branching and location of substituted LPS isoforms. The ability to detect a diverse LPS population over a wide dynamic range of expression using CE–MS enables the correlation of structural changes between bacterial strains and isogenic mutants to assign functional gene relationship.

Introduction

Gram-negative bacteria comprise both capsular and nontypable (acap-sular) strains of significant virulence and pathogenicity to humans. Lipo-polysaccharides (LPS) are the major components of the outer membrane of these human pathogens (Moxon and Maskell, 1992; Murphy and Apicella, 1997; Preston *et al.*, 1996). In *Haemophilus influenzae* and other mucosal pathogens, such as *Neisseria meningitidis* and *N. gonorrhoreae*, the antigenic LPS molecules have been associated with microbial virulence and pathogenesis (Cope *et al.*, 1990; Hood *et al.*, 1996; Kimura and Hansen, 1986; Weiser *et al.*, 1990). These outer surface components are amphipathic molecules that consist of a hydrophilic oligosaccharide core to which is linked a hydrophobic glycolipid, referred to as lipid A (Zhou *et al.*, 1994). The composition and structure of the lipid A from *H. influenzae*, *N. meningitidis*, and *N. gonorrhoreae* is typically conserved among strains and is composed of a glucosamine disaccharide to which *O*- and *N*-linked fatty acids are attached (Wilkinson, 1983).

The LPS structures of *H. influenzae* and *N. meningitidis* have been investigated in some detail, and a number of reports have described rele-vant structural features (Masoud *et al.*, 1997; Plested *et al.*, 1999; Thibault *et al.*, 1998). It has been found that the LPS of *H. influenzae* contains a common L-*glycero*-D-*manno*-heptose-containing inner-core trisaccharide unit attached to the lipid A moiety via a phosphorylated 2-keto-3-deox-yoctulosonic acid (KDO) residue (Fig. 1A) (Masoud *et al.*, 1997; Thibault *et al.*, 1998). Similarly, the LPS of *N. meningiditis* shown in Fig. 1B contains a conserved inner-core structure that contains two KDOs, two Heps, and one *N*-acetylglucosamine (GlcNAc) (Plested *et al.*, 1999). Further exten-sion in the oligosaccharide chain can take place at each heptose (Hep)

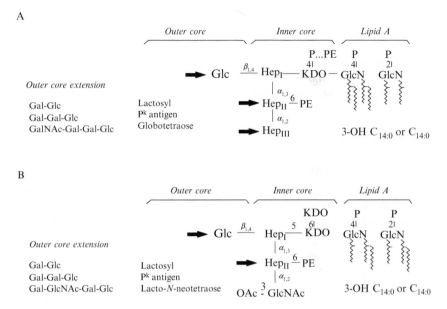

FIG. 1. Structure of the conserved inner core oligosaccharide regions of LPS from *N. meningitidis* and *H. influenzae*. (A) *Haemophilus influenzae* capsular and noncapsular strains. (B) *Neisseria meningitidis* strain.

residue. Additional heterogeneity is conferred by the incorporation of noncarbohydrate substituents such as acetate (Ac), phosphate (P), phosphoethanolamine (PE), pyrophosphoethanolamine (PPE), and phosphocholine (PC) groups as well as amino acids such as L-alanine, L-serine, L-threonine, and L-lysine (Knirel and Kochetkov, 1994). A recent survey of both typable and nontypeable strains of *H. influenzae* indicated that glycine is a common substituent of the inner core of LPS (Landerholm *et al.*, 2004; Li *et al.*, 2001).

   This gives rise to a heterogeneous population of complex glycans showing both intrastrain and interstrain variability. This structural variability is partly accounted for by a complex network of glycosyltransferases together with molecular mechanisms, such as slipped-strand mispairing that generates phase variation of terminal epitopes (High *et al.*, 1993; Jarosik and Hansen, 1994; Weiser *et al.*, 1990). This variability in the expression of different core oligosaccharides provides an important mechanism whereby bacteria can adapt to changing environmental conditions encountered during the pathogenic cycle of infection (Inzana *et al.*, 1992; Weiser and Pan, 1998).

Previous investigations have outlined the application of mass spectrometry (MS) for the structural characterization of trace-level bacterial LPS and complex glycolipids (Cope *et al.*, 1990; Gaucher *et al.*, 2000; Hood *et al.*, 1996; Kimura and Hansen, 1986; Masoud *et al.*, 1997; Moxon and Maskell, 1992; Murphy and Apicella, 1997; Plested *et al.*, 1999; Preston *et al.*, 1996; Thibault *et al.*, 1998; Weiser *et al.*, 1990; Wilkinson, 1983; Zhou *et al.*, 1994). The availability of high-resolution techniques, such as capillary electrophoresis (CE), also enabled the analysis of complex biological extracts with unparallel resolution and sensitivity (Krylov and Dovichi, 2000). Furthermore, changes in electrophoretic mobilities imparted by chain extension or structural motifs of isomeric glycolipids can be resolved by capillary electrophoresis as presented for a subpopulation of LPS obtained from *H. influenzae* and *N. meningitidis* (Li *et al.*, 2004; Schweda *et al.*, 2003). This chapter examines the analytical potentials of CE combined to electrospray mass spectrometry (CE–ESMS) for the analysis of bacterial glycolipids. The coupling of CE to ESMS facilitates the separation of closely related glycoform and isoform families based on their unique molecular conformation and ionic charge distribution as evidenced from the corresponding contour profile of mass-to-charge ratio (m/z) versus time. Structural relationships between closely related LPS can thus be visualized to facilitate the assignment of glycolipids of extended chain length and isoforms differing by the location of substituted phosphorylated functionalities (Auriola *et al.*, 1998; Kelly *et al.*, 1996; Li *et al.*, 1998, 2004; Schweda *et al.*, 2003; Thibault *et al.*, 1998). Precise location of functional groups within the glycolipid structure is facilitated through first- and second-generation product ion spectra using a hybrid quadrupole/time-of-flight (QTOF) mass spectrometer. This discussion also presents specific scanning functions facilitating the identification of biologically relevant groups and residues such as PPE, PC, and *N*-acetyl neuraminic acid (Neu5Ac) that are often present as constituents of the bacterial LPS population.

Procedures

Bacterial Strains and Growth Conditions

H. influenzae strains Rd and 375 used in this study have been previously described (Cope *et al.*, 1990; Hood *et al.*, 1999). Mutations in the *siaB* and *lic1* genes were constructed by insertion and deletion methods (Hood *et al.*, 1999; Liu *et al.*, 1996). Strains were grown at 37° on brain heart infusion agar (1% w/v) supplemented with haemin (10 ug/ml), nicotinamide adene dinucleotide (NAD) (2 ug/ml), and, when appropriate, NeuAc (25 ug/ml) or CMP-Neu5Ac (50 ug/ml). For the selection of transformants, kanamycin

(10 ug/ml) or tetracyline (50 ug/ml) was added to the growth medium. *N. meningitidis* strain BZ157 was obtained and grown as previously described in High *et al.* (1993).

Preparation and Extraction of Lipopolysaccharides

Small-scale extraction of LPS was conducted by washing the colonies from the plates and dispersing them in 1.5 mL tubes each containing 500 μL of deionized water. The cells were freeze-dried overnight, dissolved in 360 μL of deionized water, and a 40 μL aliquot of a 25 μg/mL solution of proteinase K was added to each vial. The suspended cell solutions were incubated at 37° for 90 min, and the digestion was stopped by raising the temperature to 65° for 10 min. The solutions were allowed to cool at room temperature and were subsequently freeze-dried. The cells were further digested by incubating them at 37° for 4 h in a 200 μL solution of 20 mM ammonium acetate buffer at pH 7.5 containing DNase (10 μg/mL) and RNase (5 μg/mL).

Preparation of O-Deacylated LPS

The freeze-dried and digested cells containing the free LPS were dissolved in 200 μL of hydrazine and incubated at 37° for 50 min with constant stirring to release *O*-linked fatty acids. The reaction mixtures were cooled (0°), the hydrazine was destroyed by addition of cold acetone (600 μL), and the final product was obtained by centrifugation. The pellets were washed with 2 × 600 μL of acetone, centrifuged, and then lyophilized from water. When required, enzymatic release of Neu5Ac containing LPS was achieved by incubating the *O*-deacylated LPS with $\alpha_{2,3}$ sialidase (New England Biolabs, Beverley, MA, www.neb.com) 1:50 enzyme substrate weight ratio in a buffer solution containing 50 mM sodium citrate at pH 6.0 and 100 mM NaCl.

Instrumentation

A crystal Model 310 CE instrument (ATI Unicam, Boston, MA) was coupled to an API 3000 or a Q-Star QTOF mass spectrometers (Perkin-Elmer/Sciex, Concord, Canada) via a microionspray interface (Fig. 2A). A sheath solution (isopropanol–methanol, 2:1) was delivered at a flow rate of 1 μL per minute to low dead volume tee (250 μm internal diameter (I.D.), Chromatographic Specialities, Brockville, Canada). All aqueous solutions were filtered through a 0.45 μm filter (Millipore, Bedford, MA) before use. An electrospray stainless steel needle (27 gauge) was butted against the low dead volume tee and enabled the delivery of the sheath solution to the end of the capillary column. The separation was achieved using a 90-cm

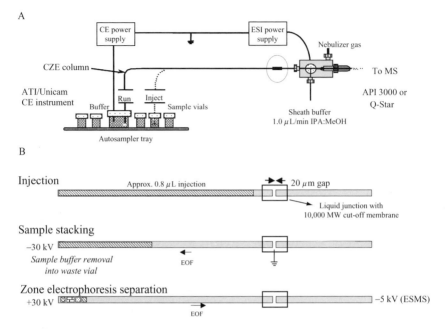

FIG. 2. Schematic description of the CE–ESMS using a sheath flow interface. (A) Diagram of the autosampler CE column and interface assembly. (B) An expanded view of the circled area showing dialysis membrane liquid junction for on-line sample stacking.

bare fused-silica capillary with a buffer of 30 mM morpholine/formic acid in deionized water, pH 9.0 with 5% methanol. For positive detection mode, a 10 mM ammonium acetate buffer, at pH 9.0, containing 5% methanol was used. A separation voltage of 30 kV was typically applied at the injection end of the capillary. The outlet of the capillary was tapered to circa 15 μm I.D. using a laser puller (Sutter Instruments, Novato, CA).

For sample stacking preconcentration experiments, a microdialysis interface junction similar to that reported elsewhere was used (Liu *et al.*, 1996; Yang *et al.*, 1999). To explain briefly, a 1.5-cm length of dialysis tubing (A/G Technology Corporation, Needham, MA) was inserted through a predrilled 1.5 mL Eppendorf tube, and epoxy was applied to around the outside of the tubing (Fig. 2B) to hold the membrane in place. The separation capillary (70 cm in length) and a tapered transfer line (15 cm in length) capillary were butted together inside dialysis tubing and sealed with epoxy. The Eppendorf tube was filled with the solution identical to the separation buffer. A platinum wire was inserted in the solution and maintained at 0 V at all the time. After the sample loading, a −30 kV voltage was applied to the inlet for typically 2 to 3 min to focus the analytes toward the injection end of the capillary while simultaneously removing the sample buffer.

Mass spectra were acquired using a dwell time of 3.0 ms per step of 1 m/z unit in full-mass scan mode. For CE–MS–MS experiments, a sample loading of approximately 25 nL was applied. The tandem MS (MS/MS) data were acquired with dwell times of 2.0 ms per step of 1 m/z unit on the triple quadrupole instrument or over 2 s/scan using the Q-Star. Collisional activation of selected precursor ions in the RF-only quadrupole collision cell was achieved using nitrogen as a target gas at energies of 80 to 90 eV (laboratory frame of reference). A mix scan function was employed for probing specific functionalities and residues such as PPE, PC, and Neu5Ac. On the triple quadrupole instrument, this scanning function involved selected ion monitoring of specific fragment ions at high orifice voltage (up to three different m/z values at OR 120 V) followed by a conventional scan at a low orifice voltage (typically m/z 500–1500, OR 30 V). Alternate positive and negative ion scans were achieved using a polarity switching function with a 1.5 s pause between ion detection modes.

Enhancement of Sensitivity for Trace-Level Identification of Bacterial Glycolipids Using Sample Preconcentration Strategies

The distribution of bacterial glycolipids not only varies in terms of their respective glycoform and isoform populations but also in the levels of LPS expression between strains of the same genus. To probe the subtle changes in LPS structure as a result of phase variation or as a consequence of site-directed mutation, it becomes important to develop sensitive analytical tools providing the required level of specificity and selectivity. To this end, the coupling of a capillary electrophoresis to ESMS provides an ideal structural tool because hydrophilic analytes can be separated and characterized based on both charge and molecular shape (see following section). The maintenance of the relatively high separation efficiencies of CE (N > 150,000 plates/m) requires that the injection volume in zone electrophoresis format be kept typically below 2% of the total column volume or 40 nL for a 1 m × 50 μm I.D. capillary (Krylov and Dovichi, 2000; Yang et al., 1999). While mass spectral identification of bacterial glycolipids can be achieved using 100 to 300 pg on-column loadings (Li et al., 1998), this sample injection constraint usually results in a concentration detection limit of low micrograms per milliliters when using electrospray ionization.

Previous investigations from this laboratory have investigated the use of on-line preconcentration techniques prior to zone electrophoresis to improve the concentration detection limits for trace-level analysis of

O-deacylated LPS (Auriola *et al.*, 1997; Li *et al.*, 1998). In transient isotachophoresis (tCITP), sample loadings of up to 1 μL of dilute *Pseudomonas aeruginosa* core oligosaccharide solution could be used without significant loss in separation performance (Auriola *et al.*, 1998). Further improvement in sample loading (up to 5 μL injection volume) was demonstrated for *O*-deacylated LPS from *H. influenzae* using a hydrophobic styrene-divinyl benzene (SDB) adsorption membrane inserted at the inlet end of the CE column (Li *et al.*, 1998). In combination with enzymatic releasing methods using DNase and proteinase K, this technique provided excellent sensitivity and enabled the identification of LPS surface antigens from as few as five bacterial colonies of *H. influenzae* strain Eagan.

The enhancement of sample loading can also be achieved using field amplification sample stacking (FASS), where the analyte zone is focused by taking advantage of the differential conductivity between the sample plug and the separation buffer (Burgi and Chien, 1991; Chien and Burgi, 1992a,b; He and Lee, 1999; Yang *et al.*, 1999; Zhang and Thormann, 1996, 1998; Zhao *et al.*, 1998). The sample is typically reconstituted in water or in a buffer of low conductivity and injected into the capillary column by hydrodynamic (on-column sample stacking) (Burgi and Chien, 1991; Chien and Burgi, 1992; He and Lee, 1999) or electrokinetic injection (head-column sample staking) (Zhang and Thormann, 1996, 1998; Zhao *et al.*, 1998). In the former method, a pressure is applied at the inlet of the capillary to inject the sample and under proper optimization can result in an injection volume corresponding to the whole separation column (Fig. 2B). Following the sample injection, the analytes are focused at the inlet of the column by applying an electric field across the sample and the electrophoresis buffer, both having different conductivity and ionic strength. At the beginning of the focusing period, the current is relatively low reflecting the weak ionic strength of the sample buffer. The current is progressively increasing as the sample buffer is removed from the capillary while the analyte is stacked against the separation buffer of high ionic strength. The focusing period is then halted when the current has reached 90% of its maximum value, at which point the electric field is reversed, and the analyte separation is allowed to proceed via zone electrophoresis (Fig. 2B).

For head-column FASS, the analytes are stacked against the boundary of the separation buffer using an extended electrophoretic injection period that prevents the introduction of a large volume of sample buffer (Zhang and Thormann, 1996, 1998). The relative proportion of individual analyte introduced in the capillary is not uniform because this injection technique favors compounds of high mobility. Quantitative analysis performed with this preconcentration technique is thus more challenging and requires proper control of the sample buffer composition and the use of an appropriate internal standard.

The application of FASS preconcentration with on-line electrospray ionization also requires a suitable interface to facilitate the sample stacking and the replenishment of the outlet electrophoresis buffer with minimal interruption or capillary movement (Zhao et al., 1998). To this end, a liquid junction containing a microdialysis membrane provides a viable solution to ensure proper exchange of electrolytes and make-up solutions for electrospray ionization. Previous reports have described a microdialysis membrane junction to introduce a post CZE separation buffer for electrochemical detection (Zhou and Lunte, 1995). This technique was subsequently used for on-line sample clean-up (Liu et al., 1996) and for capillary isoelectric focusing (CIEF) with on-line ESMS (Yang et al., 1998).

In preliminary investigations, the FASS preconcentration technique was optimized using electrophoretic conditions amenable to the separation of acidic peptides such as leu-enkephalin, glu-fibrinopeptide B, and phospho-DSIP (data not shown). The electrolyte used for focusing and zone electrophoresis consisted of 5% methanol in 50 mM morpholine pH 9.0, a buffer also compatible with the analysis of O-deacylated glycolipids (Li et al., 1998). For hydrodynamic injection of up to 1 μL, this technique provided low nanograms per milliliter detection limits with excellent linearity ($r^2 = 0.998 - 0.999$) over the dynamic range of 1 to 5000 ng/mL. Such improvement in sample loading corresponded to a 10- to 30-fold enhancement in sensitivity compared to that achievable using zone electrophoresis (25–40 nL injection).

The application of this preconcentration technique was further investigated for the analysis of O-deacylated LPS derived from isolates of small bacterial colonies. The isolation of trace-level LPS from H. influenzae was achieved using a method combining enzymatic digestions and chemical cleavages with minimal sample transfer (Li et al., 1998). Traditionally, the analysis of O-deacylated glycolipids arising from mild hydrazinolysis of intact LPS has been one of the preferred chemical methods for obtaining hydrophilic products that are amenable to electrospray ionization using aqueous buffers. Mild hydrazinolysis can be used on small amounts of material to quantitatively remove all O-linked acyl moieties though structural information on ester-linked substituents (i.e., O-acetyl groups) is lost as a result of this reaction.

The on-column FASS–CE–ESMS analysis of O-deacylated LPS from five colonies of H. influenzae 375 lic1 strain is presented in Fig. 3. In contrast to LPS from H. influenzae strain 375 (Thibault et al., 1998), the lic1 mutant is characterized by glycoforms devoid of a PC-Glc residue at

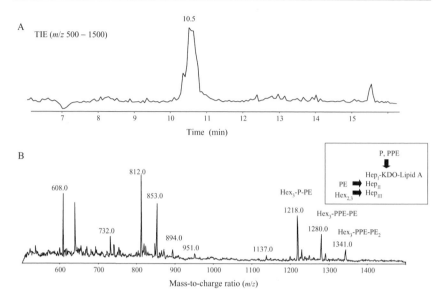

FIG. 3. Trace-level identification (TIE) of O-deacylated LPS from *H. influenzae* 375 *lic1*. (A) Total ion electropherogram. (B) Extracted mass spectrum for peak identified at 10.5 min.

Hep_I. The total ion electropherogram (TIE) for m/z 500–1500 is shown in Fig. 3A along with the extracted mass spectrum for the corresponding peak centered at 10.5 min (Fig. 3B). This spectrum is dominated by abundant multiply deprotonated ions, $[M - nH]^{n-}$, where n varies from two to four charges. The observed molecular mass for the most abundant glycolipid is 2438 Da (m/z 608, 812, 1218) in good agreement with that calculated for an O-deacylated LPS composed of Hex_3, Hep_3, KDO-P, and a PE group appended to the inner core (M_{calc}: 2437.8 Da). Further heterogeneity is conferred by the incorporation of additional PE groups giving rise to glycolipids of 2562 Da (M_{calc}: 2561.9 Da) and 2686 Da (M_{calc}: 2685.8 Da). A minor glycoform is also observed at m/z 1137 and corresponds to a truncated species containing two hexoses residues (M: 2276 Da). It is noteworthy that incorporation of PE groups can take place on Hep_{II}, Hep_{III}, or on the phosphorylated KDO of the inner core, thus accounting for a complex repertoire of structural diversity.

Identification of Glycoform and Isoform Families Using CE–ESMS

LPS glycolipids are often represented by a complex distribution of closely related glycoforms varying by the length and the site of attachment of the oligosaccharide chains. This finding in addition to the natural heterogeneity of functional groups such as P, PE, PPE, and PC present a sizable structural challenge to the analyst and a convenient mechanism for the survival of pathogenic bacteria in the host environment. Structural characterization of this diverse population of glycolipids is not only important for the further understanding of the pathogenesis processes and biosynthetic mechanisms leading to their expression but also for the development of antibodies specifically targeted toward these immunodeterminant structures.

To this end, the combination of capillary electrophoresis coupled to ESMS provides a valuable analytical tool enabling the separation of these complex biomolecules based on their respective electrophoretic mobilities. The order of migration thus reflects the intrinsic electrophoretic properties of the analyte and can be used to correlate changes (net charge and frictional drag) taking place between closely related structural analogues. This is illustrated in Fig. 4 for the separation of O-deacylated LPS from a *lex2* insert construct strain of *H. influenzae* strain Rd. In this particular example, enhanced mass spectral resolution and faster acquisition time (duty cycle of 1 s per spectrum) was achieved using a QTOF instrument. The TIE corresponding to the full-scan analysis (m/z 500–1400) is presented in Fig. 4A along with the contour intensity plots as a function of m/z versus time in the bottom panel (Fig. 4B). The contour profile shows a

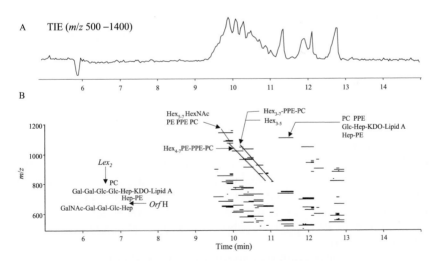

FIG. 4. CE–ES–MS (−) analysis of O-deacylated LPS from *H. influenzae* Rd *lex2* insert, (A) Total ion electropherogram. (B) Contour profile time vs. m/z. Diagonal lines indicate glycoform families. The inset shows the *H. influenzae* core oligosaccharide with arrows pointing at the site of glycan extension associated with the corresponding gene.

series of multiply deprotonated molecules from which the molecular mass of the different analytes can be calculated. One of the most prominent glycolipids observed at 9.55 min displayed $[M - 4H]^{4-}$ and $[M - 3H]^{3-}$ ions at m/z 893.29 and 1191.42 corresponding to an observed molecular mass (M_{obs}) of 3577.28 Da for the monoisotopic ^{12}C component. Based on this mass measurement and on previous structural assignments (Risberg et al., 1999), this glycolipid was assigned to a structure composed of PC-PE-PPE-Hex$_7$–HexNAc-Hep$_3$-KDO-GlcN$_2$ with two N-linked 3-OH myristic acid ($C_{14}H_{27}O_2$) and two phosphate groups (M_{calc}: 3577.20 Da).

The contour representation shown in Fig. 4B provides a valuable analytical tool to identify closely related families of glycolipids based on the appearance of diagonal lines. This phenomenon arises from the fact that progressive extension of Hex residues on the core structure results in a regular increase of molecular mass and a concurrent decrease in electrophoretic mobility. For example, the O-deacylated LPS having a molecular mass of 3577.28 Da is a member of a glycoform family extending from Hex$_5$ to Hex$_7$, all of which contains an HexNAc, a PC, and a single PPE group presumably on the KDO residue. A set of glycoforms having the same glycan distribution but lacking a terminal HexNAc residue is observed in Fig. 4B as a parallel diagonal line shifted to earlier migration time. The extensive glycosylation pattern observed here compared to the wild-type H. influenzae Rd strain (Risberg et al., 1999) is associated to the lex2 insert, which triggers a glucosyltransferase to incorporate a Glc residue next to the PC-Glc. Further extension of Gal residues to this side chain result in a P^k epitope (α–Gal(1-4)-ß-Gal (1-4)-Glc) mimicking a human blood group antigen.

Similarly, other glycoform families, such as that of PC-PPE-Hex$_{3–7}$-Hep$_3$-KDO and Hex$_{3–5}$-Hep$_3$-KDO are also observed in Fig. 4B, as indicated by the superimposed diagonal lines. Both families exhibit similar mobilities and molecular masses (ΔM: 3 Da) and thus result in a juxtaposition of their corresponding migration profiles. Distinction of these closely related glycoforms was made possible due the enhanced mass and temporal resolution of the present analytical approach. It is interesting to note that shorter O-deacylated LPS of higher mobility are also observed in Fig. 4B between 10.5 to 13.0 min and correspond to truncated glycolipids lacking a Hep$_{III}$ residue. This observation is consistent with a deficiency in the expression of the orfH gene in a subset of the bacterial population analyzed (Hood et al., 2001).

A summary of the different glycoforms observed in Fig. 4 for the lex2 insert construct strain of H. influenzae strain Rd along with the observed electrophoretic mobility for each glycolipid is presented in Table I. Mass accuracy observed on the different glycolipids was typically within ±0.1 Da

Table I
COMPOSITION OF DIFFERENT O-DEACYLATED LPS FROM H. INFLUENZAE STRAIN Rd lex2 INSERT[a]

Time (min)	$\mu_{ep} \times 10^{-4}$ (cm²V⁻¹s⁻¹)	Molecular mass (Da)		Assignment[c]						
		Observed	Calculated[b]	Hex	Hep	PC	PE	PPE	P	HexNAc
9.55	2.95	3577.28	3577.20	7	3	1	1	1	0	1
9.68	3.02	3415.24	3415.15	6	3	1	1	1	0	1
9.85	3.10	3253.16	3253.10	5	3	1	1	1	0	1
9.75	3.05	3374.23	3374.12	7	3	1	1	1	0	0
9.92	3.13	3212.14	3212.07	6	3	1	1	1	0	0
10.08	3.20	3050.07	3050.02	5	3	1	1	1	0	0
10.28	3.29	2888.06	2887.96	4	3	1	1	1	0	0
10.12	3.22	3251.13	3251.11	7	3	1	1	1	0	0
10.25	3.28	3089.09	3089.06	6	3	1	0	1	0	0
10.45	3.36	2927.08	2927.01	5	3	1	0	1	0	0
10.72	3.47	2765.03	2764.96	4	3	1	0	1	0	0
10.88	3.53	2602.96	2602.90	3	3	1	0	1	0	1
10.58	3.41	2762.01	2762.06	5	3	0	0	0	0	1
10.72	3.47	2600.00	2600.00	4	3	0	0	0	0	1
11.02	3.58	2437.91	2437.95	3	3	0	0	0	0	1
10.28	3.29	2722.98	2723.01	4	3	0	1	0	0	1
11.35	3.70	2209.80	2209.88	3	2	1	0	0	0	0
11.91	3.89	2086.80	2086.73	1	2	1	0	1	0	0
12.11	3.95	1882.68	1882.65	0	2	1	0	0	2	0
12.15	3.96	1921.74	1921.68	1	2	0	1	0	1	0
12.68	4.12	1636.68	1636.62	0	2	0	0	0	1	0
12.78	4.15	1759.67	1759.63	0	2	0	1	0	1	0

[a]Table based on Fig. 4.
[b]^{12}C monoisotopic component based on the atomic masses: ^{12}C: 12.0000, ^{1}H: 1.0078, ^{16}O: 15.9949, and ^{14}N: 14.0031.
[c]Glycans and functional groups appended to a core oligosaccharide containing Hep₃, KDO, and a lipid A composed of GlcN₂, P₂ and two N-linked 3-OH myristic acid ($C_{14}H_{27}O_2$).
Hex: hexose (Glc, Gal); GlcN: glucosamine; Hep: heptose; KDO: 3-deoxy-D-manno-2-octulosonic acid; PE: phosphoethanolamine; PC: phosphocholine.

of the predicted molecular mass. Examination of the electrophoretic mobility of related families also provided a better understanding between migration patterns and structural changes taking place between different glycoforms. In zone electrophoresis, the magnitude of the analyte mobility corresponds to the sum of the electroosmotic flow and its electrophoretic mobility, the latter being represented by the following relation:

$$\mu_{ep} = q/6\pi\eta r. \tag{1}$$

In equation (1), q is the net charge, r is the Stokes radius, and η is the viscosity. The Stokes radius is related to the molecular weight, volume, and shape of the analyte. Previous investigations on the separation of peptides and proteins have indicated that the electrophoretic mobility is proportional to $q/M^{1/3}$ or $q/M^{2/3}$, where M refers to the molecular mass (Basak and Ladisch, 1995; Cifuentes and Poppe, 1994; Hilser $et\ al.$, 1993). In the present study, the proportionality between the mobility and the molecular mass was also examined, and a plot of the electrophoretic mobility versus $1/M$ for four glycoform families is presented in Fig. 5. Excellent linearity is obtained for all glycolipids evaluated (r^2: 0.98 to 1.00), suggesting that these families share the same charge state and that variation in mobility is reflected only by their respective molecular mass. Indeed, each incremental addition of Hex residue results in a decrease of mobility of approximately $8 \times 10^{-6}\ cm^2V^{-1}s^{-1}$.

The capability of differentiating glycoform families based on their corresponding electrophoretic mobility is also an invaluable tool for the location and identification of functional groups such as P, PE, and PPE. Indeed, the natural distribution of these functionalities in glycolipid structures can give rise to a number of isoforms of identical molecular masses that differ only by the location of these substituents (Thibault $et\ al.$, 1998). This finding is illustrated in Fig. 6 for the separation of isomeric glycoforms from $N.\ meningitidis$ strain BZ157 B5+ (High $et\ al.$, 1993). Previous investigations have reported the isolation of a monoclonal antibody (MAb), designated B5, that cross-reacted with an epitope containing a phosphoethanolamine (PE) attached specifically at the 3-position on the β-chain heptose of the LPS of $N.\ meningitidis$ (High $et\ al.$, 1993). The B5 reactive LPS epitope was found to be present on 70% of all $N.\ meningitidis$ strains tested, and glycoforms that completely lack PE (L5 immunotype) did not react with MAb B5 (High $et\ al.$, 1993). The analysis shown in Fig. 6 was performed using 10 mM of ammonium acetate pH 9.0 (5% methanol) to assist in the formation of positively charged analytes. Separations conducted using morpholine as an electrolyte provided adequate resolution of isoforms and glycoforms compared to other buffers commonly used in

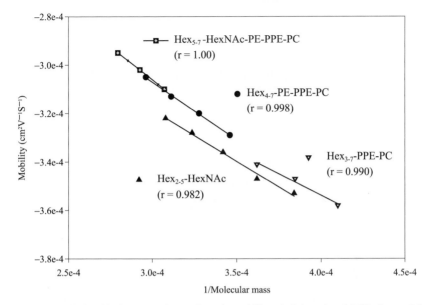

FIG. 5. Relationship between electrophoretic mobility of *O*-deacylated LPS observed in Fig. 4 and their corresponding molecular masses. Diagonal lines relate to glycoform families of related structures.

CE–ESMS experiments. However, analyses conducted with such an electrolyte are not entirely compatible with positive ion detection due to the presence of intense adduct ions corresponding to extensive addition of morpholine (1–6 molecules) to the cationic species (data not shown).

The electropherogram shown in Fig. 6A corresponds to the CE–MS/MS analysis of doubly charged precursor ions at m/z 1195.0, a predominant glycolipid composed of PE_2–HexNAc-Hex-Hep$_2$-KDO$_2$ attached to the lipid A (Fig. 1). Two distinct isoforms migrating at 11.8 and 12.6 min ($\Delta\mu_{ep}$ = 3.7×10^{-5} cm^2 V^{-1}s^{-1}) are observed in Fig. 6A. The corresponding MS/MS spectra for these two isoforms showed different fragmentation patterns (Fig. 6A and 6B). The isoform of lower mobility (Fig. 6B) is characterized by an abundant fragment ion at m/z 1313, corresponding to the neutral loss of a lipid A plus an extra PE group (neutral moiety 1075 Da). The dissociation of the core oligosaccharide fragment ion at m/z 1313 gave rise to consecutive losses of KDO residues, while fragment ions at m/z 855 and 468 were associated with the lipid A. The fragment ions at m/z 508.0, 519.0, 670.0, 711.0, and 873.0 corresponded to Hep-Hep-PE, HexNAc-Hep-PE, Hex-Hep-Hep-PE, Hep-Hep(PE)-HexNAc, and Hex-Hep-Hep (PE)-HexNAc, respectively.

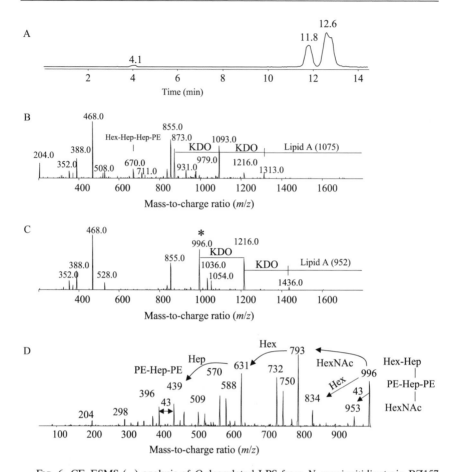

FIG. 6. CE–ESMS (−) analysis of O-deacylated LPS from N. meningitidis strain BZ157 B5+. (A) Total ion electropherogram for product ion of m/z 1195. (B) Product ion spectrum of peak migrating at 11.8 min. (C) Product ion spectrum of peak migrating at 12.6 min. (D) The second-generation product ion of fragment ion m/z 996 promoted by increasing the orifice voltage to 120 V. The structure of the substituted core oligosaccharide (Fig. 6C and D) is shown as an inset.

In contrast, the MS/MS spectrum of the isoform of higher mobility (Fig. 6C) showed a neutral loss of a lipid A at m/z 1436 (neutral moiety 952 Da) accompanied by consecutive losses of KDO residues at m/z 1216 and 996. The additional PE group was assigned to the core oligosaccharide though its exact location was still unknown. Further evidence of this structural assignment was obtained in a separate set of MS/MS experiments, where the singly charged m/z 996 formed at a high-cone voltage was selected as a precursor (Fig. 6D). This ion corresponds to the core oligosaccharide, and interpretation was simplified by the lack of fragmentation associated with the lipid A. The MS/MS spectrum of m/z 996 was dominated by simple cleavage of each glycosidic bond yielding a direct sequence assignment. The linear sequence consisting of HexNAc-(PE)Hep(PE)-Hep-Hex was confirmed by the observation of fragment ions at m/z 439, 631, and 793, respectively. Another series of fragment ions at m/z 396, 587, and 750, corresponding to a loss of the ethanolamine moiety (43 Da), also supported this assignment. Taken together, these observations enabled the identification of a Hep_{II} residue to which is attached two PE groups. The gene that encodes the phosphoethanolamine transferase (*lpt3*) adding PE to position 3 of Hep_{II} has been reported (Mackinnon et al., 2002). A separate report indicated that LPS from an inner core GlcNAc-deficient mutant (*rfaK*) of *N. meningitidis* also contained two PE groups on Hep_{II} in the major glycoform; however, the di-PE containing populations of glycoforms were not found in the parent strain from which this mutant was generated (Rahman *et al.*, 2001). An interplay between the addition of α-DGlc*p* unit at Hep_{II} in *N. meningitidis* governed by the expression of the phase variable gene lgtG and the lpt3 gene was found to be responsible for the arrangement of residues at Hep_{II} (Mackinnon *et al.*, 2002).

It is noteworthy that the occurrence of two distinct isoforms separable by their corresponding electrophoretic mobilities can be rationalized by their respective structural conformation and net charge. In the preceding example, the *O*-deacylated LPS containing a disubstituted Hep_{II} residue (Fig. 6C) is anticipated to adopt a more compact conformer compared to the glycolipid, for which the additional PE group is attached to the lipid A (Fig. 6B). More important, the presence of a PPE on one of the two phosphate groups of the lipid A contributes to a change in the number of ionizable groups, thus reducing the overall net charge by 1.

Use and Application of Diagnostic Fragment Ions for the Identification of Specific Functionalities in *O*-Deacylated LPS

The characterization of the natural diversity of glycolipid population from extracts of *O*-deacylated LPS is a challenging task and requires analytical procedures offering sensitivity, resolution, and adequate dynamic range for comprehensive structural assignment. This is particularly true not only for the identification of isoforms, as evidenced in the previous section, but also for the monitoring of trace-level glycolipids present as a small subset of a wider bacterial extract. Current methods for probing these structural changes involve gel electrophoresis and immunoblotting experiments with a number of specific monoclonal antibodies to profile the LPS epitopes (Maskell *et al.*, 1991; Moxon *et al.*, 1994; Weiser *et al.*, 1997, 1998). These methods offer excellent sensitivity for the identification of structural variants at the single colony level but rely on the availability of multiple monoclonal antibodies for adequate structural coverage. Numerous approaches are currently employed to dissect the molecular pathways involved in bacterial glycosylation and to disrupt their functions. The ability of characterizing LPS structures as a result of site-directed mutagenesis of specific gene targets facilitates *in vitro* assignment of their functions and assignment of putative glycosylation sites. The generation of gene knockouts also provides a means for producing glycosylation defects and observing their consequences of these phenotypic changes on bacterial virulence.

In an effort to provide additional sensitivity and selectivity for probing characteristic functional groups and residues present on the LPS structures, we investigated the use and application of diagnostic fragment ions in CE–ESMS experiments. Particular emphasis was placed on fragment ions specific to biologically relevant functionalities and residues such as P, PE, PPE, PC, and Neu5Ac. A list of fragment ions characteristic to these groups together with potential sites of attachment is presented in Fig. 7.

Some groups like PC have been linked to the persistence of phenotypes on the mucosal surface and to its possible evasion from innate immunity mediated by C-reactive protein (Weiser *et al.*, 1998). Furthermore, the incorporation of PC in *H. influenzae* LPS is subject to phase variation associated with the *lic1* gene (Weiser *et al.*, 1997) in which environmental PC is added to an hexose residue on the outer core region of the LPS (Risberg *et al.*, 1997; Schweda *et al.*, 2000; Weiser *et al.*, 1997). In addition to phase variation, some strains of *H. influenzae* and *N. meningitidis* have been found to mimic blood group antigens, such as P^k epitope (α–Gal(1-4)-ß-Gal (1-4)-Glc) and paragloboside (ß-Gal(1-4)-ß-GlcNAc (1-3)-ß-Gal (1-4)-Glc) (Mandrell *et al.*, 1992), which are also good acceptors for sialyl transferases. The occurrence of sialylated LPS in *H. influenzae* has been recently shown to be widespread among clinical nontypable strains (Hood *et al.*, 1999) as well as in strain Rd (Hood *et al.*, 2001), the type b strain A2 strain (Gibson *et al.*, 1993), and for a subset of LPS isolated from a clinical isolate (Thibault *et al.*, 1998). The biological importance of sialylated LPS on the pathogenicity and virulence of certain bacterial strains of both gonococci and meningococci is significant (Smith *et al.*, 1995). Most serotype groups of meningococci (B, C, W, and Y) have the ability to synthesize the substrate cytidine monophosphate Neu5Ac (CMP-Neu5Ac) that can be incorporated into capsular polysaccharides and to terminal galactose residues of LPS (Mandrell *et al.*, 1993). Human pathogens

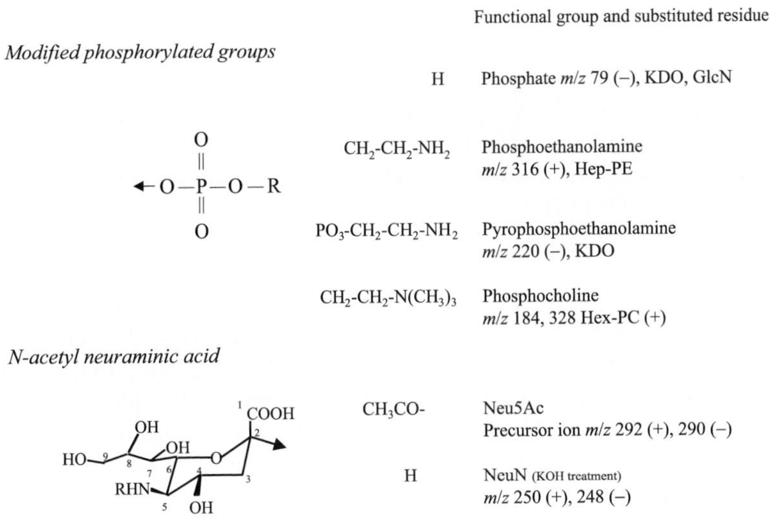

Functional group and substituted residue

Modified phosphorylated groups

	H	Phosphate m/z 79 (−), KDO, GlcN

$$\leftarrow O-\underset{\underset{O}{\|}}{\overset{\overset{O}{\|}}{P}}-O-R$$

CH_2-CH_2-NH_2		Phosphoethanolamine m/z 316 (+), Hep-PE
PO_3-CH_2-CH_2-NH_2		Pyrophosphoethanolamine m/z 220 (−), KDO
CH_2-CH_2-$N(CH_3)_3$		Phosphocholine m/z 184, 328 Hex-PC (+)

N-acetyl neuraminic acid

CH_3CO-	Neu5Ac Precursor ion m/z 292 (+), 290 (−)
H	NeuN (KOH treatment) m/z 250 (+), 248 (−)

FIG. 7. Structural determinants and functionalities observed in *H. influenzae* and *N. meningitidis* together with characteristic fragment ions specific to these groups.

deficient in endogenous sialic acid synthetase, but with active sialyl transferase genes, could incorporate such residues in their LPS structures through the use of host CMP-Neu5Ac substrate. Evidence would suggest this to be the case for *H. influenzae* strains (Hood *et al.*, 2001).

The monitoring of important virulence factors and functional groups, such as those listed in Fig. 7, can provide further insight into the biological functions of LPS and indicate, more particularly, when their occurrence is in only a small subset of the bacterial population. A previous report from this laboratory described the application of a method employing a mixed scan function with a high/low orifice voltage stepping in CE–ES–MS experiments to detect sialylated LPS from *N. meningitidis* immunotype L1 (Wakarchuk *et al.*, 1998). In the present investigation, we have further extended this method to monitor other important functional groups such as PPE and PC. Fragment ions arising from in-source dissociation (high orifice) are acquired in selected ion monitoring (SIM) mode, while multiply charged glycolipid precursors are recorded immediately after using a conventional scan (m/z 500–1500) at low orifice voltage. The SIM scan function is also performed using polarity switching on a quadrupole instrument to sequentially monitor the cation m/z 328 (Hex-PC) together with the anions m/z 220 (PPE) and m/z 290 (Neu5Ac). These combined scanning functions thus facilitate the identification of specific groups while simultaneously maximizing sample utilization.

The application of this analytical strategy is demonstrated in Fig. 8 for the analysis of O-deacylated LPS from two mutant strains of the clinical isolate *H. influenzae* 375, where site-directed mutagenesis was targeted toward the expression of the *siaB* and *lic1* genes. The wild-type strain *H. influenzae* 375 originated from a child otitis media, and LPS isolated from this sample comprised a major glycoform composed of Hex$_4$-PC-PPE-PE (Mr: 2890.4 Da) (Thibault *et al.*, 1998). A minor proportion of the LPS subpopulation contained sialylated residues (<10%), and these components were assigned to the Hex$_3$-Neu5Ac-PC-PPE-PE (Mr: 3019.5 Da) and Hex$_3$-Neu5Ac-PC-P-PE (Mr: 2896.5 Da) glycoforms (Thibault *et al.*, 1998). In comparison, the *H. influenzae* 375 *siaB* strain (Fig. 8A) showed no detectable Neu5Ac-containing O-deacylated LPS, suggesting that directed mutagenesis targeted toward this gene compromised the biosynthesis of the CMP-Neu5Ac substrate. However, other LPS glycoforms containing PE and PC functionalities remained unaffected as indicated in Fig. 8A. O-deacylated LPS from the *sia B* mutant strain exhibited a distribution of Hex$_{3-6}$ glycoforms all containing a Hex-PC but with variable incorporation of PPE and PE groups (Fig. 8A).

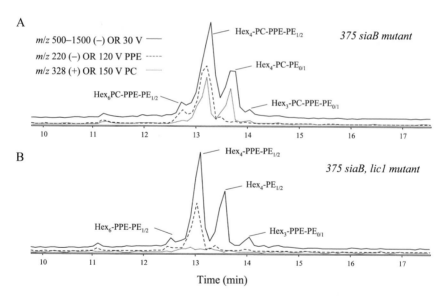

Fig. 8. Probing PC and PPE in O-deacylated LPS from *H. influenzae* 375. (A) Results for *H. influenzae* 375 *siaB* mutant strain. (B) Results for *H. influenzae* 375 *siaB/lic1* mutant strain. The total ion electropherogram (solid line) is shown together with the selected ion monitoring profile for the anion m/z 220 (dotted line) and the cation m/z 328 (dashed line).

In contrast, *O*-deacylated LPS isolated from the double mutant *siaB* and *licl* from *H. influenzae* 375 showed glycolipids lacking the Hex-PC group as evidenced in Fig. 8B from the low abundance of the *m/z* 328 (dotted line). The most abundant glycoform observed in Fig. 8B was a glycolipid Hex$_4$-PPE-PE with no PC substituent. This double mutant strain also expressed minor Hex$_{3-6}$ glycoforms, consistent with that observed for the *sia B* mutant (Fig. 8A). These CE–ESMS experiments thus confirmed that the addition of a PC group on terminal Hex residue attached to Hep$_I$ is directly associated with the *licl* gene, while the occurrence of PPE and PE groups on the LPS structures remained unaffected.

The ability to probe specific residues and functionalities can also be an important analytical tool in monitoring the incorporation of carbohydrate residues by glycosyltransferases or their removal by glycosidases. This is illustrated in Fig. 9 for the monitoring of Neu5Ac residue in *H. influenzae* 375 strain. For convenience, the reconstructed ion electropherogram for the fragment ions *m/z* 290 (Neu5Ac) is superimposed with the TIE. The wild-type *H. influenzae* 375 strain shows two sialylated components at

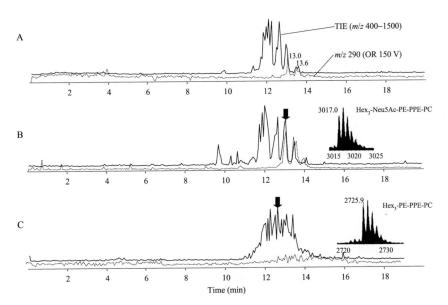

FIG. 9. Analysis of sialylated *O*-deacylated LPS from *H. influenzae* 375. (A) Results for *H. influenzae* 375 wild-type strain. (B) Strain grown in presence of CMP-Neu5Ac. (C) Strain following incubation with $\alpha_{2,3}$ sialidase. The total ion electropherogram (solid line) is shown together with the fragment anion *m/z* 290 characteristic of Neu5Ac. The arrows indicate the migration of the glycolipid shown as an inset on the right in Fig. 9B and C.

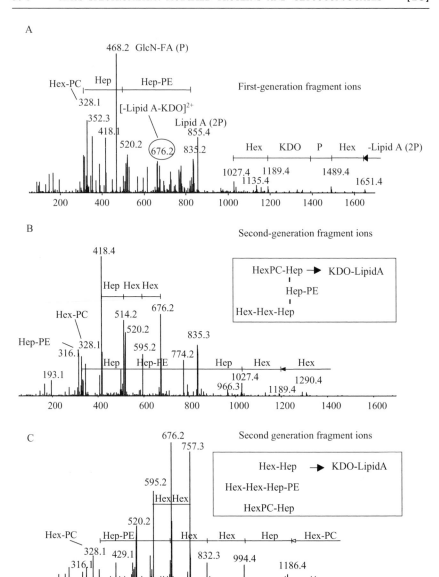

13.0 and 13.6 min (Fig. 9A) with molecular masses of 3017.0 and 2894.0 Da, respectively. These two glycolipids are consistent with the Hex$_3$-Neu5Ac-PC-P(PE)-PE glycoform, where the difference in 123 Da is associated with the incorporation of an additional PE group.

Incubation of the *H. influenzae* 375 wild-type strain in the presence of CMP Neu5Ac (Fig. 9B) gave a similar electropherogram to that grown in normal medium (Fig. 9A) though increased abundance in sialylated LPS is clearly evident in the former experiment. The reconstructed molecular mass profile of the major sialylated glycolipid, indicated by an arrow at 13.0 min, is shown as an inset in Fig. 9B. The incorporation of Neu5Ac into the LPS structures is consistent with the ability of certain strains to use environmental biosynthetic precursors as convenient surrogates when the organism lacks putative synthetases. In addition to the Hex$_3$-Neu5Ac-PC-P (PE)-PE glycoforms mentioned earlier, a third sialylated glycolipid is observed in Fig. 9B at 14.0 min. The molecular mass of this latter *O*-deacylated LPS is 3309.1 Da, consistent with a glycoform containing an additional Neu5Ac onto the Hex$_3$-Neu5Ac-PC-PPE-PE substructure. It is interesting to note that the efficient removal of the Neu5Ac residue was achieved by incubating the sialylated *O*-deacylated LPS with $\alpha_{2,3}$ sialidase (Fig. 9C) because no distinct signal was observed in the *m/z* 290 trace. The resulting incubation products were significantly enriched in the Hex$_3$-PC-PPE-PE glycoform indicated by the arrow in Fig. 9C.

It is noteworthy that sialylation of *H. influenzae* 375 strain occurs on the terminal Gal of a lactosamine. Assuming that the PC is linked to the Glc residue of Hep$_I$, the remaining glycan extension would either be a Hex$_3$ or a Hex$_2$-Neu5Ac saccharide. One could presume ambivalence in the chain elongation beyond the Hex$_2$ glycan, whereby further extension could proceed via the addition of a Hex (major pathway) or a Neu5Ac (minor pathway) residues. The extension of the oligosaccharide chain with the addition of a second Neu5Ac residue was found to take place on the terminal Neu5Ac. Further studies are under way to fully characterize the structure of these short LPS and their relationships to pathogenicity.

FIG. 10. Tandem mass spectra of PC-containing *O*-deacylated LPS from different strains of *H. influenzae*. (A) First-generation MS/MS product ion of *m/z* 868 from *H. influenzae* strain RM 118. (B) Second-generation MS/MS product ion of the doubly charged fragment ion *m/z* 676 (shown in Fig. 10A). (C) Second-generation MS/MS product ion of the doubly charged fragment ion *m/z* 757 from *H. influenzae* strain RM 318. The orifice voltage was increased to 120 V in Fig. 10B and C to promote the formation of specific fragment ions. The structure of the core oligosaccharide from *H. influenzae* RM 118 and RM 318 with the specific PC location is shown in Fig. 10B and C, respectively.

The precise location of modification or the assignment of branching site on the glycolipid structures often requires the use of MS/MS. Sequential MS, MS^n using Fourier transform ion cyclotron resonance (FTICR) mass spectrometer, has been described previously for sequence and linkage assignments in oligosaccharide glycoforms (Gaucher *et al.*, 2000). Higher order MS is sometimes required to rationalize the observation of characteristic fragment ions or to access fragmentation pathways that would be unaccessible otherwise. An example of this was presented earlier in Fig. 6 using second-generation fragment ions to obtain additional information on the branching of isomeric bacterial glycolipids from *N. meningitidis*.

Such an analytical approach is also valuable to identify the precise location of functionality, such as a PC group that may show interstrain variability. Figure 10 illustrates the application of second-generation MS/MS for assigning the location of a PC group on two strains of *H. influenzae*. The first-generation product ions of a prominent glycoform from *O*-deacylated LPS of *H. influenzae* RM118 is shown in Fig. 10A. This spectrum was obtained using on-line CE–MS/MS because flow injection analysis of the same sample proved to be unsuccessful. The corresponding MS/MS spectrum is dominated by fragment ions at *m/z* 468.2 and *m/z* 855.4 associated with the dehydrated monoglucosamine and diglucosamine *N*-acylated lipid A, respectively. Other fragment ions are observed at *m/z* 1651.4, 1489.4, 1027.4, and 1189.4 and arise from consecutive cleavages of the lipid A, Hex, and KDO (P) from the *O*-deacylated glycolipid precursor ion. Of particular interest is the observation of an abundant fragment ion at *m/z* 328 corresponding to the Hex-PC residue.

In a separate experiment, the doubly charged fragment ion *m/z* 676.2 (*O*-deacylated LPS fragment devoid of the KDO and lipid A) was promoted by increasing the orifice voltage and was selected as a precursor ion (Fig. 10B). In contrast to Fig. 10A, the interpretation of this MS/MS spectrum is facilitated through the observation of consecutive cleavages of glycosidic bonds with no fragmentation associated with the lipid A. The linear sequence consisting of PC-Hex-Hep$_I$-Hep$_{II}$(PE)-Hep$_{III}$-Hex-Hex was confirmed by the observation of fragment ions at *m/z* 328, 520, 835, 1027, and 1189, 1351, respectively. The location of this PC group differs from that observed in other *H. influenzae* strains (Risberg *et al.*, 1999; Schweda *et al.*, 2000). Indeed, the second generation of the doubly charged fragment ion *m/z* 757 (also formed at high orifice voltage) from an *O*-deacylated LPS of *H. influenzae* strain RM 318 is shown in Fig. 10C. In this case, the PC group was located on the Hex residue attached to Hep$_{III}$ as indicated by fragment ions *m/z* 355, 670, 832, 994, and 1186 and was consistent with the linear structure Hex-Hep$_I$-[Hep$_{II}$(PE)-Hex-Hex]-Hep$_{III}$-Hex-PC (Fig. 10C). The location of this PC group is similar to that observed previously for an Rd$^-$-derived mutant labeled RM 118-28, where this immunodeterminant was assigned to a Gal residue attached to Hep$_{III}$ (Risberg *et al.*, 1997). The variability in acceptor specificity and/or the involvement of multiple transferases could account for the occurrence of PC epitope on different residues.

Summary

This chapter has shown the analytical potentials of capillary electrophoresis coupled to CE–ESMS for the analysis complex glycolipids from *H. influenzae* and *N. meningitidis*. This technique provides invaluable information on the structure of trace-level glycolipids from these human pathogens. Specific enrichment methods were developed to facilitate the identification of *O*-deacylated LPS from glycolipid extracts corresponding to less than five colonies. Separation conditions enabling the detection of glycolipids as their anionic or cationic ions were described and facilitated the separation of closely related glycoforms and isoforms from different bacterial LPS.

The identification of specific functionalities present on the LPS structures, such as P, PE, PPE, and PC, is facilitated using a mixed scan function to promote the in-source formation of selected fragment ions under high orifice voltage conditions while enabling the detection of multiply charged ions using low orifice voltage. Precise location of these functional groups and the branching of oligosaccharide chains within the LPS structure can be achieved using on-line tandem mass spectrometry. The possibility of identifying glycolipid structures in terms of electrophoretic mobility and molecular mass provides unique structural maps enabling the characterization of these complex molecules into closely related families. Changes in molecular mass and electrophoretic mobilities can be visualized easily on these two-dimensional contour profiles of mass-to-charge ratio versus time and enabled the identification of families differing only by the extension of Hex residues (glycoforms) or by the position functional groups such as PPE, PE, or PC (isoforms). CE–ESMS thus provides unique analytical features that facilitate the characterization of complex distribution of bacterial LPS populations over a wide dynamic range of expression.

References

Auriola, S., Thibault, P., Sadovskaya, I., and Altman, E. (1998). Enhancement of sample loadings for the analysis of oligosaccharides isolated from *Pseudomonas aeruginosa* using transient isotachophoresis and capillary zone electrophoresis–electrospray–mass spectrometry. *Electrophoresis* **19**, 2665–2676.

Basak, S. K., and Ladisch, M. R. (1995). Correlation of electrophoretic mobilities of proteins and peptides with their physicochemical properties. *Anal. Biochem.* **226,** 51–58.

Burgi, D. S., and Chien, R.-L. (1991). Optimization in sample stacking for high-performance capillary electrophoresis. *Anal. Chem.* **63,** 2042–2047.

Chien, R.-L., and Burgi, D. S. (1992a). On-column sample concentration using field amplified in CZE. *Anal. Chem.* **64,** 489A–496A.

Chien, R.-L., and Burgi, D. S. (1992b). Sample stacking of an extremely large injection volume in high-performance capillary electrophoresis. *Anal. Chem.* **64,** 1046–1050.

Cifuentes, A., and Poppe, H. (1994). Simulation and optimization of peptide separation by capillary electrophoresis. *J. Chromatogr.* **A680,** 321–340.

Cope, L. D., Yogev, R., Mertsola, J., Argyle, J. C., McCracken, G. H., and Hansen, E. J. (1990). Effect of mutations in lipooligosaccharide biosynthesis genes on virulence of *Haemophilus influenzae* type b. *Infect. Immun.* **58,** 2343–2351.

Gaucher, S. P., Cancilla, M. T., Phillips, N. J., Gibson, B. W., and Leary, J. A. (2000). Mass spectral characterization of lipooligosaccharides from *Haemophilus influenzae* 2019. *Biochemistry* **39,** 12406–12414.

Gaucher, S. P., Morrow, J., and Leary, J. A. (2000). STAT: A saccharide topology analysis tool used in combination with tandem mass spectrometry. *Anal. Chem.* **72,** 2331–2336.

Gibson, B. W., Melaugh, W., Phillips, N. J., Apicella, M. A., Campagnari, A. A., and Griffiss, J. M. (1993). Investigation of the structural heterogeneity of lipooligosaccharides from pathogenic *Haemophilus* and *Neisseria* species and of R-type lipopolysaccharides from *Salmonella typhimurium* by electrospray mass spectrometry. *J. Bacteriol.* **175,** 2702–2712.

He, Y., and Lee, H. K. (1999). Large-volume sample stacking in acidic buffer for analysis of small organic and inorganic anions by capillary electrophoresis. *Anal. Chem.* **71,** 995–1001.

High, N. J., Deadman, M. E., and Moxon, E. R. (1993). The role of a repetitive DNA motif (5′-CAAT-3′) in the variable expression of the *Haemophilus influenzae* lipopolysaccharide epitope alpha Gal(1-4)beta Gal. *Mol Microbiol.* **9,** 1275–1282.

Hilser, V. J., Jr., Worosila, G. D., and Rudnick, S. E. (1993). Protein and peptide mobility in capillary zone electrophoresis. A comparison of existing models and further analysis. *J. Chromatogr.* **630,** 329–336.

Hood, D. W., Deadman, M. E., Allen, T., Masoud, H., Martin, A., Brisson, J.-R., Fleischmann, R., Venter, J. C., Richards, J. C., and Moxon, E. R. (1996). Use of the complete genome sequence information of *Haemophilus influenzae* strain Rd to investigate lipopolysaccharide biosynthesis. *Mol. Microbiol.* **22,** 951–965.

Hood, D. W., Makepeace, K., Deadman, M. E., Rest, R. F., Thibault, P., Martin, A., Richards, J. C., and Moxon, E. R. (1999). Sialic acid in the lipopolysaccharide of *Haemophilus influenzae*: Strain distribution, influence on serum resistance, and structural characterization. *Mol. Microbiol.* **33,** 679–692.

Hood, D. W., Cox, A. D., Gilbert, M., Makepeace, K., Walsh, S., Deadman, M. E., Cody, A., Martin, A., Mansson, M., Schweda, E. K. H., Brisson, J.-R., Richards, J. C., Moxon, E. R., and Wakarchuk, W. W. (2001). Identification of a lipopolysaccharide alpha-2,3-sialyltransferase from *Haemophilus influenzae*. *Mol. Microbiol.* **39,** 341–350.

Hood, D. W., Cox, A. D., Wakarchuk, W. W., Schur, M., Schweda, E. K. H., Walsh, S. L., Deadman, M. E., Martin, A., Moxon, E. R., and Richards, J. C. (2001). Genetic basis for expression of the major globotetraose-containing lipopolysaccharide from *H. influenzae* strain Rd (RM118). *Glycobiology* **11,** 957–967.

Inzana, T. J., Gogolewski, R. P., and Corbeil, L. B. (1992). Phenotypic phase variation in *Haemophilus somnus* lipooligosaccharide during bovine pneumonia and after *in vitro* passage. *Infect. Immun.* **60,** 2943–2951.

Jarosik, G. P., and Hansen, E. J. (1994). Identification of a new locus involved in expression of *Haemophilus influenzae* type b lipooligosaccharide. *Infect. Immun.* **62,** 4861–4867.

Kelly, J., Masoud, H., Perry, M. B., Richards, J. C., and Thibault, P. (1996). Separation and characterization of O-deacylated lipooligosaccharides and glycans derived from *Moraxella catarrhalis* using capillary electrophoresis–electrospray mass spectrometry and tandem mass spectrometry. *Anal. Biochem.* **233,** 15–30.

Kimura, A., and Hansen, E. J. (1986). Antigenic and phenotypic variations of *Haemophilus influenzae* type b lipopolysaccharide and their relationship to virulence. *Infect. Immun.* **51,** 69–79.

Knirel, Y. A., and Kochetkov, N. K. (1994). The structure of lipopolysaccharides of Gram-negative bacteria. III. The structure of O antigens: A review. *Biochemistry (Moscow)* **59,** 1325–1383.

Krylov, S. N., and Dovichi, N. J. (2000). Capillary electrophoresis for the analysis of biopolymers. *Anal. Chem.* **72,** 111R–128R.

Landerholm, M. K., Li, J., Richards, J. C., Hood, D. W., Moxon, E. R., and Schweda, E. K. H. (2004). Characterization of novel structural features in the lipopolysaccharide of nondisease-associated nontypeable *Haemophilus influenzae*. *Eur. J. Biochem.* **271,** 941–953.

Li, J., Bauer, S. H. J., Mansson, M., Moxon, E. R., Richards, J. C., and Schweda, E. K. H. (2001). Glycine is a common substituent of the inner core in *Haemophilus influenzae* lipopolysaccharide. *Glycobiology* **11,** 1009–1015.

Li, J., Cox, A. D., Hood, D. W., Moxon, E. R., and Richards, J. C. (2004). Application of capillary electrophoresis–electrospray mass spectrometry to the separation and characterization of isomeric lipopolysaccharides of *Neisseria meningitides*. *Electrophoresis* **25,** 2017–2025.

Li, J., Thibault, P., Martin, A., Richards, J. C., Wakarchuk, W. W., and van der Wilp, W. (1998). Development of on-line preconcentration method for the analysis of pathogenic lipooligosaccharides using capillary electrophoresis–electrospray mass spectrometry, application to single colony. *J. Chromatogr.* **A817,** 325–336.

Liu, C., Wu, Q., Harms, A. C., and Smith, R. D. (1996). On-line microdialysis sample cleanup for electrospray ionization mass spectrometry of nucleic acid samples. *Anal. Chem.* **68,** 3295–3299.

Mackinnon, F. G., Cox, A. D., Plested, J. S., Tang, C. M., Makepeace, K., Coull, P. A., Wright, J. C., Chalmers, R., Hood, D. W., Richards, J. C., and Moxon, E. R. (2002). Identification of a gene (*lpt-3*) required for the addition of phosphoethanolamine to the lipopolysaccharide inner core of *Neisseria meningitidis* and its role in mediating susceptibility to bactericidal killing and opsonophagocytosis. *Mol. Microbiol.* **43,** 931–943.

Mandrell, R. E., Griffiss, J. M., Smith, H., and Cole, J. A. (1993). Distribution of a lipooligosaccharide-specific sialyltransferase in pathogenic and nonpathogenic *Neisseria*. *Microb. Pathog.* **14,** 315–327.

Mandrell, R. E., McLaughlin, R., Aba Kwaik, Y., Lesse, A., Yamasaki, R., Gibson, B., Spinola, S. M., and Apicella, M. A. (1992). Lipooligosaccharides (LOS) of some *Haemophilus* species mimic human glycosphingolipids, and some LOS are sialylated. *Infect. Immun.* **60,** 1322–1328.

Maskell, D. J., Szabo, M. J., Butler, P. D., Williams, A. E., and Moxon, E. R. (1991). Molecular analysis of a complex locus from *Haemophilus influenzae* involved in phase-variable lipopolysaccharide biosynthesis. *Mol. Microbiol.* **5,** 1013–1022.

Masoud, H., Moxon, E. R., Martin, A., Krajcarski, D., and Richards, J. C. (1997). Structure of the variable and conserved lipopolysaccharide oligosaccharide epitopes expressed by *Haemophilus influenzae* serotype b strain Eagan. *Biochemistry* **36,** 2091–2103.

Moxon, E. R., and Maskell, D. J. (1992). *Haemophilus influenzae* lipopolysaccharide: The biochemistry and biology of a virulence factor. Molecular biology of bacterial infection, current status and future perspectives (C. E. Hormaeche, C. W. Penn, and C. J. Smythe, eds.). Cambridge University Press, Cambridge, United Kingdom.

Moxon, E. R., Rainey, P. B., Nowak, M. A., and Lenski, R. E. (1994). Adaptive evolution of highly mutable loci in pathogenic bacteria. *Curr. Biol.* **4**, 24–33.

Murphy, T. F., and Apicella, M. A. (1997). Nontypable *Haemophilus influenzae*: A review of clinical aspects, surface antigens, and the human immune response to infection. *Rev. Infec. Dis.* **9**, 1–15.

Plested, J. S., Makepeace, K., Jennings, M. P., Gidney, M. A., Lacelle, S., Brisson, J.-R., Cox, A. D., Martin, A., Bird, A. G., Tang, C. M., Mackinnon, F. M., Richards, J. C., and Moxon, E. R. (1999). Conservation and accessibility of an inner core lipopolysaccharide epitope of *Neisseria meningitidis*. *Infect. Immun.* **67**, 5417–5426.

Preston, A., Mandrell, R. E., Gibson, B. W., and Apicella, M. A. (1996). The lipooligosaccharides of pathogenic gram-negative bacteria. *Crit. Rev. Microbiol.* **22**, 139–180.

Rahman, M. M., Kahler, C. M., Stephens, D. S., and Carlson, R. W. (2001). The structure of the lipooligosaccharide (LOS) from the alpha-1,2-N-acetyl glucosamine transferase (rfaK (NMB)) mutant strain CMK1 of *Neisseria meningitidis*: implications for LOS inner core assembly and LOS-based vaccines. *Glycobiology* **11**, 703–709.

Risberg, A., Masoud, H., Martin, A., Richards, J. C., Moxon, E. R., and Schweda, E. K. H. (1999). Structural analysis of the lipopolysaccharide oligosaccharide epitopes expressed by a capsule-deficient strain of *Haemophilus influenzae* Rd. *Eur. J. Biochem.* **261**, 171–180.

Risberg, A., Schweda, E. K. H., and Jansson, P. E. (1997). Structural studies of the cell-envelope oligosaccharide from the lipopolysaccharide of *Haemophilus influenzae* strain RM.118–28. *Eur. J. Biochem.* **243**, 701–707.

Schweda, E. K. H., Landerholm, M. K., Li, J., Moxon, E. R., and Richards, J. C. (2003). Structural profiling of lipopolysaccharide glycoforms expressed by nontypeable *Haemophilus influenzae*, phenotypic similarities between NTHi strain 162 and the genome strain Rd. *Carbohydr. Res.* **338**, 2731–2744.

Schweda, E. K. H., Brisson, J.-R., Alvelius, G., Martin, A., Weiser, J. N., Hood, D. W., Moxon, E. R., and Richards, J. C. (2000). Characterization of the phosphocholine-substituted oligosaccharide in lipopolysaccharides of type b *Haemophilus influenzae*. *Eur. J. Biochem.* **267**, 3902–3913.

Smith, H., Parsons, N. J., and Cole, J. A. (1995). Sialylation of neisserial lipopolysaccharide: A major influence on pathogenicity. *Microb. Pathog.* **19**, 365–377.

Thibault, P., Li, J., Martin, A., Richards, J. C., Hood, D. W., and Moxon, E. R. (1998). Electrophoretic and mass spectrometric strategies for the identification of lipopolysaccharides and immunodeterminants in pathogenic strains of *Haemophilus influenzae*; application to clinical isolates. *In* "Mass Spectrometry in Biology and Medicine" (A. L. Burlingame, S. A. Carr, and M. A. Baldwin,, eds.). Humana Press, Totowa, NJ.

Wakarchuk, W. W., Gilbert, M., Martin, A., Wu, Y., Brisson, J.-R., Thibault, P., and Richards, J. C. (1998). Structure of an alpha-2,6-sialyligosaccharide from *Neisseria meningitidis* immunotype L1. *Eur. J. Biochem.* **254**, 626–633.

Weiser, J. N., Williams, A., and Moxon, E. R. (1990). Phase-variable lipopolysaccharide structures enhance the invasive capacity of *Haemophilus influenzae*. *Infect. Immun.* **58**, 3455–3457.

Weiser, J. N., Maskell, D. J., Butler, P. D., Lindberg, A. A., and Moxon, E. R. (1990). Characterization of repetitive sequences controlling phase variation of *Haemophilus influenzae* lipopolysaccharide. *J. Bacteriol.* **172**, 3304–3309.

Weiser, J. N., and Pan, N. (1998). Adaptation of *Haemophilus influenzae* to acquired and innate humoral immunity based on phase variation of lipopolysaccharide. *Mol. Microbiol.* **30,** 767–775.

Weiser, J. N., Pan, N., McGowan, K. L., Musher, D., Martin, A., and Richards, J. C. (1998). Phosphorylcholine on the lipopolysaccharide of *Haemophilus influenzae* contributes to persistence in the respiratory tract and sensitivity to serum killing mediated by C-reactive protein. *J. Exp. Med.* **187,** 631–640.

Weiser, J. N., Shchepetov, M., and Chong, S. T. (1997). Decoration of lipopolysaccharide with phosphorylcholine: A phase-variable characteristic of *Haemophilus influenzae. Infect. Immun.* **65,** 943–950.

Wilkinson, S. G. (1983). Composition and structure of lipopolysaccharides from *Pseudomonas aeruginosa. Rev. Infect. Dis.* **5,** S941–S949.

Yang, L., Lee, C. S., Hofstadler, S. A., and Smith, R. D. (1998). Characterization of microdialysis acidification for capillary isoelectric focusing–microelectrospray ionization mass spectrometry. *Anal. Chem.* **70,** 4945–4950.

Yang, Q., Tomlinson, A. J., and Naylor, S. (1999). Membrane preconcentration CE: A new approach to preconcentrating samples before separation. *Anal. Chem.* **71,** 183A–189A.

Zhang, C.-X., and Thormann, W. (1996). Head-column field-amplified sample stacking in binary system capillary electrophoresis: A robust approach providing over 1000-fold sensitivity enhancement. *Anal. Chem.* **68,** 2523–2532.

Zhang, C.-X., and Thormann, W. (1998). Head-column field-amplified sample stacking in binary system capillary electrophoresis. 2. Optimization with a preinjection plug and application to micellar electrokinetic chromatography. *Anal. Chem.* **70,** 540–548.

Zhao, Y., McLaughlin, K., and Lunte, C. E. (1998). On-column sample preconcentration using sample matrix switching and field amplification for increased sensitivity of capillary electrophoretic analysis of physiological samples. *Anal. Chem.* **70,** 4578–4585.

Zhou, D., Stephens, D. S., Gibson, B. W., Engstrom, J. J., McAllister, C. F., Lee, F. K., and Apicella, M. A. (1994). Lipooligosaccharide biosynthesis in pathogenic *Neisseria.* Cloning, identification, and characterization of the phosphoglucomutase gene. *J. Biol. Chem.* **269,** 11162–11169.

Zhou, J., and Lunte, S. M. (1995). Membrane-based on-column mixer for capillary electrophoresis/electrochemistry. *Anal. Chem.* **67,** 13–18.

Author Index

399

Subject Index

A

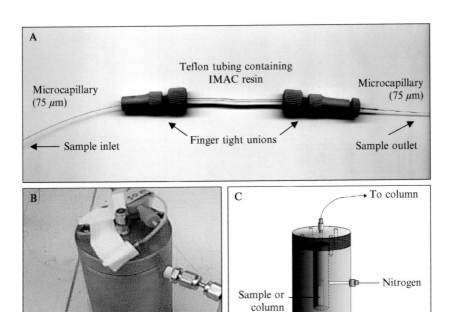

CORTHALS *ET AL.*, CHAPTER 4, FIG. 1. (A) Detailed IMAC column setup including 6 cm of Teflon tubing packed with NTA-sepharose. (B) Photo of IMAC column in use. Setup was used for identification of eNOS phosphorylation sites[16,17]. (C) Cartoon of pressure vessel and sample flow path. Table II lists the protocol for enrichment of phosphopeptides using this setup.

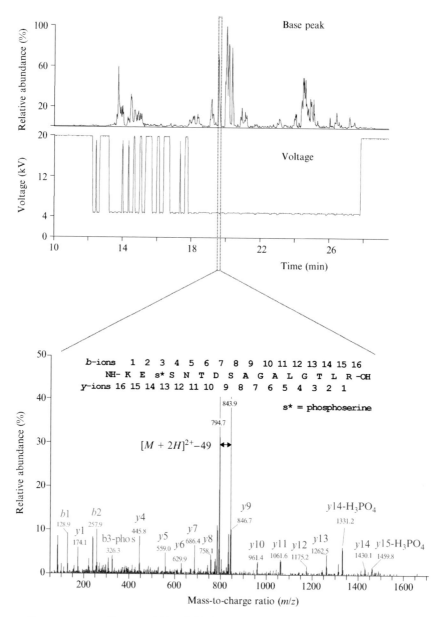

CORTHALS *ET AL.*, CHAPTER 4, FIG. 4. Peak parking in which each peak entering the mass spectrometer from a SPE–CE fractionation triggered a voltage drop, providing sufficient time for the acquisition of most abundant peptides in each fraction. The base peak trace shows the complexity of the sample despite two stages of chromatographic fractionation. The phosphopeptide identified was from the eNOS protein17.

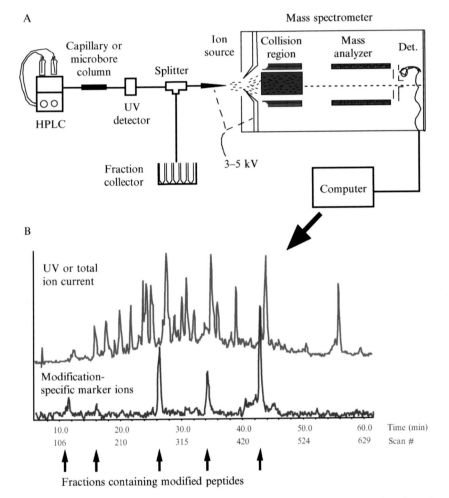

CARR *ET AL.*, CHAPTER 5, FIG. 1. Selective detection and preparative fractionation of modified peptides using marker ions during LC–ESMS. (A) Typical experimental arrangement for formation and detection of modification-specific marker ions during LC–MS analysis of proteins. (B) Representative data showing how fractions containing modified peptides are identified and collected.

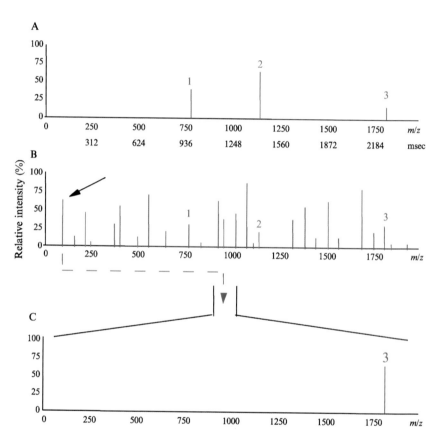

CARR *ET AL.*, CHAPTER 5, FIG. 3. Scheme illustrating precursor-ion scanning to determine the molecular masses of phosphopeptides for subsequent MS/MS analysis. (A) Q1: Normal scan (2s scan). Observed are (M – H) of all peptides. (B) Q2: Collison-induced decomposition of all ions. All peptides fragment; only phosphopeptides yield *m*/*z* 79 marker ion, which is observed only when peptide 3 fragments. (C) Q3: Selective detection of *m*/*z* 79. Observed intensity of *m*/*z* 79 was recorded at the *m*/*z* of the precursor that produced it in Q1. See text for detailed discussion.